Microbial Based Land Restoration Handbook

Plant-microbe interaction is a powerful and promising link to mitigate the various kinds of stresses like drought, salinity, heavy metals, and pathogenic effects. It is more beneficial for crop improvement and sustainable approaches for reclamation of problematic soils. Taking a multidisciplinary approach, this book explores the recent uses of plant-microbe interactions in ecological and agricultural revitalization beyond normal agriculture practices and offers practical and applied solutions for the restoration of degraded land to fulfill human needs with food, fodder, fuel, and fiber. It provides a single comprehensive platform for soil scientists, agriculture specialists, ecologists, and those in related disciplines.

Features

- Presents cutting-edge microbial biotechnology as a tool for restoring degraded lands
- Explores the aspects of sustainable development of degraded lands using microbe inspired land remediation
- Highlights sustainable food production intensification in nutrient poor lands through the innovative use of microbial inoculants
- Explains the remediation of polluted land for regaining biodiversity and achieving United Nations Sustainable Development Goals
- Includes many real-life applications from South Asia offering solutions to today's agricultural problems

This book will be of interest to professionals, researchers, and students in environmental, soil, and agricultural sciences, as well as stakeholders, policy makers, and practitioners with an interest in this field.

Microbial Based Land Restoration Handbook

VOLUME 1

Plant-Microbial Interaction and Soil Remediation

Edited by

Umesh Pankaj and Vimal Chandra Pandey

CRC Press is an imprint of the
Taylor & Francis Group, an **informa** business

First edition published 2023
by CRC Press
6000 Broken Sound Parkway NW, Suite 300, Boca Raton, FL 33487-2742

and by CRC Press
4 Park Square, Milton Park, Abingdon, Oxon, OX14 4RN

CRC Press is an imprint of Taylor & Francis Group, LLC

ISBN: 9780367702267 (hbk)
ISBN: 9780367705862 (pbk)
ISBN: 9781003147091 (ebk)

DOI: 10.1201/9781003147091

Typeset in Times
by KnowledgeWorks Global Ltd.

Dedication

This book is dedicated to our beloved mothers.

—Umesh Pankaj and Vimal Chandra Pandey

Contents

Foreword

Exponential enhanced global urbanization and industrialization have increased the exploitation of natural resources and the extravagant discharge of environmental contaminants such as heavy metals, by-products of disinfectants, polycyclic aromatic hydrocarbons, greenhouse gases, and so on, leading to air, water, and soil pollution. These environmental pollutants pose a potential health risk to human beings, along with birds and animals and have an adverse impact on crop production. Hence, we need effective and eco-friendly approaches for the remediation of such pollutants in a sustainable manner. Toxic contaminants can be cured by using some bio-based approaches like the effective use of various native microbes, plants, and enzymes. Over the past few decades, bioremediation has emerged as a potential and cost-effective strategy during the mitigation of many environmental contaminants. In this regard, plant-microbial-based remediation is the most appropriate approach to reduce the contaminants. The present book, *Microbial Based Land Restoration Handbook: Plant-Microbial Interaction and Soil Remediation,* offers such promising and potential tools to enhance plant performance with microbial intervention. The symbiotic relationships between plants and microbes have led to the alleviation of plant responses to adverse effects of harsh environmental conditions. It is generally governed by a complex molecular cross-talk between plants and microbial communities, wherein, the plants provide shelter and food to microorganisms and, in return, the microbes assist plants in water and nutrient absorption during environmental adversities. This book covers various aspects such as plant-microbial cross-talk, next-generation sequencing, metagenomics, metabolomics, bio-nanotechnology, natural resources preservation, compost-biochar-microbial and arbuscular mycorrhizal fungi-assisted phytoremediation, plant-microbial secondary metabolite-based soil remediation, and sustainable plant growth. The book is an up-to-date and timely contribution, presented in a way that is accessible to a wide-ranging audience like scientists, researchers, policy makers, practitioners, entrepreneurs, and other stakeholders alike.

I congratulate the editors, Dr. Umesh Pankaj and Dr. Vimal Chandra Pandey, in bringing out this valuable book with the help of the leading global publisher, CRC Press/Taylor & Francis. The book comprises 13 chapters covering various aspects of microbial-assisted soil remediation with contributions from the national and international scientific community.

Prof. Arvind Kumar
Vice-Chancellor
Rani Lakshmi Bai Central Agricultural University
Jhansi, India

Preface

Restoring degraded land is imperative for recuperating ecosystem services, that is, conserving the diversity of flora and fauna, enabling nutrient cycling, and meeting the demands of food, feed, and fiber for present and future generations. In this regard, microbial bio-inoculants are acknowledged as promising candidates for restoring degraded and contaminated lands. But recent microbial and biotechnological progressions, comprising the development of effective microbial consortia and designing plants with particular microbial partners, are opening new prospects in bioremediation technology. This book provides insights into promising ways to harness plant-microbial interactions for remediation and restoration of degraded land in an eco-friendly and sustainable manner. Biotechnological progressions such as genomics, metabolomics, and proteomics can be used for restoring the degraded and contaminated lands for environmental welfare. Modified (site-specific, pollutant-specific, and cost-effective) approaches are crucial for effective restoration programs. The present book, *Microbial Based Land Restoration Handbook, Volume 1: Plant-Microbial Interaction and Soil Remediation* covers various microbial bio-inoculant applications in degraded lands and their role in plant growth and productivity for sustainable agriculture. This volume is exclusively based on microbial remediation of various polluted soils and their interactions with plants. The book is highly useful for the scientific community, including students, policy makers, and agriculture professionals.

The book consists of 13 chapters contributed by experts from around the globe. The first chapter describes the microbial-assisted phytoremediation method for heavy metals containing industrial discharge and the interactions among microbes, plants and metal-containing soil. Chapter 2 throws light on the interaction of plant-associated microbes in the rhizosphere that encourage plant health and growth via a variety of mechanisms. Chemical signals play an important role in forming close relationships between plants and microbes in this interaction. The secondary metabolites from microbes and plant root exudates are the chemical agents that untangle this interaction in the rhizosphere. Chapter 3 covers the information related to the beneficial effects of biochars on soil health by enhancing carbon sequestration, soil fertility, transformation, and immobilization of emerging contaminants like heavy metals, polyaromatic hydrocarbons (PAHs), and so on. These characteristics of biochar provide a favorable environment for plant growth promotion under contaminated soil conditions. Moreover, biochar has been observed to improve soil microbial activities as well and acts as a biostimulant to the indigenous bacterial population. Chapter Four focuses on the meta-transcriptomics approaches used to identify the active microbial gene expression/regulation during stress tolerance in plants. Chapter 5 provides information on bio-nanoparticle applications for sustainable management of insect pests and diseases and minimizing the harmful effects of dangerous synthetic pesticides. Chapter 6 is a detailed discussion on the metagenomics study of microbial diversity in contaminated sites and considered one of the potential tools to remove various harmful contaminants from the environment. Chapter 7 emphasizes the development of PGPR formulation with compatible promising strains that provide strength to plants under abiotic stressed conditions. Chapter 8 summarizes the biological methods for the rehabilitation of contaminated soils. Chapter 9 describes the new research achievements in the field of microbial pesticides, also called biopesticides, that are effective, safe and eco-friendly integrated pest management strategies. Chapter 10 refers to the various plastic degrading microorganisms that are capable of degrading plastics to a lesser or greater extent, which can be exploited to solve the plastic problem. Chapter 11 deals with the source, nature, and persistence of pollutants emitted from the coal mining industry and the efficiency of plant-microbe synergy for the rehabilitation of the impacted land. Chapter 12 focuses on the microbial metabolism of indigenous microorganisms that can be exploited for degradation remediation of hazardous chemical pesticides toward sustainable agriculture. Chapter 13 outlines the plant-soil-microbial interaction for organic production of oilseed crops.

Umesh Pankaj
Vimal Chandra Pandey

Acknowledgments

We wish to extend our sincere thanks to Irma Shagla Britton (Senior Editor), Rebecca Pringle (Editorial Assistant) and Shannon Welch (Editorial Assistant), CRC Press/Taylor & Francis for their whole-hearted support, guidance, and coordination to carry out this project. We would like to thank all the contributors for their excellent chapter contributions. We would like to thank all the reviewers for their time and expertise to review the chapters of this book. We are thankful to Prof. Arvind Kumar, Vice-Chancellor Rani Lakshmi Bai Central Agricultural University, Jhansi, for writing the Foreword at short notice. Finally, diction is not enough to express our deepest sense of unbounded gratitude to our respected parents and family for their support, love, and encouragement, for which we will remain indebted to them throughout our lives.

Editors

Dr. Umesh Pankaj is presently working as a Teaching cum Research Associate in Microbiology at Rani Lakhsmi Bai Central Agricultural University, Jhansi, Uttar Pradesh, India. His doctoral degree was awarded in field of Agricultural Microbiology from Jawaharlal Nehru University, New Delhi and research work carried out from CSIR-Central Institute of Medicinal and Aromatic Plants, Lucknow, India. His area of research is mainly focused on plant-microbe interaction with special emphasis on the phytoremediation and restoration of salt-affected soil using halotolerant bio-inoculants like plant growth promoting microbes and arbuscular mycorrhiza fungi. Currently, he is working on biofertilizers development and their use for sustainable development. He is also engaged in teaching for the courses at post-graduate and undergraduate level. Dr Pankaj has published 12 research articles in peer reviewed international journals and 5 book chapters to his credit. He has received Young Microbiologist Award from Agro-Environmental Development Society, Uttar Pradesh. He is a life member of various scientific societies.

Dr. Vimal Chandra Pandey featured in the world's top 2% scientists curated by Stanford University, United States. Dr. Pandey is a leading researcher in the field of environmental engineering, particularly phytomanagement of polluted sites. His research focuses mainly on the remediation and management of degraded lands, including heavy metal-polluted lands and post-industrial lands polluted with fly ash, red mud, mine spoil, and others, to regain ecosystem services and support a bio-based economy with phytoproducts through affordable green technology such as phytoremediation. Dr. Pandey's research interests also lie in exploring industrial crop-based phytoremediation to attain bioeconomy security and restoration, adaptive phytoremediation practices, phytoremediation based biofortification, carbon sequestration in waste dumpsites, climate resilient phytoremediation, fostering bioremediation for utilizing polluted lands, and attaining the United Nations Sustainable Development Goals. His phytoremediation work has led to the extension of phytoremediation beyond its traditional application. He is now engaged to explore profitable phytoremediation with least risk, low input and minimum care. Dr. Pandey worked as a CSIR-Pool Scientist and DS Kothari Postdoctoral Fellow at Babasaheb Bhimrao Ambedkar University, Lucknow; Consultant at the Council of Science and Technology, Uttar Pradesh and DST-Young Scientist at CSIR-National

Botanical Research Institute, Lucknow. He is the recipient of a number of awards / honors / fellowships and is a member of the National Academy of Sciences India. Dr. Pandey serves as a subject expert and panel member for the evaluation of research and professional activities in India and abroad for fostering nature sustainability. He has published over 100 scientific articles/book chapters in peer-reviewed journals/books. Dr. Pandey is also the author and editor of several books published by Elsevier, with several more forthcoming. He is associate editor of *Land Degradation and Development* (Wiley); editor of *Restoration Ecology* (Wiley); associate editor of *Environment, Development and Sustainability* (Springer); associate editor of *Ecological Processes* (Springer Nature); academic editor of *PLoS ONE* (PLOS); advisory board member of *Ambio* (Springer); editorial board member of *Environmental Management* (Springer); *Discover Sustainability* (Springer Nature) and *Bulletin of Environmental Contamination and Toxicology* (Springer). He is also a guest editor for many reputed journals.

Contributors

Faissal Aziz
Laboratory of Water
Biodiversity and Climate Change
Semlalia Faculty of Sciences
Cadi Ayyad University
Marrakech, Morocco

Mounir EL Achaby
Department of Materials Science and Nano-Engineering (MSN)
Mohammed VI Polytechnic University
Ben Guerir, Morocco

Khalid Aziz
Laboratory of Materials and Biotechnology and Environment (LBME)
Faculty of Sciences
Ibn Zohr University
Agadir, Morocco

Laila Midhat
National Centre for Research and Study on Water and Energy (CNEREE)
Cadi Ayyad University
Marrakech, Morocco

Majida Lahrouni
Laboratory of Bioactive, Health and Environment
Faculty of Science and Techniques
Moulay Ismail University
Errachidia, Morocco

Brahim Oudra
Laboratory of Water
Semlalia Faculty of Sciences
Cadi Ayyad University
Marrakech, Morocco

Namo Dubey
CSIR–Institute of Himalayan Bioresource Technology
Palampur, India
and
Academy of Scientific and Innovative Research (AcSIR)
Ghaziabad, India

Umesh Pankaj
College of Horticulture and Forestry
Rani Lakshmi Bai Central Agricultural University
Jhansi, India

Vimal Chandra Pandey
Babasaheb Bhimrao Ambedkar University (BBAU)
Lucknow, India

Kunal Singh
CSIR–Institute of Himalayan Bioresource Technology
Palampur, India

Pratibha Tripathi
Division of Plant Production and Protection
CSIR–Central Institute of Medicinal and Aromatic Plants
Lucknow, India

Puja Khare
Division of Plant Production and Protection
CSIR–Central Institute of Medicinal and Aromatic Plants
Lucknow, India

Romit Seth
Department of Horticulture Science
Plants for Human Health Institute-NC State University
Kannapolis, USA
Department of Biotechnology
CSIR–Institute of Himalayan Bioresource Technology (CSIR-IHBT)
Palampur, India
and
Department of Agricultural Biotechnology
CSK HP-Agricultural University
Palampur, India

Ram Kumar Sharma
Department of Biotechnology
CSIR–Institute of Himalayan Bioresource Technology (CSIR-IHBT)
Palampur, India

Rakesh Yonzone
College of Agriculture (Extended Campus)
Uttar Banga Krishi Viswavidyalaya
Majhian, India

M. Soniya Devi
College of Agriculture
Rani Lakshmi Bai Central Agricultural University
Jhansi, India

Bimal Das
College of Agriculture (Extended Campus)
Uttar Banga Krishi Viswavidyalaya
Majhian, India

Tony Kipkoech Maritim
Biotechnology Department
CSIR-Institute of Himalayan Bioresource Technology (CSIR-IHBT)
Palampur, India
and
Academy of Scientific and Innovative Research (AcSIR)
Ghaziabad, India
and
Division of Tea Breeding and Genetic Improvement
KALRO–Tea Research Institute
Kericho, Kenya

Abhishek Bhandawat
Biotechnology Department
CSIR-Institute of Himalayan Bioresource Technology (CSIR-IHBT)
Palampur, India
and
Division of Agri-Biotechnology
National Agri-Food Biotechnology Institute
Mohali, India

Vyoma Mistry
C.G. Bhakta Institute of Biotechnology
Uka Tarsadia University, Gopal-Vidyanagar, Maliba Campus
Surat, India

Abhishek Sharma
C.G. Bhakta Institute of Biotechnology
Uka Tarsadia University, Gopal-Vidyanagar, Maliba Campus
Surat, India

Elmehdi Ouatiki
Polyvalent Laboratory in Research and Development
Sultan Moulay Slimane University
Témara, Morocco

Suman Singh
Department of Botany
University of Lucknow
Lucknow, India

Laila Mandi
Laboratory of Water
Semlalia Faculty of Sciences
Cadi Ayyad University
Marrakech, Morocco

Sucheta Singh
Department of Molecular Biology and Biotechnology
CSIR–National Botanical Research Institute
Lucknow, India

Naaila Ouazzani
Laboratory of Water
Semlalia Faculty of Sciences
Cadi Ayyad University
Marrakech, Morocco

Ashutosh Awasthi
College of Agriculture Sciences
Teerthanker Mahaveer University
Moradabad, India

Abdessamad Tounsi
Polyvalent Laboratory in Research and Development
Sultan Moulay Slimane University
Témara, Morocco

Lahcen Ouahmane
Laboratory of Microbial Biotechnologies, Agro-Sciences and Environment (BioMAgE)
Semlalia Faculty of Sciences
Cadi Ayyad University
Marrakesh, Morocco

Soumia Amir
Polyvalent Laboratory in Research and Development
Sultan Moulay Slimane University
Témara, Morocco

Usha
College of Agriculture
Rani Lakshmi Bai Central Agricultural University
Jhansi, India

Mohamed Radi
Laboratory of Microbial Biotechnologies, Agro-Sciences and Environment (BioMAgE)
Faculty of Sciences Semlalia
Cadi Ayyad University
Marrakesh, Morocco

Vijay Kumar Mishra
College of Agriculture
Rani Lakshmi Bai Central Agricultural University
Jhansi, India

Rashmi Rawat
Department of Botany and Microbiology
Hemvati Nandan Bahuguna Garhwal University
Srinagar, Uttarakhand, India

Ekta Mishra
Department of Environmental Studies
The Maharaja Sayajirao University of Baroda
Baroda, India

Meena
Department of Botany and Microbiology
Hemvati Nandan Bahuguna Garhwal University
Srinagar, Uttarakhand, India

Shilpi Jain
Department of Environmental Studies
The Maharaja Sayajirao University of Baroda
Baroda, India

Shahenaz Jadeja
Department of Environmental Studies
The Maharaja Sayajirao University of Baroda
Baroda, India

Kulasekaran Ramesh
ICAR–Indian Institute of Oilseeds Research
Hyderabad, India

A. Aziz Qureshi
ICAR–Indian Institute of Oilseeds Research
Hyderabad, India

Disha Mishra
CSIR–Central Institute of Medicinal and Aromatic Plants
Lucknow, India

Radhey Shyam
Bihar Agriculture University
Sabour, India

Sankari Meena K.
ICAR–Indian Institute of Oilseeds Research
Hyderabad, India

P. Ratna Kumar
ICAR–Indian Institute of Oilseeds Research
Hyderabad, India

1 Plant-Microbe Consortium for Heavy Metal Removal from Contaminated Soil

Faissal Aziz
Cadi Ayyad University, Marrakech, Morocco

Laila Midhat
National Centre for Research and Study on Water and Energy (CNEREE), Marrakech, Morocco

Mounir EL Achaby
Mohammed VI Polytechnic University, Ben Guerir, Morocco

Khalid Aziz
Ibn Zohr University, Agadir, Morocco

Majida Lahrouni
Moulay Ismail University, Errachidia, Morocco

Brahim Oudra
Cadi Ayyad University, Marrakech, Morocco

CONTENTS

DOI: 10.1201/9781003147091-1

1.1 INTRODUCTION

Environmental pollution has been a major concern of society since the 1960s (Hlavackova 2005). Among the different types of pollution (organic and/or mineral), the problem posed by pollution due to heavy metals (HMs) is unique. It is linked to the specificity of the contamination, often multi-element, and to the chemical characteristics of the soil/pollutant system. The abundance of HMs in soils is particularly problematic because of their non-biodegradability compared to organic pollution and their acute toxicity (Remon et al. 2005; Baycu et al. 2014). Although some HMs are essential to life (e.g. zinc and copper) at low concentrations, they are toxic at high doses. In contrast, others are toxic even at very low doses (e.g. lead and cadmium) (Hlavackova 2005).

Indeed, HMs present in soils result from the inheritance of the geochemical background on the one hand, and the accumulation of anthropogenic inputs on the other. The metals naturally occurring in soil originate from the rock on which the soil has formed and the sediment supply. Anthropogenic contributions can be of different types: Activities linked to agricultural practices, industrial activities or even urban activities linked to city development and road networks.

From the time of the industrial revolution, metallurgical industries and mining activity have been the main sources of HMs released into the environment (Adriano et al. 2001; Singh et al. 2010). The excessive use of HMs for many decades led to the accumulation of a hundred tons of tailings waste usually left without treatment (Hakkou et al. 2009). These wastes present the principal environmental risk because the HM contents are considered toxic substances that are non-biodegradable and consequently persist for a long time (Remon et al. 2005; Baycu et al. 2014). Furthermore, it's required to consider their possible generation of acid mine drainage (Hakkou et al. 2008; Avila et al. 2012; Esshaimi et al. 2013; El Amari et al. 2014). In addition, few preventive measures are taken during the operational period in mines.

To our knowledge, few studies have been carried out to determine the concentration of HMs in mining residues and their impact on both surrounding soil and water resources (El Gharmali et al. 2004; El Adnani et al. 2007; Aziz et al. 2014). When HMs are present as freely bound forms such as soluble, exchangeable and adsorbed fractions, they tend to be easily moved and disseminated. However, metals associated with organic ligands or crystal lattices are not considered easily separated or mobilized. Thus, it is necessary to remove toxic elements from tailings using efficient methods such as sustainable restoration strategies to rehabilitate these contaminated areas.

Several methods have been used to remedy this problem over the last few decades. Conventional physical and chemical remediation methods may prove highly effective, but they have the drawback of being usually expensive and difficult to implement. They may also disturb soils' physical, chemical and biological structure (McGrath et al. 2001). Therefore, remediation options have turned to biological approaches considered alternative solutions to soil engineering; this strategy is based on the properties and functioning of specific plants and root-associated microorganisms. Metallophytes (metal-tolerant plants) are specific plant species that have the ability to survive and reproduce on metal-enriched soils without suffering from toxicity.

Thus, they represent a potential opportunity to be applied for phytoremediation such as phytostabilization or phytoextraction (Boularbah et al. 2006; Alford et al. 2010; Batty and Dolan 2013). Nowadays, about 500 plant species (0.2% of angiosperms) have been identified as hyperaccumulators of HMs (Krämer 2010).

Phytoremediation is a plant-based approach; it is a low-cost technology that involves the use of plants to extract and remove pollutants from the soil. However, this new technology has some disadvantages, such as small biomass productivity, differential tolerance to specific contaminants, climate limitations and the long time requirement of this process (Mani et al. 2015). To overcome these problems, scientists have resorted to supporting phytoremediation by other biological techniques via their combination with symbiotic microorganisms, such as rhizobia bacteria and/or mycorrhizal fungi (Singh et al. 2019). This microbial-assisted phytoremediation method enhances the plant's capacity to remove a huge quantity of pollutants.

The aim of this chapter is to contribute to a good understanding of the metallic pollutant's generation and its spread in soils. Furthermore, it examines bioremediation, from the mechanism by which microorganisms uptake those pollutants, then the role of plants as defined by phytoremediation. Finally, it reports the microbial-assisted phytoremediation method and presents some case studies that combine both techniques to enhance the bioremediation performance on HMs removal.

1.2 HEAVY METALS SOIL CONTAMINATION: A CHALLENGING ISSUE

Anthropic activities have not changed the overall quantity of metallic and metalloid elements on Earth but have principally changed their availability as well as their chemical form and distribution. The most common sources of contamination are mining activities, the metallurgical and iron and steel industries, fertilization using chemical and organic fertilizers, pesticides, emissions from factories, municipal waste reception centers, sludge purification and wastewater (Stepniewska and Bucior 2001; Douay et al. 2008).

In general, the principal sectors producing elements of metals are mining activities and associated industries. These practices lead to highly contaminated soils due to the dissemination of HMs by the wind and/or runoff of residues. Generally, HM-contaminated soils have a low amount of organic matter as well as limited levels of total nitrogen and assimilable phosphorus (Boularbah et al. 2006; Avila et al. 2012; Baycu et al. 2014).

We can distinguish two types of contamination following HM inputs (Bliefert and Perraud 2001):

- **Local contaminations** are linked to one or more localized source(s), well-identified and often very close (a few meters to a few kilometers). These are usually massive inflows, often associated with mining, industrial facilities and other facilities, both in operation and after closure.
- **Diffuse contaminations** result mainly from atmospheric deposition, which cannot be linked to one or more identifiable point source(s), as in agriculture (spreading of fertilizers, amendments, wastewater treatment plants, phytosanitary treatments, etc.).

Soils contain different types of compounds that can interact with contaminants. These interactions vary depending on the physicochemical properties of soils. Trace metal elements (TMEs) are linked to different soil constituents and are present in various chemical forms. They are distributed between the solid phase and the liquid phase of soil. Their amount in the soil solution represents a tiny percentage of the total pollutants. TMEs are mostly retained in the solid soil fraction, at which level they are spread in diverse organic and mineral parts (Hooda 2010). They are found in an exchangeable form in clays and organic matter, in the form of complexes associated with organic molecules, included in the crystalline structure, adsorbed on particles of oxides or hydroxides of Fe, Al and Mn or maintained in the fragments of living organisms (Kabata-Pendias and Pendias 2001).

Due to their high perseverance in the environment, HMs in soil can be converted and stored in the bodies of animals and/or human beings via the food chain (Kabata-Pendias and Mukherjee 2007). Consequently, they could be accumulated in the vital organs of the human body, causing numerous serious health disorders, such as destroying the immune system, Alzheimer's diseases, impaired psychosocial abilities, and disabilities associated with malnutrition (Bhargava et al. 2017). Therefore, if not controlled adequately, these tailings pose a major danger for the environment and a great threat to the health of human populations living in these mining areas.

These tailings contain large quantities of untreated toxic HMs that could be released into the environment in the form of particle-bound or aqueous solutions (El Adnani et al. 2007; El Amari et al. 2014), and stored in high amounts in soils and tailings (Boularbah et al. 2006; Hakkou et al. 2008; Esshaimi et al. 2013). Furthermore, HMs, particularly in high concentrations, have negative effects on physiological and biochemical systems in plants, as well as on associated soil microorganisms (Lenart-Boron and Wolny-Koladka 2015).

It should be noted that it is not so much the otherwise normal presence of metals in the soil that is problematic but rather the fact that they can be mobilized and, therefore, can reach an exposed population. Knowing the mechanisms of mobilization and transfer plays an equally important role in this regard as identifying the presence of a contaminant at a given location. This, therefore, amounts to identifying how a site is polluting (or potentially polluting) rather than being polluted.

When metallic pollutants are present in the soil, regardless of how they were introduced (air, water, deposit), all physical, chemical and biological mechanisms contribute to their transformation. This can result in metals' partial or total immobilization in soil or their entrainment by water toward the water tables. Thus, soil pollution is directly linked to the degradation of soil quality and water pollution: it presents a danger to all living organisms in the soil and the aquatic environment.

1.2.1 Overview of the Issue of Abandoned Mines

Abandoned mines have become a global issue, especially in countries with a long mining tradition. Indeed, it presents a significant challenge for all environmental stakeholders in mining projects.

In Canada, almost 10,000 abandoned sites (including more than 1,000 in the province of Quebec) have been identified. In addition, more than 550,000 sites are classified as abandoned in the United States, and more than 1,000 sites in Sweden, 10,000 sites in Great Britain, 5,500 in Japan and more than 689 sites in France (Van Zyl et al. 2002; Bussière 2008).

In Africa, several countries represent a continent with strong mining activity and as much as 30% of the world reserves of strategic minerals for the world economy: for example, 30% of bauxite, 60% of manganese, 75% of phosphates, 85% of platinum, 80% of chromium, 60% of cobalt, 30% of titanium, 75% of diamonds and almost 40% of gold. However, more than 6,000 mines have been abandoned in the southern region following the economic downfall (Chevrel et al. 2003).

The different stages associated with mining activity generate varying degrees of risks for the environment and society in general, particularly for neighboring populations. Acid Mine Drainage "AMD" is one of the principal environmental issues of the mining industry, especially in areas where the economically exploited minerals have pyrite or associated sulfides (Hakkou et al. 2008; Valente et al. 2011). AMD appears after a chemical reaction between oxygen, water and sulfurous minerals (pyrite and pyrrhotite), and is almost always present in waste rock piles and tailings ponds, with the help of acidophilic bacteria of the Thiobacillus family. Once exposed to air and water, sulfurous minerals can turn into sulfuric acid and attack rocks, drastically lowering the pH to values below 4, causing acidic and charged leachate to form HMs. Under a pH below 4.5, an iron-oxidizing bacterium *Thiobacillus ferrooxidans* catalyzes the oxidation of Fe^{2+} (ferrous iron) to Fe^{3+} (ferric iron) (Akcil and Koldas 2006; El Amari et al. 2014).

The decrease in pH can lead not only to the dissolution of metals present in mining wastes (Cu, Fe, Zn, etc.), but it may also induce the release of metals present in the sediments of streams, rivers and lakes, thus increasing the toxicity of the environment where the phenomenon occurs (Akcil and Koldas 2006). The acid generation is explained by several factors such as pH, temperature, the oxygen content of the phase gas, the degree of saturation with water, the chemical activity of ferric iron, the exposed surface of the sulfide and the microbial activity (Akcil and Koldas 2006). The effect of AMD extends far beyond the ore mining and tailings storage sector. AMD is transported to soils, rivers and aquifers. In addition, AMD impacts landscapes—the visual impact of rust-colored deposits over several kilometers of watercourses or the bare appearance of waste stocks or mine tailings is stark.

The acidity associated with potentially harmless pollutants such as various HMs (Cu, Cd, Pb, Co, Hg, As, etc.) can seriously affect the surrounding ecosystems.

Mining activity can be a source of sediment and soil pollution over large areas (Zhang et al. 2012; Khalil et al. 2013). It presents at abnormal concentrations (different from those naturally present), on the surface or in the first few meters of soil. Contamination and/or alteration of the soil by mining activity can be linked to different factors: (1) soil removal for exploration and subsoil exploitation, (2) disposal of tailings and waste rock piles or (3) contamination from tailings generated during the various stages of mining activity. Moreover, spills and leaks of materials and deposits of pollutants' wind-blown dust may also contribute to soil pollution. However, high concentrations of HMs are generally the greatest risk to soils and lead to vegetation degradation (Zhao et al. 2012; El Amari et al. 2014).

1.3 HEAVY METAL UPTAKE MECHANISMS BY MICROORGANISMS

1.3.1 Interactions between Bacteria and Metals in the Environment

Certain metals are essential for microorganisms (K, Na, Mg, Ca, etc.) and participate in the functioning of cells. For example, metalloenzymes require a metal ion as a central atom in some catalytic reactions essential for functioning. In addition, to producing energy, many microorganisms use metal compounds as electron donors for metabolism, chemotropism and electron acceptors during anaerobic respiration. Therefore, it is usual to find metals on the surface of bacteria. Indeed, metal ions are an important component of bacterial walls (Beveridge, 1989). Divalent ions are strongly present in the walls of two bacteria (*B. subtilis* 168 and *B. licheniformis* NCTC 6346). Although magnesium (Mg^{2+}) and calcium (Ca^{2+}) appear to be the most abundant metals, other metals in lesser quantities have also been observed. These metals are strongly bound to teichoic acids and peptidoglycan. Metal ions can also affect the phospholipids and lipopolysaccharides' organization outside the membrane and even affect the binding forces between different membrane parts.

On the other hand, certain metallic elements (Al, Ag, Cd, Sn, Hg, etc.) are toxic to organisms. However, despite the presence of toxic ions, some microorganisms are able to survive in polluted environments. This ability may be the result of mechanisms induced or intrinsic to individuals (Gadd 1992). For microorganisms, Gadd (1992) defines tolerance as the faculty to cope with the toxic effects of metals using the intrinsic properties of the organisms. In parallel, he describes resistance as the faculty to survive toxic metals through detoxification pathways in direct response to metal exposition in the environment. In the remainder of this chapter, we will first focus on the general reactivity of bacteria toward metals and then on the various adaptive mechanisms developed by bacteria to protect themselves from metal toxicity.

Microorganisms, in particular bacteria, can interact with metals via different mechanisms (Ledin 2000). Metals can be transformed by oxidation/reduction or alkylation processes. These changes generally modify the toxic effect and mobility of the parent metal. Metals can also be accumulated by passive adsorption phenomena (independent of metabolism) or by active transport within the cell (dependent on metabolism). The involvement of extracellular polymeric substances (EPS) released by microorganisms in a natural or artificial environment has also been observed. EPS are made up of a large variety of high-molecular-weight macromolecules such as polysaccharides, proteins, nucleic acids and phospholipids and some low-molecular-weight non-polymeric molecules. EPS play an important role in cell adhesion, forming cell aggregates (biofilms, sludges, biogranules, etc.) and protecting cells from environmental attacks (Bhaskar and Bhosle 2006; Pal and Paul 2008). EPS, under their chemical characteristics, are potentially metal fixing.

It has also been observed that the microorganism's substances, such as organic compounds or sulfides, modify the solubility and, therefore, the mobility of the metal. In addition, through their participation in biogeochemical cycles, microorganisms modify the characteristics of organic matter in their environment, which can

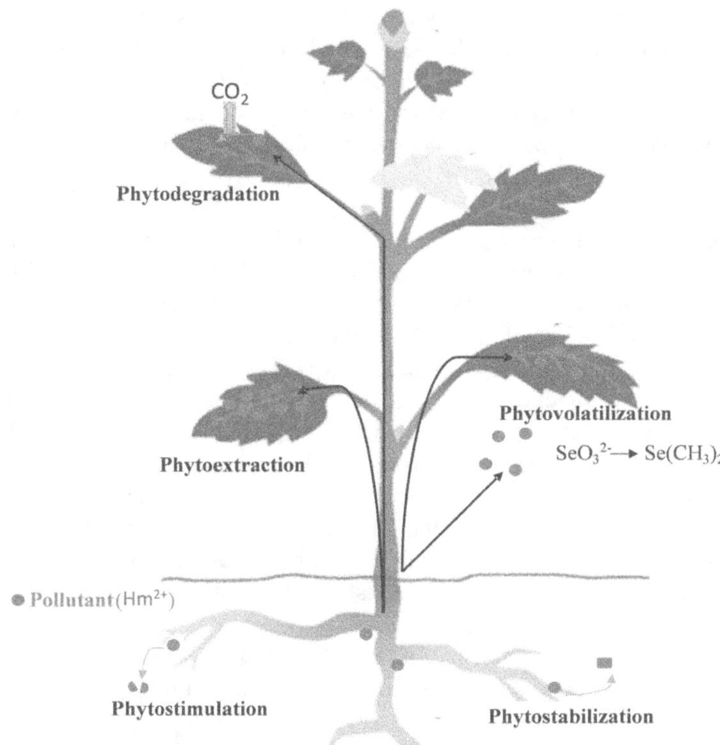

FIGURE 1.1 Schematization of interactions between metals and bacteria. **S** corresponds to the reactive groups that exist on the bacterial wall. **Hm^{2+}** corresponds to a metallic cation. **Org** corresponds to organic compounds.

modify the behavior of metals via chelation or complexation mechanisms. Finally, bacteria can indirectly influence metals' mobility through modifications of the medium, for example, by acidification. These different mechanisms are summarized in Figure 1.1.

Passive or active mechanisms control the exchanges between bacterial cells and metals (Chang 1997; Haferburg and Kothe 2007).

In metabolically active cells, absorption occurs in two phases. An initial phase of fast assimilation followed by a slower active absorption phase depends on the metabolism and the metals considered. Unfortunately, most studies of metal biosorption in bacterial cells have been performed without taking into account the effect of the physiological activity of cells.

1.3.2 Survival Strategies of Bacterial Resistance against Metals

Metals with no biological function are generally tolerated at low concentrations, while essential metals are accepted at higher concentrations (Bruins et al. 2000).

The latter participate in the metabolic functioning of cells as constituents of enzymes or structural constituents (example of the membrane). Thus, the speciation and concentration of the metal can state if it is useful or harmful to the cell (Sigg et al. 2001); therefore, control of internal concentrations, or homeostasis, is necessary. This is why bacteria have developed different defence strategies to protect themselves from metal toxicity:

- Many bacteria are now known for their ability to excrete metals through efflux systems. This pathway is categorized by a strong likeness to the substrate and makes it possible to maintain low metal concentrations in the cytosol (Mergeay et al. 1987; Nies and Silver, 1995; Nies, 2003). One of the most known is the bacterium *Cupriavidus metallidurans* CH34, which has been the subject of numerous studies. Three major families of proteins responsible for efflux in microorganisms (Nies 2003) have been described:
    - In the proteins of the resistance-nodulation-cell division (RND) family, the first finding was the CzcCBA protein of the bacterium *C. metallidurans* (Saier et al. 1994) in which the plasmid pMOL30 permits resistance to cobalt, zinc and cadmium through an efflux mechanism. Equally, the protein permits resistance to cobalt and nickel achieved by the plasmid pMOL28 (Nies, 2003).
    - The "cation diffusion facilitator" (CDF) protein family, in which the main substrates are zinc, cobalt, nickel, cadmium and iron. The concentration gradient controls these efflux systems. All the CDF proteins in bacteria are concerned with the resistance to Zn. This is the case with the CzcD protein in *C. metallidurans* CH34. In *E. coli*, the ZitB protein has been demonstrated to be a part of the resistance mechanism to Zn; the presence of Zn stimulates the expression of this protein. The proteins of the CDF family are generally responsible for the efflux of metals present in the cytoplasm (Nies 1999).
    - The third large family involved in the efflux of metals is that of "P-type ATPases," which form a large family of active transporters whose energy comes from the hydrolysis of ATP. The first protein described was CadA, which is involved in the efflux of Cd, discovered in *Staphylococcus aureus* (Nies 2003).
- Microorganisms can produce and release organic (EPS) or inorganic (metabolites) substances to the outside which are likely to modify the mobility of metals either by immobilizing them (precipitation, adsorption) or by (re-)mobilizing them (solubilization). This mechanism is often described as "bioweathering" or bioleaching (Gadd 2009; Gilmour and Riedel 2009; Uroz et al. 2009).
- While toxic metals have entered the cell and cannot be released through efflux systems, many organisms have developed cytosolic sequestration mechanisms to protect themselves. It has been shown in many organisms resistant to metals that internal compounds such as, for example, granules of polyphosphates or thiol groups (containing sulfur), are able to sequester large quantities of metal cations (Finlay et al. 2000; Gadd 2000; Pagès et al.

2007). HM bioaccumulation and subsequent storage in inert forms allow the cell to decrease its toxicity. This is, for example, the case with *C. metallidurans* CH34, which reduces selenite into elemental selenium (red) and accumulates as nodules in the cytoplasm (Roux et al. 2001; Sarret et al. 2005).
- Extracellular precipitation occurs when microorganisms produce or release substances that can react with soluble metals to produce an insoluble metallic compound. Inorganic metabolites such as sulfate ions, carbonate or phosphate derived mainly from respiratory metabolism can precipitate toxic metal ions. The formation of metal sulfides by sulfate-reducing bacteria (anoxic sediments, poorly aerated soils) is, for example, one of the best-known microbiological immobilization processes (Ledin 2000).
- Finally, metals can be biotransformed by redox mechanisms (e.g. Fe and Mn) linked to cellular respiration or alkylation (e.g. Hg). These transformations are very important for certain bacteria (in particular sulfate-reducing bacteria) and affect the metal's bioavailability, mobility and toxicity (depending on its speciation). Toxic metals can also be transformed into less toxic or even non-toxic forms by oxidation or enzymatic reduction. Many prokaryotes can use metals present in different oxidation states (Cr, Mn, Fe, Co, Cu, As or Se) as electron donors or acceptors (Ledin 2000).

To sum up, various studies have highlighted the productive interaction between trace metals and microorganisms. Metal complexes diffuse from the external medium to the body surface through the mass transport mechanism. They are often dynamic, able to dissociate (dissociation) and reassociate (complexation) before reaching the biological surface.

In the biological membrane, each metal reacts with one or more sensitive sites through an adsorption mechanism. After adsorption, an internalization phase is often but not necessarily observed. Each metal flux involved can be limiting for the entire assimilation process. The magnitude of these processes varies depending on the chemical characteristics of compounds, the nature of the organism and the surrounding physicochemical conditions such as pH, metal or ligand concentrations and membrane potential. These mechanisms can directly influence the metals physicochemistry in solution.

The study of the metal distribution at a cellular scale could highlight the crucial roles of these mechanisms. Furthermore, few studies have taken into account the effect of the physiological state of cells, while the establishment of active resistance systems can greatly modify the take-up of metals by bacteria (efflux, overproduction of EPS, bioleaching, biotransformation, etc.).

1.4 PHYTOREMEDIATION

Plants have been a source of energy (since wood was burned) much longer than they have been appreciated for their aesthetic appearance. However, it was only very recently (since the last decade) that researchers have measured the importance of their ability to store, degrade or mineralize environmental pollutants and, in particular,

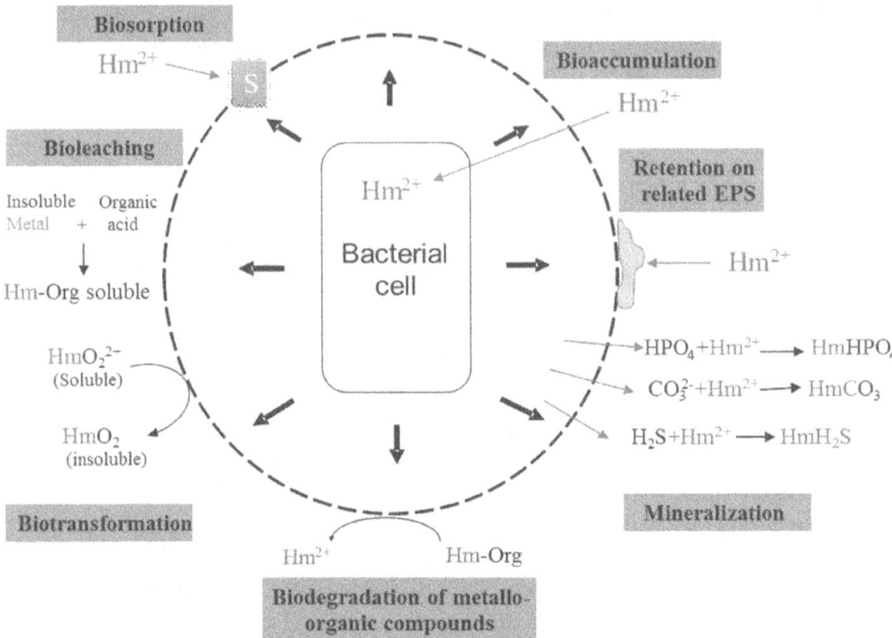

FIGURE 1.2 Summary scheme of the different phytoremediation processes (**Hm^{2+}** corresponds to a cation metallic).

soil pollutants. Thanks to different processes (Figure 1.2), plants can purify the soil *in situ*. This is called phytoremediation.

In situ treatment of soils and sediments is possible thanks to the ability of plants to be able to either fix pollutants (and thus avoid any groundwater migration), known as **phytostabilization**, or to reduce their concentrations in the soil by different processes:

- **Phytoextraction**: The accumulation of inorganic compounds, particularly TMEs, in aerial plant tissues.
- **Rhizodegradation or phytostimulation**: The biodegradation of organic compounds by microorganisms in the rhizosphere.
- **Phytovolatilization**: The transformation of the contaminant at the plant level into a volatile compound such as mercury or selenium.
- **Phytodegradation**: The transformation or metabolization of complex organic molecules by plant metabolism.

The choice of phytoremediation technique depends on the nature of the contaminant and its fate in the plant itself or in the soil and the extent of the contaminated area.

Indeed, inorganic pollutants such as TMEs, metalloids and radioactive molecules will be stabilized by phytostabilization or eliminated by phytoextraction. While organic pollutants such as hydrocarbons and pesticides will be degraded or eliminated by phytodegradation, rhizodegradation or phytovolatilization (Leung et al. 2013).

Phytoremediation adapts well to certain types of contamination, such as surface contamination of soils (<5 m depth) and moderately hydrophobic pollutants. But plants can also be used to store or degrade certain pollutants in rivers or marshes and thus participate in the decontamination of polluted water. This process is called phytofiltration. Because phytoremediation is still in development, it is not yet widely used. Still, it is low-cost compared to conventional remediation methods (incineration and washing of the soil) gives it a particular interest.

Phytoremediation is, therefore, a promising solution for soil decontamination. The gentle technique can be applied *in situ* on a wide variety of polluted soils (soils, agricultural fields, brownfields, excavated sediments, etc.) in rural and urban areas. Judged a priori to be more in line with sustainable development challenges than traditional on-site and off-site treatment techniques (excavation, thermal desorption, incineration). These techniques have a positive impact on the functions and structure of the soil (Origo et al. 2012; Singh et al. 2020). However, some drawbacks arise, such as the slowness of action (years to clean up an industrial site partially) and the implementation of this system which requires a lot of time compared to physicochemical processes, even using fast-growing plants. In addition, this solution has limits: these are due to the size of plants and roots, which do not allow cleaning up the deep horizons, or the phytotoxicity of certain pollutants. These factors interfere with establishing this technology (Leung et al. 2013). To partially overcome these problems, one of the solutions would be the addition of biological amendments based on microorganisms: this method is called microbial-assisted phytoremediation.

1.5 PLANT-MICROBE SOIL INTERACTION FOR METAL DETOXIFICATION

In the rhizosphere, plants secrete substances (chemicals or signals) to select and communicate with their root microflora (Bulgarelli et al. 2013). The selected microbes may also trigger a functional signal (e.g. microbial chemotaxis and colonization) to establish an efficient associative symbiosis with plant roots (Ma et al. 2016). This process depends on the chemical composition and the concentrations of root exudates secreted and the physicochemical characteristics of rhizospheric soil (Hartmann et al. 2009; Chaparro et al. 2013).

Root exudates are the organic chemicals that healthy and intact roots release into the soil. They play a crucial role in phytoremediation by enhancing the ability of plants to adapt, tolerate and survive under metal stress conditions. Diverse mechanisms are the key to these important characteristics such as allelopathic functions and detoxification process (adsorption, chelation, transformation and inactivation) (Luo et al. 2014). Moreover, they are credited with having a major influence on the structure of rhizosphere bacterial communities (Kim et al. 2010).

Natural chelating agents such as organic acids, citric acid and oxalic acid can significantly enhance both translocation and bioaccumulation of metals and contribute to their immobilization through the formation of stable metal complexes in soil. Therefore, they can be considered an efficient tool to enhance phytoremediation (phytoextraction and phytostabilization) strategies (Zhao et al. 2005).

Generally, the interactions between plants roots and their associated microbes are highly dynamic. These interactions via various signaling molecules play a significant role in maintaining the growth of both partners. Flavonoids are known as the important key signaling components in various plant microorganisms' interactions. They play a significant role in arbuscular mycorrhizal fungi (AMF) spore germination, hyphal growth, differentiation and root colonization in AMF-plant interaction (Mandal et al. 2010). Moreover, specific flavonoids induce nod gene expression, rhizobial chemotaxis and bacterial growth (Bais et al. 2006). However, some free-living microorganisms in soil can also release various signaling molecules, such as volatile organic compounds (VOCs), Nod factors, Myc factors, microbe-associated molecular patterns (MAMPs) and exopolysaccharides (Goh et al. 2013; Ma et al. 2016), which modify or/and alter the chemical composition of root exudates as well as plant physiology. They can establish communication with plants and trigger plant defence and growth promotion mechanisms (Bailly and Weisskopf 2012).

Plant-microbial communication is a crucial process to characterize rhizospheric soil. The metal-resistant beneficial microorganisms especially bacteria and AMF are usually used as bio-inoculants to affect metabolic functions and membrane permeability of root cells to enhance the establishment, growth and development of remediating plants in polymetallic contaminated soils.

1.5.1 Bacteria-Addicted Phytoremediation

Rhizospheric soil hosts a larger community of microorganisms that can affect contaminants' mobilities, especially TMEs. They can introduce H^+ ions in the rhizosphere and create an acidic environment nearby, which modifies metal speciation and enhance their bioavailability in soil. To remediate and restore contaminated soils, assisted phytoremediation was found to be an effective, eco-friendly and feasible technique in the case of many organic and/or inorganic contaminants. Assisted phytoremediation could be enhanced using certain microbes, including photosynthetic and lactic acid bacteria, fungi, yeast and actinomycetes. When they were used in combination, results were very successful (Ma et al. 2015). Different microbe-based phytoremediation techniques could be applied, including appropriate rhizobial microflora with a specific plant. Each plant species has a distinct group of associated microorganisms. The interactions between microorganisms and root exudates in the soil rhizosphere have been recognized as a critical component of plant growth in phytoremediation. Using microorganisms is more efficient than chemical amendments because the microbial metabolites generated in rhizospheric soil are less toxic and biodegradable (Rajkumar et al. 2012).

Plant-microbe-metal interactions highlight the importance of plant-microbe interactions in the biogeochemical cycling of metals and their application in phytoremediation. To obtain the best performance of phytoremediation technologies, it is essential to consider an appropriate combination of plants and microbes applied. Microorganisms can enhance phytoremediation efficiency in different ways: increase plant growth and biomass, decrease metal availability in soil and facilitate metal translocation from soil to plant tissues (Rajkumar et al. 2012). Moreover, they can tolerate considerably high concentrations of metals and evolve resistance strategies

(Ma et al. 2015). In addition, microorganisms can rapidly alter their cell surface and proliferate in plants' roots, namely root colonization. In this process, cell-surface biopolymers play a significant role, including proteins, glycoproteins, glycopeptidolipids and other macromolecular metabolites (Wang et al. 2016).

Metal-resistant mechanisms have been widely investigated for their potential to improve plant growth, alleviate metal toxicity and immobilize/mobilize/transform metals in soil, which may help to develop new microbe-assisted phytoremediation and restoration strategies. When plants are exposed to high metal concentrations, the stress stimulates the plants' inter-linked physiological and molecular mechanisms in adapting to stressful conditions. Among the mechanisms involved in plant metal tolerance are plant cell wall binding, active transport of ions into cell vacuoles, intracellular complexation with peptide ligands [phytochelatins (PCs) and metallothioneins (MTs)], as well as sequestration of metal-siderophore complexes in root apoplasm or soils (Miransari 2011).

Inoculation of efficient fungi in compatible host-microorganism-site combination can significantly contribute to the successful application of phytoremediation technologies. AMF, particularly those isolated from metalliferous sites, play an important role in improving plant establishment and nutrient uptake, reducing metal-induced toxicity, changing metal availability through alteration of soil pH (Rajkumar et al. 2012) and affecting metal translocation (Rajkumar et al. 2012; Jiang et al. 2016; Ma et al. 2016).

We notice that soil properties, such as pH, temperature, water content and redox potential, present a significant impact on microbial growth and, consequently, on the success of microorganism-assisted phytoremediation.

1.5.2 Mycoremediation

Mycoremediation is a biological technique that uses fungi to remove toxic pollutants (HMs, polycyclic aromatic hydrocarbons, agricultural effluents, pharmaceutical wastes and phthalates, dyes, and detergents, etc.) from the environment. It has advantages over other conventional and bioremediation methods. Mycoremediation is an eco-friendly, non-invasive, cost-effective and practical solution that reduces or transforms toxic pollutants into non-toxic or harmless forms (Perelo 2010). Mushrooms, macro-fungi, are among the most important natural sources of mycoremediation (Kapahi and Sachdeva 2017). Mushrooms have fruiting bodies that grow out of a mass of mycelium. They present a special chemical composition that provide them with particular and present diverse functional characteristics such as anti-oxidant, anti-cancer, immunostimulatory, anti-inflammatory and anti-diabetic therapeutic properties (Synytsya et al. 2009), as well as a great ability to accumulate high metal concentrations in their bodies. These properties make them suitable to be used as an effective biosorption tool for removing toxic metals (Das 2005; Kapahi and Sachdeva 2017).

Furthermore, fungal biomass presents an excellent tolerance towards adverse environmental conditions such as diverse pH and temperature (Yazdani et al. 2010; Salman et al. 2014). A great deal of research has proved that the mycoremediation process using indigenous fungal isolates could be an effective tool to promote the immobilization of trace elements, decrease metal leaching from the industrial soil and serve as a promising technology towards a biological refinement of soils (Kapahi and Sachdeva 2017; Khan et al. 2019; Akhtar and Mannan 2020).

1.6 ADVANTAGES AND LIMITS OF PHYTOREMEDIATION

Phytoremediation is one of the biological techniques for the restoration of HMs contaminated soils. It uses metal hyperaccumulator plants which remove metals from soils and accumulate them in their tissues (Mahar et al. 2016). Phytoremediation uses the natural capacity of plants and their associated microorganisms to biodegrade or mineralize pollutants (Evangelou and Deram 2014). Phytoremediation uses the natural capacity of plants and associated microorganisms to contain, biodegrade or mineralize pollutants (Pilon-Smits 2005; Evangelou and Deram 2014). Phytoremediation is used more and more nowadays because it helps maintain soil structure since it does not require any excavation (EPA 2012). In addition to being recognized as an economical choice compared to conventional decontamination techniques, phytoremediation is widely accepted by the general public since few impacts are associated with it. Several other advantages are attributed to this decontamination technique, including the quantity of residues generated by phytoremediation. However, when conventional technologies are used, the volume of material to be buried or incinerated is higher than if phytoremediation is used (reduction of more than 95%) (Forget 2004; Ghosh and Singh 2005). In addition, the use of these phytotechnologies applies to a wide variety of contaminated sites. Whether it is for organic or inorganic contamination, soil contamination or groundwater contamination, phytoremediation is now an option to consider. Not to mention that the presence of vegetation on a site helps to reduce or prevent erosion and provides a visual benefit to the landscape (Vishnoi and Srivastava 2008). The energy used to decontaminate the sites where phytoremediation techniques take place is provided by the sun, which is beneficial for the environment since traditional techniques instead opt for dirty energy. For example, in excavation, it is the gasoline that drives the machinery and which unfortunately pollutes to a great degree and creates noise disturbances. In addition to using no fossil fuels during decontamination, phytoremediation generally has a positive impact on the environment. Indeed, plants improve air quality and have the ability to sequester greenhouse gases (GHGs) (ITRC 2009).

This bioprocess is considered more in line with the challenges of sustainable development than physicochemical techniques, which despite their speed, lead to the alteration of the biological functions of the soil. In addition to being ecological, phytoremediation has the advantage of being less expensive and easier to implement than physicochemical processes. It is also a well-suited technique for *in situ* treatment of large areas and with good integration into the landscape.

Among the microorganisms that influence the processes, rhizospheric include AMFs, which are ubiquitous in the most natural and anthropogenic soils. More than 80% of plants species live in symbiosis with these fungi. AMFs provide many benefits to the host plant, including improving growth due to improved water and mineral nutrition, as well as enhancing tolerance mechanisms to biotic and abiotic stresses (Clark and Zeto 2000; Singh et al. 2019).

However, the effectiveness of this phytotechnology is limited due to the phytotoxicity and/or the low bioavailability of certain pollutants. Although this method has been the subject of much research since the 1990s, knowledge remains patchy and the *in situ* application of this technology is still poorly understood. The optimization of this method requires an adequate choice of plant species, and in particular, the use of plants tolerant to pollutants and promoting the growth of rhizospheric microbial populations with degrading capacity. Indeed, the roots stimulate microbial activity and modify the physicochemical properties of the rhizosphere.

One of the limitations is the contact between the rhizosphere and the contaminants present in the medium to be decontaminated (EPA 2012). The ability of plants to reach a certain depth through their roots depends on the plant species and geomorphological and climatic conditions (EPA 2012). For example, some tree species, such as the poplar, have roots that can potentially reach a depth of 15 feet (ft) in soils, while shrubs will be shallower (EPA 2012). Finally, phytoremediation must be restricted to sites with shallow contamination and whose concentrations are relatively low so that plants can grow properly to capture all contaminants (Ghosh and Singh 2005). These contaminants absorbed by plants can also pose a potential risk to the environment since they can enter the food chain if animals ingest contaminated plants. Several studies have shown that some animals and insects do not consume contaminated plants because they taste bad (Chaney et al. 2000). The growth rate of the plants will also influence phytoremediation since it may take several years to reach an acceptable level of decontamination.

Many plants species are known to accumulate high concentrations of metals in their tissues. However, are they well adapted to the targeted environment because the choice of plants in phytoremediation should be well considered (Singh et al. 2020). That said, it is best not to opt for plants that are not native to the site where *in situ* decontamination takes place and to avoid invasive ones. These precautions will help maintain the biodiversity already in place (Ghosh and Singh 2005).

The most common method of disposing of plants is controlled incineration. This way of doing things results in the production of ash, which can be reused in industrial processes if it contains metals. As a result, sometimes phytoextraction is combined with the production of biomass and its commercial use as an energy source, resulting in economic benefits and reducing the quantity of metals to be managed during processing (Ghosh and Singh 2005). This is the case with some willow species that are felled so that their biomass can be used as a source of energy.

1.7 CONCLUSION AND FUTURE PROSPECTS

Microbial-assisted phytoremediation (bio-phytoremediation) could be considered a suitable technique for the rehabilitation of contaminated soils. It constitutes a possible avenue to address the realities of contaminated soils. It's time to take advantage of this emerging technology and take it to the next level. In other words, it is time to make bio-phytoremediation competitive with conventional remedial techniques. Some recommendations mainly focus on the possible improvements that stakeholders can make to promote phytoremediation in the management of contaminated soils.

Thus, the following recommendations support phytoremediation as a contaminated soil rehabilitation project:

- That the legal framework favor the use of *in situ* treatments, such as phytoremediation.
- That documents, directives or guides can be developed to provide a framework for phytoremediation and its management.
- That a database be set up to make accessible information on plants used in phytoremediation and projects and trials already carried out.
- That large-scale trials to improve the performance of phytoremediation and to prove its effectiveness on different sites be encouraged.
- That research and development be supported in order to develop knowledge in fields relating to phytoremediation and biotechnologies.
- That the strengths of phytoremediation be brought to the fore in order to form a significant advantage for the actors who decide on the technology of soil rehabilitation. We are talking about the reuse of metals extracted from plants that have commercial value, minimal costs associated with this technology, beneficial effects on human health and the environment, etc.
- That the methods of managing plant residue be further developed to reduce this waste and minimize its costs. In addition, there is a need to regulate this waste management and the disposal of contaminants using rules or directives.

In short, the recommendations drawn up in the light of this work are not exhaustive. Still, they will help guide the analysis and research and development surrounding this technology in future years. Ultimately, these recommendations will encourage the decision makers in these countries to opt for an emerging but green technology that meets the principles of SD.

REFERENCES

Adriano D.C. 2001. *Trace elements, in the terrestrial environments: Biogeochemistry, bioavailability, and risks of metals*, second ed. New York, NY, Springer Verlag, p. 867.

Akcil A., Koldas S. 2006. Acid mine drainage (AMD): Causes, treatment and case studies. *J. Clean Prod.* 14:1139–1145.

Akhtar N., Amin-ul Mannan M. 2020. Mycoremediation: Expunging environmental pollutants. *Biotechnol. Rep.* 26:e00452. doi:10.1016/j.btre.2020.e00452.

Alford E.R., Pilon-Smits E.A.H., Paschke M.W. 2010. Metallophytes a view from the rhizosphere. *Plant Soil.* 337:33–50.

Avila M., Perez G., Esshaimi M., Mandi L., Ouazzani N., Brianso J., Valiente M. 2012. Heavy metals contamination and mobility at the mine area of Draa Lasfar (Morocco). *Open Environ. Pollut. Toxicol. J.* 3:141–146.

Aziz F., Ouazzani N., Mandi L. 2014. Assif El Mal River: Source of human water consumption and a transfer vector of heavy metals. *Desalination and Water Treatment.* 52:13–15. DOI: 10.1080/19443994.2013.807085.

Bailly A., Weisskopf L. 2012. The modulating effect of bacterial volatiles on plant growth: Current knowledge and future challenges. *Plant Signal. Behav.* 7:79–85. doi: 10.4161/psb.7.1.18418.

Bais H.P., Weir T.L., Perry L.G., Gilroy S., Vivanco J.M. 2006. The role of root exudates in rhizosphere interactions with plants and other organisms. *Annu. Rev. Plant Biol.* 57:233–266. doi: 10.1146/annurev.arplant.57.032905.105159.

Batty L.C., Dolan C. 2013. The potential use of phytoremediation for sites with mixed organic and inorganic contamination. *Crit. Rev. Environ. Sci. Technol.* 43(3):217–259.

Baycu G., Tolunay D., Ozden H., Csatari I., Karadag S., Agba T., Rognes S.E. 2014. An abandoned copper mining site in Cyprus and assessment of metal concentrations in plants and soil. *Int. J. Phytoremed.* 17(7):622–631.

Beveridge T.J. 1989. Role of cellular design in bacterial metal accumulation and mineralization. *Annu. Rev. Microbiol.* 43:147–171.

Bhargava P., Gupta N., Vats S., Goel R. 2017. Health issues and heavy metals. *Austin J. Environ. Toxicol.* 3(1):1018.

Bhaskar P.V., Bhosle N.B. 2006. Bacterial extracellular polymeric substances (EPS): a carrier of heavy metals in the marine food chains. *Environ. Int.* 32:191–198.

Bliefert C., Perraud R. (Eds.). 2001. *Chemistry of environment, air, water, soil, waste.* Paris, Publisher-Bruxelles:De Boeck.

Boularbah A., Schwartz C., Bitton G., Aboudrar W., Ouhammou A., Morel J.L. 2006. Heavy metal contamination from mining sites in South Morocco: 2. Assessment of metal accumulation and toxicity in plants. *Chemosphere.* 63:811–817.

Bruins M., Kapil M., Oehme F.W. 2000. Microbial resistance to metals in the environment. *Ecotoxicol. Environ. Saf.* 45:198–207.

Bulgarelli D., Schlaeppi K., Spaepen S., Ver Lorenvan Themaat E., SchulzeLefert P. 2013. Structure and functions of the bacterial microbiota of plants. *Annu. Rev. Plant Biol.* 64:807–838. doi: 10.1146/annurev-arplant-050312- 120106.

Bussière B. 2008. Les sites miniers abandonnés au Québec: empreinte sur le territoire, problématique environnementale et options pour la restauration. Présentation aux 4 à 6 de la Chaire Des jardins 9 avril 2008, Montréal.

Chaney R.L., Brown S.L., Li Y.M., Angle J.S., Stuczynski T.I., Daniels W.L., Henry C.L., Siebielec G., Malik M., Ryan J.A., Crompton H. 2000. Progress in risk assessment for soil metals, and in situ remediation and phytoextraction of metals from hazardous contaminated soils. In USOEPA's Conference Phytoremediation: State of the Science Conference, Boston, MA, May 1–2, 2000.

Chang J. 1997. Biosorption of lead, copper and cadmium by biomass of *Pseudomonas aeruginosa PU21. Water Research.* 31:1651–1658.

Chaparro J.M., Badri D.V., Bakker M.G., Sugiyama A., Manter D.K., Vivanco J.M. 2013. Root exudation of phytochemicals in Arabidopsis follows specific patterns that are developmentally programmed and correlate with soil microbial functions. *PLoS ONE.* 8:e55731. doi: 10.1371/journal.pone. 0055731.

Chevrel S., Courant C., Cottard F., Coetzee H. 2003. Contribution de la très haute résolution spatiale à l'évaluation des risques environnementaux liés à l'arrêt des activités minières. Exemple du bassin aurifère du Witwatersrand (Afrique du Sud). Actes du Colloque "Après-mine 2003", Nancy, 5–7 février 2003. CD ROM.11 p.

Das N. 2005. Heavy metals biosorption by mushrooms. *Indian J Natl Prod Resour.* 4:454–459.

Douay F., Roussel H., Pruvot C., Loriette A., Fourrier H. 2008. Assessment of a remediation technique using the replacement of contaminated soils in kitchen gardens nearby a former lead smelter in Northern France. *Sci. Total Environ.* 401(1-3): 29–38. doi: 10.1016/j.scitotenv.2008.03.025.

El Adnani M., Rodriguez-Maroto J.M., Sbai M.L., Loukili I.L., Nejmeddine A. 2007. Impact of polymetallic mine (Zn, Pb, Cu) residues on surface water, sediments and soils at the vicinity (Marrakech, Morocco). *Environ. Technol.* 28:969–985.

El Amari K., Valera P., Hibti M. 2014. Impact of mine tailings on surrounding soils and ground water: Case of Kettara old mine, Morocco. *J. Afr. Earth Sci.* 100:437–449.

El Gharmali A., Rada A., El Adnani M., Tahlil N., El Meray M., Nejmeddine A. 2004. Impact du drainage minier acide sur les écosystèmes aquatiques superficiels dans la région de Marrakech, Maroc. *Environ. Technol.* 25:1431–1442.

EPA, United States Environmental Protection Agency 2012. A Citizen's Guide to Phytoremediation. 2 p. (EPA 542OFO12O016).

Esshaimi M., Ouazzani N., El Gharmali A., Berrekhis F., Valiente M., Mandi L. 2013. Speciation of heavy metals in the soil and the tailing in the Zinc-Lead Sidi Bou Othmane abandoned mine. *J. Environ. Earth Sci.* 3(8):138–146.

Evangelou M.W.H., Deram A. 2014. Phytomanagement: A realistic approach to soil remediating phytotechnologies with new challenges for plant science. *Int. J. Plant Biol. Res.* 2(4): 1023.

Finlay J.A., Allan V.J., Conner A., Callow M.E., Basnakova G., Macaskie L.E. 2000. Phosphate release and heavy metal accumulation by biofilm-immobilized and chemically coupled cells of a *Citrobacter* sp. pre-grown in continuous culture. *Biotechnol. Bioeng.* 63(1):87–97.

Forget D. 2004. Réhabilitation des sols. In ETS. École de technologie supérieure (ETS). https://cours.etsmtl.ca/ctn626/innov_fiche_cemrs_200409b_fr.pdf (Page consultée le 20 février 2013).

Gadd G.M. 1992. Metals and microorganisms: A problem of definition. *FEMS Microbiol. Lett.* 79:197–203.

Gadd G.M. 2000. Bioremedial potential of microbial mechanisms of metalmobilization and immobilization. *Curr. Opin. Biotechnol.* 11:271–279.

Gadd G.M. 2009. Metals, minerals and microbes: geomicrobiology and bioremediation. *Microbiology.* 156:609–643.

Ghosh M., Singh S.P. 2005. A review on phytoremediation of heavy metals and utilization of its byproducts. *Appl. Ecol. and Environ. Res..* 3(1):1–18.

Gilmour C., Riedel G. 2009. *Biogeochemistry of Trace Metals and Mettaloids. In: Gene E. Likens, Vol. 2, Encyclopedia of Inland Waters.* Amsterdam, Elsevier, pp. 7–15.

Goh H.H., Sloan J., Malinowski R., Fleming A. 2013. Variable expansion expression in Arabidopsis leads to different growth responses. *J. Plant Physiol.* 171:329–339. doi: 10.1016/j.jplph.2013.09.009.

Haferburg G., Kothe E. 2007. Microbes and metals: interactions in the environment. *J. Basic Microbiol.* 47:453–467.

Hakkou R., Benzaazoua M., Bussière B. 2008. Acid mine drainage at the abandoned Kettara mine (Morocco): 1 environmental characterization. *Mine Water Environ.* 27:145–159.

Hakkou R., Benzaazoua M., Bussière B. 2009. Laboratory evaluation of the use of alkaline phosphate wastes for the control of acidic mine drainage. *Mine Water Environ.* 28:206.

Hartmann A., Schmid M., van Tuinen D., Berg G. 2009. Plant-driven selection of microbes. *Plant Soil.* 321: 235–257. doi: 10.1007/s11104-008-9814-y.

Hlavackova P. 2005. Evaluation du comportement du cuivre et du zinc dans une matrice de type sol à l'aide de différentes méthodologies. Thèse de doctorat, Institut National des Sciences Appliquées de Lyon.

Hooda P.S. (Ed.). 2010. *Trace elements in soils.* Willey, Blackwell Publishing Ltd. DOI:10.1002/9781444319477

Interstate Technology & Regulatory Cooperation Work Group (ITRC) 2009. *Phytotechnology Technical and Regulatory Guidance and Decision Trees, Revised.* Washington, DC, ITRC and Phytotechnologies Teams, 187 p. PHYTOO3. Online link: https://connect.itrcweb.org/HigherLogic/System/DownloadDocumentFile.

Jiang Q.Y., Zhuo F., Long S.H., Zhao H.D., Yang D.J., Ye Z.H. et al. 2016. Can arbuscular mycorrhizal fungi reduce Cd uptake and alleviate Cd toxicity of *Lonicera japonica* grown in Cd-added soils? *Sci. Rep.* 6:21805. doi: 10.1038/srep21805.

Kabata-Pendias A., Mukherjee A.B. 2007. *Trace elements from soil to human.* Berlin, Heidelberg, Springer.

Kabata-Pendias A., Pendias H. 2001. *Trace elements in soils and plants.* 3rd Edition, Boca Raton, CRC Press, 403 pp.

Kapahi M., Sachdeva S. 2017. Mycoremediation potential of *Pleurotus* species for heavy metals: A Review. *Bioresour. Bioprocess.* 4(1):32. doi:10.1186/s40643-017-0162-8.

Khalil K., Hanich L., Bannari A., Zouhri L., Pourret O., Hakkou R. 2013. Assessment of soil contamination around an abandoned mine in a semi-arid environment using geochemistry and geostatistics: Prework of geochemical process modeling with numerical models. *J. Geochem. Explor.* 125:117–129.

Khan I., Aftab M., Shakir S., Ali M., Qayyum S., Rehman M.U. et al. 2019. Mycoremediation of heavy metal (Cd and Cr)–polluted soil through indigenous metallotolerant fungal isolates. *Environ. Monit. Assess.*191(9). doi:10.1007/s10661-019-7769-5.

Kim S., Lim H., Lee I. 2010. Enhanced heavy metal phytoextraction by *Echinochloa crus-galli* using root exudates. *J. Biosci. Bioeng.* 109:47–50. doi: 10.1016/j.jbiosc.2009.06.018.

Krämer U. 2010. Metal hyperaccumulation in plants. *Annu. Rev. Plant Biol.* 61:517–534.

Ledin M. 2000. Accumulation of metals by microorganisms—Processes and importance for soil systems. *Earth-Science Reviews.* 51:1–31.

Lenart-Boron A., Wolny-Koladka K. 2015. Heavy metal concentration and the occurrence of selected microorganisms in soils of a steelworks area in Poland. *Plant Soil Environ.* 61(6):273–278.

Leung H.-M., Wang Z.-W., Ye Z.-H., Yung K.-L., Peng X.-L., Cheung K.-C. 2013. Interactions between arbuscular mycorrhizae and plants in phytoremediation of metal-contaminated soils: A Review. *Pedosphere.* 23:549–563. doi:10.1016/S1002-0160(13)60049-1

Luo Q., Sun L., Hu X., Zhou R. 2014. The variation of root exudates from the hyperaccumulator *Sedum alfredii* under cadmium stress: metabonomics analysis. *PLoS ONE.* 9:e115581. doi: 10.1371/journal.pone.01 15581.

Ma Y., Oliveira R.S., Freitas H., Zhang C. 2016. Biochemical and molecular mechanisms of plant-microbe-metal interactions: Relevance for phytoremediation. *Front. Plant Sci.* 7:918. doi:10.3389/fpls.2016.00918.

Ma Y., Rajkumar M., Rocha I., Oliveira R.S., Freitas H. 2015. Serpentine bacteria influence metal translocation and bioconcentration of *Brassica juncea* and *Ricinus communis* grown in multi-metal polluted soils. *Front. Plant Sci.* 5:757. doi: 10.3389/fpls.2014.00757.

Mahar A., Wang P., Ali A., Awasthi M.K., Lahori A.H., Wang Q., Li R., Zhang Z. 2016. Challenges and opportunities in the phytoremediation of heavy metals contaminated soils: A review. *Ecotoxicol. Environ. Saf.* 126:111–121. doi: 10.1016/j.ecoenv.2015.12.023.

Mandal S.M., Chakraborty D., Dey S. 2010. Phenolic acids act as signaling molecules in plant-microbe symbioses. *Plant Signal. Behav.* 5:359–368. doi:10.4161/psb.5.4.10871.

Mani D., Kumar C., Patel N.K. 2015. Integrated micro-biochemical approach for phytoremediation of cadmium and zinc contaminated soils. *Ecotoxicol. Environ. Saf.* 111:86–95.

McGrath S.P., Zhao F.J., Lombi E. 2001. Plant and rhizosphere process involved in phytoremediation of metal-contaminated soils. *Plant Soil.* 232(1/2):207–214.

Mergeay M., Nies D.H., Schlegel H.G., Charles P., Gerits J., 1987. A. eutrophus CH34 is a facultative chemolitotroph with plasmid-bound resistance to heavy metal. *J. Bact.* 162:328–334.

Miransari M. 2011. Soil microbes and plant fertilization. *Appl. Microbiol. Biotechnol.* 92:875–885. doi: 10.1007/s00253-011-3521-y.

Nies D.H. 1999. Microbial heavy-metal resistance. *Appl. Microbiol. Biotechnol.* 51:730–750.

Nies D.H. 2003. Efflux-mediated heavy metal resistance in prokaryotes. *FEMS Microbiology Reviews.* 27:313–339.

Nies D.H., Silver S. 1995. Ion efflux systems involved in bacterial metal resistances. *J. Ind. Microbiol.* 14:186–199.

Origo N., Wicherek S., Hotyat M. 2012. Réhabilitation des sites pollués par phytoremédiation. *La Rev. électronique en Sci. l'environnement électronique.* (12): 2–13.

Pagès D., Sanchez L., Conrod S., Gidrol X., Fekete A., Schmitt-Kopplin P., Heulin T., Achouak W. 2007. Exploration of intraclonal adaptation mechanisms of *Pseudomonas brassicacearum* facing cadmium toxicity. *Environ. Microbiol.* 9:2820–2835.

Pal A., Paul A.K. 2008. Microbial extracellular polymeric substances: Central elements in heavy metal bioremediation. *Indian J. Microbiol.* 48:49–64.

Perelo L.W. 2010. Review: In situ and bioremediation of organic pollutants in aquatic sediments. *J. Hazard. Mater.* 177(1–3):81–89. doi:10.1016/j.jhazmat.2009.12.090.

Rajkumar M., Sandhya S., Prasad M.N.V., Freitas H. 2012. Perspectives of plant-associated microbes in heavy metal phytoremediation. *Biotechnol. Adv.* 30:1562–1574. doi: 10.1016/j.biotechadv.2012.04.011.

Remon E., Bouchardon J.L., Cornier B., Guy B., Leclerc J.C., Faure O. 2005. Soil characteristics, heavy metal availability and vegetation recovery at a former metallurgical landfill: Implications in risk assessment and site restoration. *Environ. Pollut.* 137: 316–323.

Roux M., Sarret G., Pignot-Paintrand I., Fontecave M., Coves J. 2001. Mobilization of selenite by *Ralstonia metallidurans* CH34. *Appl. Environ. Microbiol.* 67:769–773.

Saier M.H., Tam R., Reizer A., Reizer J. 1994. Two novel families of bacterial membrane proteins concerned with nodulation, cell division and transport. *Mol. Microbiol.* 11: 841–847.

Salman H.A., Ibrahim M.I., Tarek M.M., Abbas H.S. 2014. Biosorption of heavy metals—A Review. *J. Chem. Sci. Technol.* 3(4):74–102.

Sarret G., Avoscan L., Carriere M., Collins R., Geoffroy N., Carrot F., Coves J., Gouget B. 2005. Chemical forms of selenium in the metal-resistant bacterium *Ralstonia metallidurans* CH34 exposed to selenite and selenate. *Appl. Environ. Microbiol.* 71:2331–2337.

Sigg L., Behra P., Stumm W. 2001. *Chimie des milieux aquatiques chimie des eaux naturelles et des interfaces dans l'environnement*, 3 ed. Paris, Science Sup, Dunod.

Singh A., Sharma R.K., Agrawal M., Marshall F.M. 2010. Health risk assessment of heavy metals via dietary intake of foodstuffs from the wastewater irrigated site of a dry tropical area of India. *Food Chem. Toxicol.* 48(2):611.

Singh G., Pankaj U., Ajayakumar P.V., Verma R.K. 2020. Phytoremediation of sewage sludge by *Cymbopogon martinii* (Roxb.) Wats. var. motia Burk. grown under soil amended with varying levels of sewage sludge. *Int. J. Phytoremediation.* 22(5):540–550.

Singh G., Pankaj U., Chand C., Verma R.K. 2019. Arbuscular mycorrhizal fungi-assisted phytoextraction of toxic metals by *Zea mays L.* from tannery sludge. *Soil Sediment Contam.: An Internat. J.* 28(8):729–746.

Stepniewska, Z., Bucior, K. 2001. Chromium contamination of soils, waters, and plants in the vicinity of a Tannery Waste Lagoon. *Environ. Geochem. Hlth.* 23: 241–245. https://doi.org/10.1023/A:1012247230682

Synytsya A, Mickova K, Synytsya A, Jablonsky I, Spevacek J, Erban V. 2009. Glucans from fruit bodies of cultivated mushrooms *Pleurotus ostreatus* and *Pleurotus eryngii*: Structure and potential prebiotic activity. *Carbohydr Polym.* 76:548–556. doi:10.1016/j.carbpol.2008.11.02.

Uroz S., Calvaruso C., Turpault M.-P., Frey-Klett P., 2009. Mineral weathering by bacteria: Ecology, actors and mechanisms. *Trends Microbiol.* 17:378–387.

Valente T.M., Ferreira M.J., Leal Gomes C. 2011. Application of fuzzy logic to qualify the environmental impact in abandoned mining sites. *Water Air Soil Pollut.* 217(1–4):302–315.

Van Zyl D., Sassoon M., Digby C., Fleury A.M., Kyeyune S. 2002. Mining for the future. Paper prepared for Mining, Minerals and Sustainable Development (MMSD), Appendix C: Abandoned Mines, 19 p.

Vishnoi S.R., Srivastava P.N. 2008. Phytoremediation: Green for environmental clean. In Proceedings of Taal 2007: The 12th World Lake Conference, (pp. 1016–1021), Jaipur, India

Wang W.F., Zhai Y.Y., Cao L.X., Tan H.M., Zhang R.D. 2016. Endophytic bacterial and fungal microbiota in sprouts, roots and stems of rice (*Oryza sativa* L.). *Microbiol. Res.* 188:1–8. doi: 10.1016/j.micres.2016. 04.009.

Yazdani M., Chee K.Y., Faridah A., Soon G.T. 2010. An in vitro study on the adsorption, absorption and uptake capacity of Zn by the bioremediator *Trichodermaatro viride. Environ Asia.* 3:53–59.

Zhang W.H., Huang Z., He L.Y., Sheng X.F. 2012. Assessment of bacterial communities and characterization of lead-resistant bacteria in the rhizosphere soils of metal-tolerant *Chenopodium ambrosioides* grown on lead–zinc mine tailings. *Chemosphere.* 87:1171–1178.

Zhao H., Xia B., Qin J., Zhang J. 2012. Hydrogeochemical and mineralogical characteristics related to heavy metal attenuation in a stream polluted by acid mine drainage: A case study in Dabaoshan Mine, China. *J. Environ. Sci.* 24:979–989.

Zhao J., Davis L.T., Verpoort R. 2005. Elicitor signal transduction leading to production of plant secondary metabolites. *Biotechnol. Adv.* 23:283–333. doi: 10.1016/j.biotechadv.2005.01.003.

2 Role of Plant-Microbial Secondary Metabolites in Stress Mitigation

Current Knowledge and Future Directions

Namo Dubey and Kunal Singh
CSIR, Palampur, India
and
AcSIR, Ghaziabad, India
Umesh Pankaj
Central Agricultural University, Jhansi, India
Vimal Chandra Pandey
Babasaheb Bhimrao Ambedkar University (BBAU)
Lucknow, India

CONTENTS

DOI: 10.1201/9781003147091-2

2.1 INTRODUCTION

Plant-microbial interactions are vital, as they are crucial to the establishment and protection of plant- based ecosystems (Jacoby et al. 2017). These interactions affect plant health and growth through multiple modes of actions. It is therefore necessary to understand the bidirectional interaction between plant and microbes for the conservation of the soil ecosystem. Currently, there is limited knowledge of plant-microbial secondary metabolites and the underlying mechanisms for the plant-microbe interaction. Microbial signaling compounds are bioactive metabolites, produced in the late idiophase (Gonzalez et al. 2003). These metabolites are classified as secondary metabolites that are not necessary for development and growth (Pang et al. 2021) but are incorporated in (1) adapting to adverse climatic conditions; (2) providing resistance against pathogens; (3) inducing signaling compounds to interact among biota; and (4) altering interactions for mutual gain. In 1891, A. Kossel introduced the term "secondary," which implies that primary metabolites are present in every living cell; the secondary metabolites are not an essential for individual's life and present incidentally. A distinctive characteristic of secondary metabolites, apart from being structurally complex, is that they cause effects at very low concentrations and often considered as chemical messengers (Linares et al. 2006).

The pathways, such as shikimate, peptide, carbohydrate, polyketide synthase, β-lactam synthetic and the non-ribosomal polypeptide synthase have all been identified in the biosynthesis of secondary metabolites. A major class of secondary metabolites belongs to ribosomal peptides, non-ribosomal peptides, polyketides, terpenes, volatile organic compounds, pyrazines and phenylpropanoids. Since they are soluble in water, these secondary metabolites have a higher degree of polarity and functionality (Tyc et al. 2017). In the natural environment, synthesis of microbial secondary metabolites parallel to their dispensability, along with the conservation of encoding genes in microbial genomes, indicate, contrary to earlier thinking, that they play important functions for the producer's survival. In fact, the synthesis and secretion of microbial secondary metabolites by rhizospheric microbes can better their ecological fitness in a variety of ways, including improving nutrient availability, protecting against environmental stress, killing or repelling predators and also by replacing competitors between microbial communities (Curl and Truelove 2012). According to Demain and Fang (2000), this ecological fitness is imparted via secondary metabolite participation in multiple activities including (1) metal transporting; (2) chemical signals against other organisms; (3) symbiotically intermediate in between microbes and plants; (4) differentiation effectors. However, despite the significance of the ecological roles played by soil microbes and the biological functions of their metabolites, the majority of research on interspecies contact in nature has been focused on plant-insect associations (Braga et al. 2016). Fortunately, prospects seem to be improving, and the scientific community is becoming more interested in researching microbe-microbe-plant chemical interactions. Over the last 15 years, research on microbial secondary metabolites has gained significant attention. Indeed, as of December 2020, a non-exhaustive search of the PubMed database (https://pubmed.ncbi.nlm.nih.gov/) reveals that the number of scientific publications on this subject has increased from 21 in 2000 to over 172 in 2020. Henceforth, in the following sections, we will analyze some of the most recent advances concerning plant-microbial secondary metabolites and how they are responsible for preserving nutrient availability, plant productivity, soil fertility and biocontrol of phytopathogens.

2.2 PLANT-MICROBE INTERACTIONS

A fundamental concern in the area of rhizospheric research has been the nature of plant-microbe interactions. Depending on how complex molecular signals are exchanged, associations between microorganisms and plants can be favorable, negative or neutral. Many microbial phytopathogens have been characterized to date including: *Alternaria solani*, *Phytophthora infestans*, *Sclerotium* sp., *Xanthomonas oryzeae* and *Pseudomonas syringeae*, just to name a few. Similarly, rhizospheric microbes such as mycorrhizal fungi, nitrogen-fixing bacteria, plant growth promoting rhizobacteria (PGPR), microorganisms with biocontrol activity, root endophytic fungi and mycoparasitic fungi have been extensively studied in recent years for their beneficial effects on plant growth (Beneduzi et al. 2012). PGPRs are classified as group of rhizobacteria with plant growth promoting activity and consist of many genera such as *Bacillus*, *Pseudomonas*, *Acinetobacter*, *Enterobacter*, *Burkholderia*, *Arthrobacter* and *Paenibacillus* (Sasse et al. 2018). As a result, the "second plant genome" refers to complex microbial species associated with plants (Berendsen et al. 2012). These advancement in plant-beneficial interactions with microbes has led to commercialization of few microbial inoculum (either single or as consortium) such as *Asticcacaulis*, *Burkholderia*, *Chitinophaga*, *Ensifer*, *Lysobacter*, *Pseudomonas* and *Arthrobacter* (Johns et al. 2016).

Plant beneficial associations or communications are nearly divided into three groups. In the first group, the interaction of plant-related microorganisms is responsible for a significant improvement in nutritional status. In this instance, though most do not interact directly with the plants, their effects on the abiotic and biotic factors of the soil have an impact on plant health (Pankaj et al. 2020). The second type of interactions involves associations between microbes and plants that are accountable for fostering plant growth and health by releasing hormone-regulating metabolites that modulate multiple phytohormones. The third type includes interactions in which communities of microbes indirectly trigger plant growth by restricting phytopathogen growth through the process of antibiosis, inducing systemic resistance (ISR) and competition. Such microbes are considered biocontrol agents. Thus, the interaction of plant microbes in the rhizosphere is critical for maintaining soil health.

2.3 PLANT SECONDARY METABOLITES: THEIR ROLE IN MICROBIAL INTERACTIONS

A broad range of structurally diverse compounds including primary (proteins, lipid, carbohydrates) and secondary (phytohormones, flavonoids, soluble sugars, organic acids phenols, etc.) metabolites are synthesized by plants (Badri and Vivanco 2009; Jacoby et al. 2021) which play a role in attracting microorganisms. Plant secondary metabolites (PSM) are divided into several broad molecular families based on their

TABLE 2.1
A Few Plant Secondary Metabolites and Their Mode of Action

Secondary Metabolite	Primary Metabolite Precursor	Mechanistic Action on Microbes	Reference
Benzoxazinoids	Chorismate	Chemo-attractant	Neal et al. 2012
Glucosinolates	Tryptophan, phenylalanine and methionine	Isothiocyanate-mediated enzyme inactivation	Aires et al. 2009
Coumarins	Phenylalanine	Disruption of transcription, Quorum sensing and disruption of biofilm formation	Yang et al. 2016
Camalexin	Tryptophan	Disruption of membrane integrity	Rogers et al. 1996
Flavonoids	Phenylalanine	Induction of nod gene expression in *Rhizobium*	Jacoby et al. 2021
Triterpenes	Isopentenyl pyrophosphate and squalene	Membrane integrity disruption	De León et al. 2010
Strigolactones	Isopentenyl pyrophosphate and β-carotene	Hyphal branching induction in mycorrhizal fungi	Akiyama et al. 2005

biosynthetic pathways: Terpenes, phenolics, steroids, flavanoids and alkaloids that are involved in plant growth, defence response signaling, plant-microbe mutual benefit and stress responses. Many PSMs even impacts on human health and agriculture development, contributing significantly to the economy. Table 2.1 provides the role of a few such metabolites with their role in plant-microbe interaction.

The bulk of plant secondary metabolites are secreted as root exudates that are one of the two main components of rhizosphere, along with root microbiome. The "rhizosphere" is the layer of soil covering a plant's root that is affected and inhabited by microorganisms (Hiltner 1904) while the population of all root or rhizosphere-dwelling microorganisms is referred to as the root microbiome (Lundberg et al. 2012). Increased nutrient levels due to PSM exudation impact directly on the density and composition of microbial community of the soil within the rhizosphere and mediate association between the plant root and the microbes. As compared to bulk soil, the rhizosphere is correlated with greater microbial abundance and activity, but lower diversity. This suggests that plant-derived metabolites have significant roles in the nature and evolution of microbiome assemblage in the rhizosphere (Hacquard et al. 2017). However, the relationship is bidirectional in between the associated microbiome and plant exudates and helps respective communities in a dynamically evolving environment. For example, root exudates discharge many carbon (C)-based compounds into the soil, which are then used by rhizospheric microorganisms to mobilize phosphorus (P) and nitrogen (N) and make them accessible to plants. Studies have also shown that root exudates play an essential role in the production of beneficial microbial communities near the root systems, which in turn provide defence to plants against both abiotic and biotic stress (Olanrewaju et al. 2019).

2.4 MICROBIAL SECONDARY METABOLITES: THEIR ROLE IN PLANT IMMUNE RESPONSES

Plants' defensive mechanisms are triggered in response to microbe-associated metabolites that serve as signaling molecules in stressful situations (Vimal et al. 2017). As Figure 2.1 illustrates, these mechanisms and signaling molecules include microbe-associated molecular patterns (MAMPs), pathogen-associated molecular patterns (PAMPs), peptidoglycans, lipopolysaccharides, and pattern recognition receptors (PRRs) (Boller and Felix 2009). Plants also produce signaling molecules like salicylic, jasmonic acid and ethylene to enable the plant defence pathways after detecting the unwanted microbes (Yi et al. 2014). Phenolic compounds provide immunity to plants against major phytopathogens including bacteria, fungi and oomycetes. In one such study, when pea plants inoculated with *B. subtilis* and *P. aeruginosa*, no disease symptoms were observed as opposed to mock treated plants with a substantial rise in phenolic compounds under treatment (Jain et al. 2015). In the shoots of inoculated pea plants, the development of salicylic acid, an essential marker for systemic acquired resistance (SAR) was also found to be increased (Jain et al. 2015). The benzoxazinoid metabolite (2, 4-dihydroxy-7-methoxy-2H-1, 4-benzoxazin-3(4H)-one) acts against pathogenic fungi and aphids, provides immunity to plants (Ahmad et al. 2011) and has been reported in many cereals including maize. This property is utilized by *Azospirillum* by modulating the benzoxazinoid

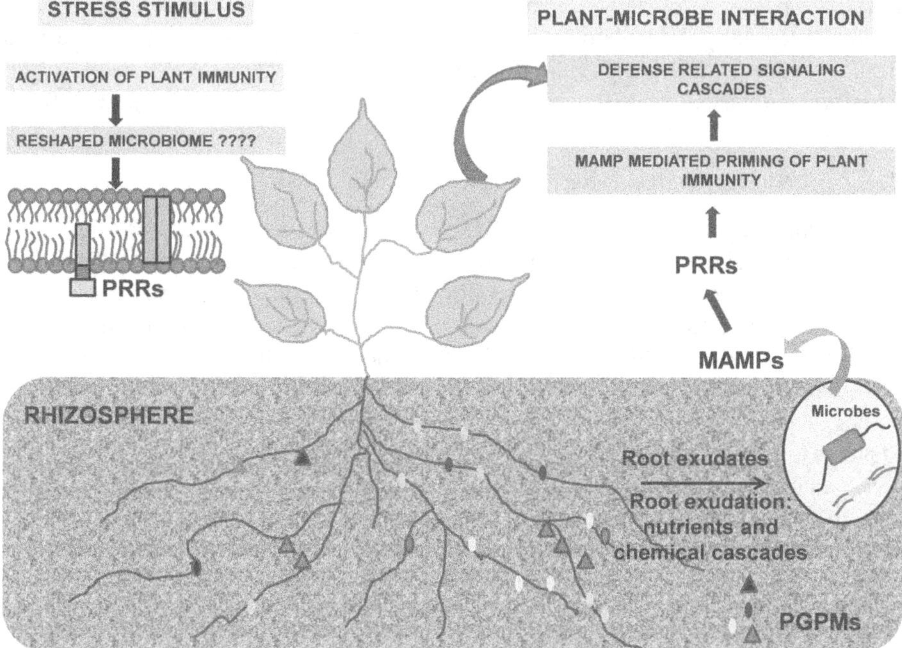

FIGURE 2.1 A pictorial representation of plant-microbe interaction against stress stimulus.

Abbreviations: PRRs, pattern recognition receptors; MAMPs, microbe-associated molecular patterns.

contents leading to better plant immunity after application (Chamam et al. 2013). *Bacillus simplex* in soybean roots enhanced the content of phytosphingosine, lactic acid, melibiose and noradrenaline-suppressing nematode penetration (Kang et al. 2020). Some microbial secondary metabolites, such as nod factors, AHLs, acetoin (bacterial VOCs), pyoverdine, DAPG (2,4-diacetylphloroglucinol) and polyamine amide, have a dual function and may boost plant growth while also increasing plant defence and immunity to abiotic and biotic stresses (Rosier et al. 2018).

2.4.1 Plant-Microbe Association in Biological Controls

Microorganisms are good candidate to be used as biocontrol agents because they provide an environmentally safe approach by providing defensive immunity against phytopathogens. The key modes of biocontrol activity in PGPR's are production of antibiotics (Raaijmakers and Mazzola 2012), iron chelating siderophores and multiple volatile compounds such as hydrogen cyanide. For instance, *Agrobacterium radiobacter* strain K84, a well-known biocontrol agent, secretes agrocin 84 (an antibiotic), which leads to control *Agrobacterium tumefaciens.* Another mode applied by beneficial microbes surrounding the plant root is that they outcompete harmful microorganisms for nutrients in the rhizospheric region restricting their colonization along the root's surface. Such competition can reduce the phytopathogen's ability to survive hence making it hard to spread. Few PGPR produced siderophores which starve the pathogen by sequestering much of the available Fe^{3+} in the rhizospheric region, eventually contributing to pathogen suppression. Siderophores derived from *Pseudomonas* have exhibited antifungal activity against *Sclerotium rolfsii, Aspergillus flavus, Aspergillus niger* and *Aspergillus oryzae* (Manwar et al. 2004).

2.4.2 Induced Systemic Resistance

Signal transduction mechanisms regulate plant-microbe interactions, allowing plants to prioritize defence pathway in response to stress (Dubey and Singh 2018). Via a signaling network mediated by the phytohormones such as jasmonic acid (JA) and salicylic acid (SA) molecular communication between the two organisms takes place. ISR signal transduction molecules ethylene and JA have been identified, and SA aggregation has been observed in systemic acquired resistance, instigated by pathogen attack (Van-Loon and Van-Strien 1999). PGPR activate a defense response in plants to reduce the harmful effects of phytopathogens via induced systemic resistance. Plants have been shown to develop systemic resistance to fungal, bacterial, and viral diseases when exposed to fluorescent Pseudomonads and other PGPRs. ISR is a network of organized signaling pathways controlled and primarily powered by plant hormones that exchange signaling elements that provides a high level of defense (Olanrewaju et al. 2017). Bacterial components like 2,4-diacetylphloroglucinol, CLPs, siderophores, and volatiles like 2,3-butanediol and acetoin can also activate ISR pathways in plants. Plants with a favorable PGPR in association will react more rapidly to pathogen infection by triggering plant defense response, as seen in the *Pseudomonas putida* 89 B27 strains, which have shown resistance to *Colletotrichum orbiculare* (Wei et al. 1991). The action of the *Pseudomonas fluorescens* strain WCS417r inhibited the development of *Fusarium oxysporum* f.sp. *dianthi*, which causes vascular wilts in Arabidopsis, and suppressed *Pseudomonas syringae* pv. *tomato* and *Alternaria brassicicola* with radish necrotic spots in Arabidopsis (Hoffland et al. 1996). ISR-mediated, PGPR-induced resistance has been linked to the PR (pathogenesis related) proteins accumulation and the activation of phenylpropanoid pathway genes. Similarly, Phenylalanine ammonia lyase (PAL) is an important enzyme for the biosynthesis of phytoalexins and phenolics. When *Solanum lycopersicum* and *Capsicum annuum* plants are inoculated with *Pseudomona fluorescence* in response to attack with *Fusarium oxysporum* f.sp. *lycopersici* and *Colletotrichum capsici*, ISR pathway is induced in these plants by the aggregation of PAL and PR proteins. ISR elicitation in sugar beet by *Bacillus pumilus* and *Bacillus mycoa* is linked to elevated peroxidase activity (Kloepper et al. 2004). In plants, endophytes can also mediate biotic stress tolerance by the synthesis of reactive oxygen species (ROS). In one such study, in endophyte-treated plants ROS production gets improved along with antioxidants such as phenols and flavonoids (Ogbe et al. 2020).

2.5 PLANT-MICROBIAL INTERACTION FOR NUTRIENT AVAILABILITY AND UPTAKE

Appropriate nutrient supply is essential for growth of both plants and microorganisms. Microbial associations that mediate nutrient availability and acquisition are affected by a variety of reactions in the rhizosphere zone, including physical and chemical reactions (Agrios 2005; Khan et al. 2015; Pankaj et al. 2019). All nutrients are dynamically sourced and sunk in rhizospheric soil (Nitrogen, Phosphorous, Potassium, Iron, and Sulfur) but different soil microorganisms use different pathways in soil environments to acquire, cycle, and ensure mineral supply to plants. Endophytic fungi have been shown to increase the acquisition of macronutrients such as potassium (K^+) and sulfur (S) (Kafle et al. 2018) in the model legume *Medicago truncatula*. K^+ deficiency is a common problem in soybeans and can cause stunted growth and decreased yields under drought stress.

Bacterial and fungal endophytes are found almost everywhere in vascular plants (Hardoim et al. 2015). There are wide range of endophytes, and it is not rare for certain plants to host various endophytic species as root endophytic fungi and Arbuscular mycorrhizal fungi (AMF) have been discovered in large numbers on colonized pine and oak roots (Chadha et al. 2014). In the relationship between host and endophytes; host selectivity, specificity, and preference often play an essential role. Endophytic fungi associate with plant roots in a variety of ways, depending on the environment and organisms. For example in tree species, the development of mantle (fungal sheath) and Hartig net (network of inward growing hyphae) is a way for fungal root endophytes and plants to associate. Endophytes *Phialocephala finlandica* (*C. finlandica*) create a rough Hartig net that is confined to the mantle and epidermis of birch roots (Wilcox and Wang 1987). Other instances of endophytic association include *Trichoderma*, *Piriformospora indica* and other dark septate endophytes (DSE). Endophytic plant growth promoting fungi (PGPF) are those that exist in the host plant roots and are helpful. These PGPFs are crop-specific due to the difference in environment and soil inconsistency and their significance on plants is studied in a limited way and needs to be explore further. PGPF may enhance plant growth via phosphate solubilization, auxin (IAA) and siderophore production (Malla et al. 2004). For example, *Trichoderma virens* synthesizes the compounds associated with auxin and enhances the growth of the *Arabidopsis thaliana* (Contreras-Cornejo et al. 2009). A major advancement of microbial-based mechanisms of nitrogen, phosphorus and iron uptake are discussed below.

2.5.1 Nitrogen Availability and Uptake

Nitrogen is an essential component for plant growth and health (Jetten 2008). While nitrogen is plentiful in gaseous form, it is not readily accessible to plants until it is converted to ammonia (Baas et al. 2014). Interactions between symbiotic and non-symbiotic organisms are primarily responsible for biological nitrogen fixation. Symbiotic microbes like *Frankia*, *Rhizobium*, and free-living soil bacteria such as *Azospirillum*, *Azotobacter*, *Cyanobacteria*, *Klebsiella* and *Clostridium* are examples of nitrogen-fixating microorganisms. In the symbiotic Rhizobium-legume association, the plant provides fixed carbon such as organic acids after photosynthesis, which the bacteria requires for multiplication and growth, while continuing to fix nitrogen in the nodules. Nodules are nitrogen-fixing symbiotic locations that develop after a string of encounters between rhizobium and leguminous plants. Most diazotrophic (endophytic) bacteria have the capacity to fix nitrogen. In addition, these bacterial endophytes associate with rhizobia bacteria and may strengthen soybean plant root nodulation (Subramanian et al. 2014).

It has been observed that microbial secondary metabolites play a role in nitrogen uptake and provide a defense against adverse conditions. For instance, GA1 and GA3 isolated from *Azospirillum brsilense* and *Azospirillum lipoferum* improved nitrogen uptake in stress conditions (Ghorai et al. 2020). Dual inoculation of a salt-tolerant bacterial endophyte and rhizobia has been shown to produce synergistic responses and improve soybean plant fitness when exposed to salt (Egamberdieva et al. 2016). Root exudates contain flavonoid compounds, which are produced by leguminous plants to fuel their growth and recruit rhizobial cells by inducing nod gene transcription, are essential signals in Rhizobium's symbiotic relationship. The insertion of Nod genes in stems, which are implicated in the synthesis/formation of lipo-chitooligosaccharide signals known as Nod factors, determines flavonoid concentration. Other compounds in exudates, in addition to flavonoids, play a significant role in deciding the microbe composition in the plant rhizosphere.

2.5.2 Phosphate Solubilization

After nitrogen, phosphorus (P) is the second most important macronutrient for plant growth promotion. Like nitrogen, phosphate is also found in immobilized form in the soil as inorganic and organic form (Khan et al. 2009). Because of its high-level reactivity with certain metal complexes such as Fe, Ca and Al which contribute to soil precipitation, phosphorus is not easily accessible to plants. Phosphate-solubilizing microbes solubilize insoluble phosphate forms by the release of mineral dissolving molecules or compounds, including siderophores, hydroxyl ions (OH^-), protons, organic acids and CO_2 (Rodriguez and Fraga 1999) resulting in a decrease of pH. Gluconic acid and 2-ketogluconic acid are among the major organic acids which are synthesized by phosphate-solubilizing microorganisms. These organic acids are responsible for the microbial cells acidification and the release of the phosphate ions in their surroundings. As a result, rhizosphere pH and phosphorous availability are inversely linked. The release of enzymes such as phytases, phosphatases, phosphonatases and CP lyases is the second process through which microbes increase phosphate supply. Finally, when the substrate degrades, phosphate is released and this process is known as biological phosphate mineralization (BPM). Henceforth, phosphate-solubilizing microorganisms (PSM) are fungi and bacteria that enable plants to mobilize phosphate in insoluble forms. *Bacillus* is one of the most important phosphate solubilizer bacteria among them. In addition, *Azotobacter, Rhizobium, Azospirillum, Mycobacterium, Agrobacterium, Achromobacterium, Erwinia, Enterobacter, Flavobacterium, Escherichia, Pseudomonas* and *Serratia* are very efficient at converting unavailable complex phosphate ions into accessible inorganic phosphate ions (Swarnalakshmi et al. 2020). *Penicillium, Trichoderma, Aspergillus* and *Rhizoctonia solani* are the major phosphate solubilizers among fungi (Jacobs et al. 2002).

The rock phosphate can also be solubilized by *Arthrobotrys oligospora* (a nematophagous fungus) (Duponnois et al. 2006). In return for reduced biomass, the other main mutualist groups in the phosphorus cycle, the mycorrhizal fungi, are major suppliers of phosphorus to plants. AMF improves plant phosphate nutrition by increasing phosphorus supply through their hyphae's large surface area and high-affinity absorption of phosphorus mechanisms (Begum et al. 2019). PSM is a biofertilizer that solubilizes fixed soil phosphate which results in improved crop yields. For instance, many observations have shown the importance of endophytic microorganisms as biocontrol agents and biofertilizers. The *Pantoea* sp., an endophyte from the peanut root nodule, for example, has been observed with possible solubilizing action (Yadav et al. 2018). Similar to organic phosphorus chelation and redox modifications, endophytic actinomycetes have been shown to play a significant role in the solubilization of phosphate and increasing its supply to plants (Singh and Dubey 2018). In another case study, two bacterial endophytic strains of *Acinetobacter* sp. ACMS25 as well as *Bacillus* sp. derived from the *Phyllanthus amarus* medicinal plant, PVMX4 also were implicated as PSM. The behavior of significant plant growth promotion, for *P. amarus* was found to be dependent on these two strains. It was observed that under *in vitro* salt stress conditions, these strains supported a higher vigor index, percent germination and plant biomass (Joe et al. 2016).

2.5.3 Iron Uptake

Iron (Fe) is a necessary micronutrient for virtually all organisms, including plants, fungi and bacteria, and belongs to the transition group of elements. While it aids photosynthesis, respiration and nitrogen fixation, plants and microbes find it difficult to attain sufficient quantities due to its immobilized nature in soil as insoluble ferric hydroxide or iron oxides. Alas, many times even in iron-rich soils, iron is a limiting factor for plant growth since its availability to organisms is restricted due to rapid ferrous oxidation (Fe^{2+}) to the insoluble ferric state (Fe^{3+}). In rhizospheric microorganisms, specific mechanisms for iron assimilation have evolved through the formation of low-molecular weight, Fe-chelating compounds or molecules identified as siderophores, which transfer iron to their cells. The siderophore complex in the bacterial membrane oxidizes Fe^{2+} to Fe^{3+}, which is then introduced into the cell by endophytes through a gating mechanism. Soluble metal concentration rises when siderophores adhere to the metal surface (Rajkumar et al. 2010). Plants use a variety of pathways to uptake iron from bacterial siderophores after heavy metal residues, such as iron chelates, have been removed to specifically absorb siderophore-Fe complexes (Schmidt 1999). In a wide range of bacterial genera, including *Aeromonas, Rhizobium*, *Pseudomonas, Serratia* and *Bacillus*, siderophore production has been cited. When the plant was inoculated with *Pseudomonas* strain GRP3 for iron feeding, the findings revealed that the plants have shown a decline in chlorotic symptoms after 45 days, while the content of chlorophyll was increased as compared to the mock or control strain (Sharma et al. 2003). There are some endophytic actinomyces such as *Streptomyces* sp. mhcr0816, *Streptomyces* sp. GMKU 3100, *Nocardia* sp. and *Streptomyces* sp. UKCW/B that have been observed to synthesize siderophores. Similarly, under nickel stress, *Streptomyces acidiscabies* E13 has been identified as a great siderophore producer that helps *Vigna unguiculata* to grow better (Sessitsch et al. 2013).

Siderophores have been found to be enabled both directly and indirectly in the improvement of plant health and productivity. Plants benefit from increased siderophore-mediated Fe supply and they select beneficial siderophore-producing microbes to colonize their roots in order to improve iron uptake as well as availability (Ahmed and Holmström 2014). Few microbial siderophores have been documented to alleviate iron deficiency-induced chlorosis in dicot (Yehuda et al. 2000). Plant-derived Fe-phytosiderophore complexes, on the other hand, tend to be a rich source of iron for bacteria. Therefore, plant or microbial origin Fe-chelating compounds are highly stable and soluble for a wide range of pH and are uniformly essential since they assist in the delivery of Fe to plants and soil as a nutrient (Marschner and Crowley 1998).

2.6 PHYTOHORMONE PRODUCTION AND REGULATION BY MICROBES IN RHIZOSPHERE

Phytohormones are produced by microorganisms residing in rhizospheric region within the root zone. In addition, they promote plant growth by enhancing the length and density of root hair. This growth in root hair expands the surface region, which enhances the plant's ability to retain water and mineral nutrients (Choi and Cho 2019). Many phytohormones are produced by microbes, viz. abscisic acid (ABA), auxins, cytokinins and gibberellins via secondary metabolite pathway. These hormones induce tolerance against various stresses including abiotic and biotic stress. Few such microbes with their respective hormone and other secondary metabolite production along with function are provided in Table 2.2. These rhizospheric microorganisms are also known as PGPR because they directly enhance the plant growth

TABLE 2.2
Various Types of Secondary Metabolites Secreted by a Diverse Group of Plant Beneficial Microbes and Their Role in Stress Mitigation

Name of the Secondary Metabolite	Producer Microorganism	Function	Reference
Abscisic acid	*Azospirillum* sp.	Acts against water stress; acts against drought stress	Casanovas et al. 2002; Cohen et al. 2009
Salicylic acid	*A. brasilense, P. aeruginosa, Burkholderia cepacia, Promicromonospora* sp., *Acinetobacter calcoaceticus*	Decreases the harsh effects of salinity; acts against biotic stress; acts against drought stress	Syeed and Khan 2010; Saikia et al. 2006; Tortora et al. 2011; Kang et al. 2014
Indole acetic acid	*P. aurantiaca, P. extrmorientalis, Azospirillum brasilense, B. amyloliquefaciens*	Acts against salt stress; Acts against drought stress	Egamberdieva 2009
Cytokinin	*B. subtilis, P. fluorescens*	Acts against drought stress; acts against biotic stress	Liu et al. 2013; Grobkinsky et al. 2016
GA4	*Serratia nematodiphila*	Acts against cold stress	Kang et al. 2015
Strigolactones	*Arbuscular mycorrhiza*	Acts against drought stress	Ruiz-Lozano et al. 2016
2R,3R-butanediol	*P. chlororaphis*	Acts against drought stress	Cho et al. 2008
Brefeldine A	*Phoma medicaginis*	Antibacterial activity	Weber et al. 2004
Phomoenamide	*Phomopsis* sp. PSU-D15	Antimycobacterial effect	Rukachaisirikul et al. 2008
Naphthoquinone spiroketal	*Chaetomium acuminate*	Allelochemical activity	Macías-Rubalcava et al. 2008
Naphthalene	*Muscodor vitigenus*	Use in insecticides antimicrobials, anti-helminthics and vermicides	Strobel et al. 2001
Altersetin	*Alternaria* spp.	Antibacterial effect	Hellwig et al. 2002
Aciphyllene, 2-butanone and 2-methyl furan	*Muscodor albus*	Antibiotic effect	Atmosukarto et al. 2005

and development. The manufacture of ACC deaminase and multiple compounds is another approach of plant growth stimulation by PGPR. The primary abiotic stress-related phytohormone is ABA whose level rises in response to environmental stimuli and triggers diverse signaling and gene expression pathways (Finkelstein 2013). It controls the expression of various stress responsive genes that prevent the stress-related agglutination of other proteins due to low-temperature stress (Verslues et al. 2006) and dehydrin vesicles associated with lipid-proteins that shield plants from various other abiotic stresses including drought, heat, and salt stress (Sreenivasulu et al. 2012). The ABA level was found to be increased by *Azospirillum* sp. in wheat and maize under drought stress by inducing stomata closure thus providing resistance against osmotic stress (Casanovas et al. 2002). The stomata closure and callose deposition in the cell wall of plants assisted by PGPR against phytopathogens are controlled by two different ABA- independent and dependent pathways (Cohen et al. 2015). The water content is maintained under drought stress in maize plants, when inoculated with *Azospirillum lipoferum* (Cohen et al. 2009).

Salicylic acid is mostly involved in biotic stress, but also plays a role under high salinity, resulting in an increase in osmoprotectants, sugar and proline and reduces the effects of salt stress. *Burkholderia cepacia*, *Promicromonospora* sp., and *Acinetobacter calcoaceticus* inoculated cucumber plants produced more salicylic acid and endogenous hormones in response to drought and salinity stress (Kang et al. 2014). Salicylic acid boosts the immune response of plants against pathogen invasion by inducing pathogen-related genes. Salicylic acid-induced resistance to the phytopathogen *Rhizoctonia solani* is augmented by *Pseudomonas aeruginosa* (Saikia et al. 2006). Similarly, inoculating *Cucumis sativus* with *Acinetobacter calcoaceticus* has been shown to dramatically increase endogenous gibberellic acid (GA) which mediate plant antioxidant status and reduce the negative effects of osmotic stress. Exogenous application of GAs boosts production of photosynthetic pigment, seed germination, and iron absorption in plants, as well as providing resilience to salt stress (Iqbal et al. 2011).

Auxins are vital components of plant growth and stress tolerance as they carry the capacity to modulate many stress responsive genes and intervene via crosstalk between different stress responses. *Pseudomonas extremorientalis* and *Pseudomonas aurantiaca* produce Indole Acetic Acid (IAA), thus alleviating seed dormancy and reducing salt stress in wheat. *Bacillus amyloliquefaciens* and *Azospirillum brasilense* reduced the negative effects on wheat caused by drought (Kasim et al. 2013). Similarly, cytokinins stimulate cell division, root initiation, root growth, root elongation and other physiological responses in plants. Inoculation with *Bacillus subtilis* increases cytokinins and ABA levels in *Platycladus orientalis*, which decreases the impact of water stress (Liu et al. 2013). *Pseudomonas syringae* infection can be controlled by *Pseudomonas fluorescens* in the Arabidopsis plant due to an increase cytokinin activity (Grobkinsky et al. 2016).

2.7 SECONDARY METABOLITES FROM MARINE MICROFLORA

Marine fungi, which coexist with marine flora and fauna, supposedly develop a wide variety of terpenes with striking structural diversity and useful pharmacological properties. Peribysins and phomactins have mostly been described as common marine terpenes extracted from fungi. Three novel pimarane diterpenes and diaporthin B were discovered in *Apostichopus japonicas* (a fungus associated with sea cucumber) (Deshmukh et al. 2018). A bisabolane type of aspergiterpenoid-A and sesquiterpenoid fermented by *Aspergillus* sp. was recovered from the *Xestospongia testudinaria* (Li et al. 2012). Also from *Aspergillus insuetus* (OY-207), three novel compounds, insuetolides A-C, (E)-6-(40-hydroxy-20-butenoyl)-strobilactone-A and meroterpenoids have been identified. Seven phenalenone derivatives were obtained from *Coniothyrium cereale* (an algal endophytic fungus) (Elsebai et al. 2011). Aspergillin A is an aromatic polyketide extracted from *Aspergillus versicolor.* Secondary metabolites including depsidones, diaryl ether, aspergillusidones A-C and aspergillus ether A have been isolated from *Aspergillus unguis.* The chlorogentisyl alcohol, an alcohol-containing class of secondary metabolite, was discovered in marine *Aspergillus* genus in the red alga *Hypnea saidana.* Kjer et al. (2010), on the other hand, provided a thorough description of the methods for isolating and cultivating fungi associated with different marine species (algae, sponges).

2.8 QUORUM SENSING

Mutual interactions of microbes in the rhizosphere determine the ultimate microbial population (intra- and interspecies). These interactions between microbes occur via multiple signaling molecules or compounds that allow bacteria to sense their environment. For bacterial cell-cell contact, the quorum sensing (QS) process is well-known. This procedure entails the development and identification of signaling compounds and molecules, which permit bacterial populations to jointly induce gene expression (Hawver et al. 2016). In multiple bacterial species with diverse molecular architectures, different types of quorum-sensing networks are present based on auto inducers or signaling molecules. In order to sense the ecological niche, spatial diffusion and population distribution, these metabolites are secreted inside the cell. Gram-negative bacteria have a lot of auto inducers of N-acyl-homoserine lactone (AHL), which promote a lot of plant-microbe interactions, whereas gram-positive bacteria only have QS signals. The plant-associated bacteria that synthesize AHL are *Sinorhizobium*, *Rhizobium*, *Pseudomonas*, *Pantoea*, *Agrobacterium*, *Xanthomonas* and *Erwinia*. Various bacteria may synthesize the same AHLs with similar structures and functions, implying population crosstalk, and it is apparent that QS through AHLs is more widespread among plant-related bacteria than the other non-plant-associated bacteria of soil (Elasri et al. 2001). AHL has been shown in many studies to promote root length or utilize AHL priming for increased resistance (Zhao et al. 2020). The application of a mixture of AHL molecules modulates the expression of many defense-related genes and imparts resistance to *Pseudomonas syringae*, a foliar pathogen (Shrestha et al. 2020). Previous studies have shown that resistance to salt stress has increased after the treatment with AHL like compound oxo-C6-HSL (Zhao et al. 2020). The cytoskeleton remodeling, aggregation of PR proteins, and phytohormone reaction were both altered in 3-oxo-C8-homoserine lactone (HSL) inoculated *Arabidopsis thaliana* (Miao et al. 2012). The inoculation of 3-oxo-C10-HSL also causes adventitious root development and up-regulation of auxin-responsive genes in *Vigna radiata* (Bai et al. 2012). Systemic resistance was induced in *A. thaliana* when infected with the *P. syringae* and *Golovinomyces orontii* through AHL molecules. When *A. thaliana* was infected with *P. syringae*, oxo-C14-HSL was induced, resulting in increased phenolics, cell wall lignification, callose deposition and salicylic acid formation (Schenk et al. 2012). Yet, the exact mechanism of AHL-induced resistance is still unknown.

2.9 CONCLUSION AND FUTURE PROSPECTS

This chapter throws light on the interaction of plant-associated microbes in the rhizosphere that encourages plant health and growth via a variety of mechanisms. Chemical signals play an important role in forming close relationships between plants and microbes in this interaction. The secondary metabolite from microbes and plant root exudates are the chemical agents that untangle this interaction in the rhizosphere. These interactions facilitate nutrient transfer (by fixing nitrogen, mobilizing and immobilizing micro and macro-nutrients), plant growth factors, phytopathogen thwarting and cross-talking between pathogenic and beneficial microorganisms, all of which enhance plant development, plant growth, soil fertility and crop yield. The importance of signaling molecules and quorum sensing in the rhizosphere, especially with regard to PGPMs, has drawn attention to the microbe-plant and microbe-microbe interaction-based research. A variety of factors has been established that initiate such communication in the rhizosphere, but there are still many characteristics that need to be studied to further explain plant-microbe associations at the environmental level for agricultural use. The advent of current "omics" technologies (transcriptomics, metabolomics, proteomics, etc.) will certainly improve the multi-trophic communication analysis. In such a scenario, a better understanding of the interactions between plants and microbes, as well as the interactions among microbes would open new vistas for the improvement of agriculture, ultimately offering a positive outlook for environmental stability.

ACKNOWLEDGMENT

The authors are thankful to the Director, CSIR–Institute of Himalayan Bioresource Technology, Palampur, for providing the necessary facilities. We acknowledge the financial support from project OLP0042 and MLP0168, funded by CSIR, India. Namo Dubey gratefully acknowledges the University Grant Commission (UGC), Government of India for providing this fellowship. The CSIR-IHBT publication number for this manuscript is 9567.

REFERENCES

Agrios, G.N. 2005. Plant pathology, *End of the English version,* 5th ed. Academic Press, New York, 922.

Ahmad, S., Veyrat, N., Gordon-Weeks, R. et al. 2011. Benzoxazinoid metabolites regulate innate immunity against aphids and fungi in maize. *Plant Physiology* 157(1):317–327.

Ahmed, E., Holmström, S.J. 2014. Siderophores in environmental research: Roles and applications. *Microbial Biotechnology* 7(3):196–208.

Aires, A., Mota, V.R., Saavedra, M.J. et al. 2009. Initial *in-vitro* evaluations of the antibacterial activities of glucosinolate enzymatic hydrolysis products against plant pathogenic bacteria. *Journal of Applied Microbiology* 106(6):2096–2105.

Akiyama, K., Matsuzaki, K.I., Hayashi, H. 2005. Plant sesquiterpenes induce hyphal branching in arbuscular mycorrhizal fungi. *Nature* 435(7043):824–827.

Atmosukarto, I., Castillo, U., Hess, W.M., Sears, J., Strobel, G. 2005. Isolation and characterization of *Muscodor albus* I-41.3s, a volatile antibiotic producing fungus. *Plant Science* 169(5):854–861.

Baas, P., Markewitz, D., Knoepp, J.D., Mohan, J.E. 2014. Nitrogen cycling heterogeneity: An approach for plot scale assessments. *Soil Science Society of American Journal* 78:S237–S247.

Badri, D.V., Vivanco, J.M. 2009. Regulation and function of root exudates. *Plant, Cell and Environment* 32(6):666–681.

Bai, M.Y., Fan, M., Oh, E., Wang, Z.Y. 2012. A triple helix–loop–helix/basic helix–loop–helix cascade controls cell elongation downstream of multiple hormonal and environmental signaling pathways in Arabidopsis. *Plant Cell* 24(12):4917–4929.

Begum, N., Qin, C., Ahanger, M.A. et al. 2019. Role of arbuscular mycorrhizal fungi in plant growth regulation: implications in abiotic stress tolerance. *Frontiers in Plant Science* 10:1068.

Beneduzi, A., Ambrosini, A., Passaglia, L.M. 2012. Plant growth-promoting rhizobacteria (PGPR): Their potential as antagonists and biocontrol agents. *Genetics and Molecular Biology* 35(4):1044–1051.

Berendsen, R.L., Pieterse, C.M., Bakker, P.A. 2012. The rhizosphere microbiome and plant health. *Trends in Plant Science* 17(8):478–486.

Boller, T., Felix, G. 2009. A renaissance of elicitors: Perception of microbe-associated molecular patterns and danger signals by pattern-recognition receptors. *Annual Review of Plant Biology* 60:379–406.

Braga, R.M., Dourado, M.N., Araújo, W.L. 2016. Microbial interactions: Ecology in a molecular perspective. *Brazilian Journal of Microbiology* 47:86–98.

Casanovas, E.M., Barassi, C.A., Sueldo, R.J. 2002. Azospiriflum inoculation mitigates water stress effects in maize seedlings. *Cereal Research Communications* 30(3):343–350.

Chadha, N., Mishra, M., Prasad, R., Varma, A. 2014. Endophytic fungi: Research update. *Journal of Biology and Life Science* 5(2):135–158.

Chamam, A., Sanguin, H., Bellvert, F. et al. 2013. Plant secondary metabolite profiling evidences strain-dependent effect in the Azospirillum–*Oryza sativa* association. *Phytochemistry* 87:65–77.

Cho, S.M., Kang, B.R., Han, S.H. et al. 2008. 2R, 3R-butanediol, a bacterial volatile produced by *Pseudomonas chlororaphis* O6, is involved in induction of systemic tolerance to drought in *Arabidopsis thaliana. Molecular Plant-Microbe Interactions* 21(8):1067–1075.

Choi, H.S., Cho, H.T. 2019. Root hairs enhance Arabidopsis seedling survival upon soil disruption. *Scientific Reports* 9(1):1–10.

Cohen, A.C., Bottini, R., Pontin, M., Berli, F.J. et al. 2015. *Azospirillum brasilense* ameliorates the response of *Arabidopsis thaliana* to drought mainly via enhancement of ABA levels. *Physiologia Plantarum* 153(1):79–90.

Cohen, A.C., Travaglia, C.N., Bottini, R., Piccoli, P.N. 2009. Participation of abscisic acid and gibberellins produced by endophytic *Azospirillum* in the alleviation of drought effects in maize. *Botany* 87(5):455–462.

Contreras-Cornejo, H.A., Macías-Rodríguez, L., Cortés-Penagos, C., López-Bucio, J. 2009. *Trichoderma virens*, a plant beneficial fungus, enhances biomass production and promotes lateral root growth through an auxin-dependent mechanism in arabidopsis. *Plant Physiology* 149(3):1579–1592.

Curl, E. A., Truelove, B. 2012. *The rhizosphere. Advanced Series in Agricultural Science*, Springer-Verlag, Vol. 15, 288.

De León, L., López, M.R., Moujir, L. 2010. Antibacterial properties of zeylasterone, a triterpenoid isolated from *Maytenus blepharodes*, against *Staphylococcus aureus. Microbiological Research* 165(8):617–626.

Demain, A.L., Fang, A. 2000. The natural functions of secondary metabolites. *Advances in Biochemical Engineering/Biotechnology* 69:1–39.

Deshmukh, S.K., Prakash, V., Ranjan, N. 2018. Marine fungi: A source of potential anticancer compounds. *Frontiers in Microbiology* 8:1–24.

Dubey, N., Singh, K. 2018. Role of NBS-LRR proteins in plant defense. *Molecular Aspects of Plant-Pathogen Interaction*, Springer-Nature, 115–138.

Duponnois, R., Kisa, M., Plenchette, C. 2006. Phosphate-solubilizing potential of the nematophagous fungus *Arthrobotrys oligospora*. *Journal of Plant Nutrition and Soil Science* 169(2):280–282.

Egamberdieva, D. 2009. Alleviation of salt stress by plant growth regulators and IAA producing bacteria in wheat. *Acta Physiologiae Plantarum* 31(4):861–864.

Egamberdieva, D., Jabborova, D., Berg, G. 2016. Synergistic interactions between *Bradyrhizobium japonicum* and the endophyte *Stenotrophomonas rhizophila* and their effects on growth, and nodulation of soybean under salt stress. *Plant and Soil* 405(1):35–45.

Elasri, M., Sandrine, D., Philippe, L. et al. 2001. Acyl-homoserine lactone production is more common among plant-associated *Pseudomonas* spp. than among soilborne *Pseudomonas* spp. *Applied and Environmental Microbiology* 67(3):1198–1209.

Elsebai, M.F., Kehraus, S., Lindequist, U. et al. 2011. Antimicrobial phenalenone derivatives from the marine-derived fungus *Coniothyrium cereale*. *Organic and Biomolecular Chemistry* 9(3):802–808.

Finkelstein, R. 2013. Abscisic acid synthesis and response. *The Arabidopsis Book* 11:e0166.

Ghorai, A.K., Patsa, R., Jash S., Dutta, S. 2020. Microbial secondary metabolites and their role in stress management of plants. *Biocontrol Agents and Secondary Metabolites*, 1st ed. Elsevier Science, 283–319.

Gonzalez, J.B., Fernandez, F.J., Tomasini, A. 2003. Microbial secondary metabolites production and strain improvement. *Indian Journal of Biotechnology* 2(3):322–333.

Grobkinsky, D.K., Tafner, R., Moreno V.M. et al. 2016. Cytokinin production by *Pseudomonas fluorescens* G20-18 determines biocontrol activity against *Pseudomonas syringae* in Arabidopsis. *Scientific Reports* 6(1):1–11.

Hacquard, S., Spaepen, S., Garrido-Oter, R., Schulze-Lefert, P. 2017. Interplay between innate immunity and the plant microbiota. *Annual Review of Phytopathology* 55:565–589.

Hardoim, P. R., Van Overbeek, L. S., Berg, G., Pirttilä, A. M., Compant, S., Campisano, A., Döring, M., Sessitsch, A. 2015. The hidden world within plants: ecological and evolutionary considerations for defining functioning of microbial endophytes. *Microbiology and Molecular Biology Reviews* 79(3): 293–320.

Hawver, L.A., Jung, S.A., Ng, W.L. 2016. Specificity and complexity in bacterial quorum-sensing systems. *FEMS Microbiology Reviews* 40(5):738–752.

Hellwig, V., Grothe, T., Mayer-Bartschmid, A.N.K.E. et al. 2002. Altersetin, a new antibiotic from cultures of endophytic *Alternaria* spp.: Taxonomy, fermentation, isolation, structure elucidation and biological activities. *Journal of Antibiotics* 55(10):881–892.

Hiltner, L. 1904. About recent experiences and problems in the field of soil bacteriology and with special attention to the foundation and broche. *Arbeitskräfte Deutschland Landwirtschaftliche Gesamtfläche* 8: 59–78.

Hoffland, E., Hakulinen, J., Van Pelt, J.A. 1996. Comparison of systemic resistance induced by avirulent and nonpathogenic *Pseudomonas* species. *Phytopathology* 86(7):757–762.

Iqbal, N., Rahat, N., Iqbal, M. et al. 2011. Role of gibberellins in regulation of source–sink relations under optimal and limiting environmental conditions. *Current Science* 100(7): 998–1007.

Jacobs, H., Boswell, G.P., Ritz, K., Davidson, F.A., Gadd, G.M. 2002. Solubilization of calcium phosphate as a consequence of carbon translocation by *Rhizoctonia solani*. *FEMS Microbiology Ecology* 40(1):65–71.

Jacoby, R., Peukert, M., Succurro, A., Koprivova, A., Kopriva, S. 2017. The role of soil microorganisms in plant mineral nutrition—current knowledge and future directions. *Frontiers in Plant Science* 8:1617.

Jacoby, R.P., Koprivova, A., Kopriva, S. 2021. Pinpointing secondary metabolites that shape the composition and function of the plant microbiome. *Journal of Experimental Botany* 72(1):57–69.

Jain, A., Singh, A., Singh, S., Singh, H.B. 2015. Phenols enhancement effect of microbial consortium in pea plants restrains *Sclerotinia sclerotiorum*. *Biological Control* 89:23–32.

Jetten, M.S. 2008. The microbial nitrogen cycle. *Environmental Microbiology* 10 (11):2903–2909.

Joe, M.M., Devaraj, S., Benson, A., Sa, T. 2016. Isolation of phosphate solubilizing endophytic bacteria from *Phyllanthus amarus* Schum & Thonn: Evaluation of plant growth promotion and antioxidant activity under salt stress. *Journal of Applied Research on Medicinal and Aromatic Plants* 3(2):71–77.

Johns, N.I., Blazejewski, T., Gomes, A.L., Wang, H.H. 2016. Principles for designing synthetic microbial communities. *Current Opinion in Microbiology* 31:146–153.

Kafle, A., Garcia, K., Peta, V., Yakha, J., Soupir, A., Bücking, H. 2018. Beneficial plant microbe interactions and their effect on nutrient uptake, yield, and stress resistance of soybeans. In *Soybean-Biomass, Yield and Productivity,* IntechOpen, London, UK.

Kang, S.M., Khan, A.L., Waqas, M., You, Y.H. et al. 2015. Gibberellin-producing *Serratia nematodiphila* PEJ1011 ameliorates low temperature stress in *Capsicum annuum* (L.). *European Journal of Soil Biology* 68:85–93.

Kang, S.M., Khan, A.L., Waqas, M., You, Y.H., et al. 2014. Plant growth-promoting rhizobacteria reduce adverse effects of salinity and osmotic stress by regulating phytohormones and antioxidants in *Cucumis sativus*. *Journal of Plant Interactions* 9(1):673–682.

Kang, W.S., Chen, L.J., Wang, Y.Y. et al. 2020. *Bacillus simplex* treatment promotes soybean defence against soybean cyst nematodes: A metabolomics study using GC-MS. *PloS One* 15(8):e0237194.

Kasim, W.A., Osman, M.E., Omar, M.N., Abd El-Daim, I.A., Bejai, S., Meijer, J. 2013. Control of drought stress in Wheat using plant growth-promoting bacteria. *Journal of Plant Growth Regulation* 32(1):122–130.

Khan, K., Pankaj, U., Verma, S.K., Gupta, A.K., Singh, R.P., Verma, R.K. 2015. Bioinoculants and vermicompost influence on yield, quality of *Andrographis paniculata*, and soil properties. *Industrial Crops and Products* 70:404–409.

Khan, M. S., Zaidi, A., Wani, P. A. 2009. Role of phosphate solubilizing microorganisms in sustainable agriculture-a review. *Sustainable agriculture*, Springer-Dordrecht, 551–570.

Kjer, J., Debbab, A., Aly, A.H., Proksch, P. 2010. Methods for isolation of marine-derived endophytic fungi and their bioactive secondary products. *Nature Protocols* 5(3):479–490.

Kloepper, J.W., Ryu, C.M., Zhang, S. 2004. Induced systemic resistance and promotion of plant growth by *Bacillus* spp. *Phytopathology* 94(11):1259–1266.

Kossel A. 1891. Archives of analytical physiology. *Physiology Abteilung, Berlin University Laboratory, Germany* 181–186.

Li, D., Xu, Y., Shao, C.L., Yang, R.Y. et al. 2012. Antibacterial bisabolane-type sesquiterpenoids from the sponge-derived fungus *Aspergillus* sp. *Marine Drugs* 10(12):234–241.

Linares, J.F., Gustafsson, I., Baquero, F., Martinez, J.L. 2006. Antibiotics as intermicrobial signaling agents instead of weapons. *Proceedings of the National Academy of Sciences* 103(51):19484–19489.

Liu, F., Xing, S., Ma, H., Du, Z., Ma, B. 2013. Cytokinin-producing, plant growth-promoting rhizobacteria that confer resistance to drought stress in *Platycladus orientalis* container seedlings. *Applied Microbiology and Biotechnology* 97(20):9155–9164.

Lundberg, D.S., Lebeis, S.L., Paredes, S.H. et al. 2012. Defining the core Arabidopsis thaliana root microbiome. *Nature* 488(7409):86–90.

Macías-Rubalcava, M.L., Hernández-Bautista, B.E., Jiménez-Estrada, M. et al. 2008. Naphthoquinone spiroketal with allelochemical activity from the newly discovered endophytic fungus *Edenia gomezpompae*. *Phytochemistry* 69(5):1185–1196.

Malla, R., Prasad, R., Kumari, R. et al. 2004. Phosphorus solubilizing symbiotic fungus: *Piriformospora indica. Endocytobiosis Cell Research* 15(2):579–600.

Manwar, A.V., Khandelwal, S.R., Chaudhari, B.L., Meyer, J.M., Chincholkar, S.B. 2004. Siderophore production by a marine *Pseudomonas aeruginosa* and its antagonistic action against phytopathogenic fungi. *Applied Biochemistry and Biotechnology* 118(1–3):243–252.

Marschner, P., Crowley, D.E. 1998. Phytosiderophores decrease iron stress and pyoverdine production of *Pseudomonas fluorescens* PF-5 (pvd-inaZ). *Soil Biology and Biochemistry* 30(10–11):1275–1280.

Miao, C., Liu, F., Zhao, Q., Jia, Z., Song, S. 2012. A proteomic analysis of *Arabidopsis thaliana* seedling responses to 3-oxo-octanoyl-homoserine lactone, a bacterial quorum-sensing signal. *Biochemical and Biophysical Research Communications* 427(2):293–298.

Neal, A.L., Ahmad, S., Gordon-Weeks, R., Ton, J. 2012. Benzoxazinoids in root exudates of maize attract *Pseudomonas putida* to the rhizosphere. *PLoS One* 7(4):35498.

Ogbe, A.A., Finnie, J.F., Van Staden, J. 2020. The role of endophytes in secondary metabolites accumulation in medicinal plants under abiotic stress. *South African Journal of Botany* 134:126–134.

Olanrewaju, O. S., Glick, B. R., Babalola, O. O. 2017. Mechanisms of action of plant growth promoting bacteria. *World Journal of Microbiology and Biotechnology* 33(11): 1–16.

Olanrewaju, O.S., Ayangbenro, A.S., Glick, B.R., Babalola, O.O. 2019. Plant health: Feedback effect of root exudates-rhizobiome interactions. *Applied Microbiology and Biotechnology* 103(3):1155–1166.

Pang, Z., Chen, J., Wang, T., Gao, C., Li, Z., Guo, L., Xu, J., Cheng, Y. 2021. Linking plant secondary metabolites and plant microbiomes: A Review. *Frontiers in Plant Science* 12:300.

Pankaj, U., Singh, D.N., Mishra, P., Gaur, P., Vivekbabu, C.S., Shanker, K., Verma, R.K. 2020. Autochthonous halotolerant plant growth promoting rhizobacteria promote bacoside A yield of *Bacopa monnieri* (L) Nash and phytoextraction of salt-affected soil. *Pedosphere* 30(5): 671–683.

Pankaj, U., Singh, D.N., Singh, G., Verma, R.K. 2019. Microbial inoculants assisted growth of *Chrysopogon zizanioides* promotes phytoremediation of salt affected soil. *Indian Journal of Microbiology* 59(2):137–146.

Raaijmakers, J.M., Mazzola, M. 2012. Diversity and natural functions of antibiotics produced by beneficial and plant pathogenic bacteria. *Annual Review of Phytopathology* 50:403–424.

Rajkumar, M., Ae, N., Prasad, M.N.V., Freitas, H. 2010. Potential of siderophore-producing bacteria for improving heavy metal phytoextraction. *Trends in Biotechnology* 28(3):142–149.

Rodriguez, H., Fraga, R. 1999. Phosphate solubilizing bacteria and their role in plant growth promotion. *Biotechnology Advances* 17(4–5):319–339.

Rogers, E.E., Glazebrook, J., Ausubel, F.M. 1996. Mode of action of the *Arabidopsis thaliana* phytoalexin camalexin and its role in Arabidopsis–pathogen interactions. *Molecular Plant-Microbe Interactions* 9(8):748.

Rosier, A., Medeiros, F.H.V., Bais, H.P. 2018. Defining plant growth promoting rhizobacteria molecular and biochemical networks in beneficial plant–microbe interactions. *Plant and Soil* 428(1–2):35–55.

Ruiz-Lozano, J.M., Aroca, R., Zamarreño, Á.M., et al. 2016. Arbuscular mycorrhizal symbiosis induces strigolactone biosynthesis under drought and improves drought tolerance in lettuce and tomato. *Plant, Cell & Environment* 39(2):441–452.

Rukachaisirikul, V., Sommart, U., Phongpaichit, S., Sakayaroj, J., Kirtikara, K. 2008. Metabolites from the endophytic fungus *Phomopsis* sp. PSU-D15. *Phytochemistry* 69(3):783–787.

Saikia, R., Kumar, R., Arora, D.K., Gogoi, D.K., Azad, P. 2006. *Pseudomonas aeruginosa* inducing rice resistance against *Rhizoctonia solani*: Production of salicylic acid and peroxidases. *Folia Microbiologica* 51(5):375–380.

Sasse, J., Martinoia, E., Northen, T. 2018. Feed your friends: Do plant exudates shape the root microbiome? *Trends in Plant Science* 23(1):25–41.

Schenk, S.T., Stein, E., Kogel, K.H., Schikora, A. 2012. Arabidopsis growth and defense are modulated by bacterial quorum sensing molecules. *Plant Signaling and Behavior* 7(2):178–181.

Schmidt, W. 1999. Mechanisms and regulation of reduction-based iron uptake in plants. *New Phytologist* 141(1):1–26.

Sessitsch, A., Kuffner, M., Kidd, P., Vangronsveld, J., Wenzel, W.W., Fallmann, K., Puschenreiter, M. 2013.The role of plant-associated bacteria in the mobilization and phytoextraction of trace elements in contaminated soils. *Soil Biology and Biochemistry* 60:182–194.

Sharma, A.; Johri, B.N., Sharma, A.K., Glick, B.R. 2003. Plant growth-promoting bacterium *Pseudomonas* sp. strain GRP3 influences iron acquisition in mung bean (*Vigna radiata* L. Wilzeck). *Soil Biology and Biochemistry*, 35(7):887–894.

Shrestha, A., Grimm, M., Ojiro, I., Krumwiede, J., Schikora, A. 2020. Impact of quorum sensing molecules on plant growth and immune system. *Frontiers in Microbiology* 11:1545.

Singh, R., Dubey, A.K. 2018. Diversity and applications of endophytic actinobacteria of plants in special and other ecological niches. *Frontiers in Microbiology* 9:1767.

Sreenivasulu, N., Harshavardhan, V.T., Govind, G., Seiler, C., Kohli, A. 2012. Contrapuntal role of ABA: Does it mediate stress tolerance or plant growth retardation under long-term drought stress? *Gene* 506(2):265–273.

Strobel, G.A., Sears, J., Dirkse, E., Markworth, C. 2001. Volatile antimicrobials from *Muscodoralbus*, a novel endophytic fungus. *Microbiology* 147(11):2943–2950.

Subramanian, P., Kim, K., Krishnamoorthy, R., Sundaram, S., Sa, T. 2014. Endophytic bacteria improve nodule function and plant nitrogen in soybean on co-inoculation with *Bradyrhizobium japonicum* MN110. *Plant Growth Regulation* 76(3):327–332.

Swarnalakshmi, K., Yadav, V., Tyagi, D., Dhar, D.W., Kannepalli, A., Kumar, S. 2020. Significance of plant growth promoting rhizobacteria in grain legumes: Growth promotion and crop production. *Plants* 9(11):1596.

Syeed, S., Khan, N.A. 2010. Physiological aspects of salicylic acid-mediated salinity tolerance in plants. *Plant Stress* 1:39–46.

Tortora, M.L., Díaz-Ricci, J.C., Pedraza, R.O. 2011. *Azospirillum brasilense* siderophores with antifungal activity against *Colletotrichum acutatum. Archives of Microbiology* 193(4):275–286.

Tyc, O., Song, C., Dickschat, J.S., Vos, M., Garbeva, P. 2017. The ecological role of volatile and soluble secondary metabolites produced by soil bacteria. *Trends in Microbiology* 25(4):280–292.

Van-Loon, L.C., Van-Strien, E.A. 1999. The families of pathogenesis-related proteins, their activities, and comparative analysis of PR-1 type proteins. *Physiological and Molecular Plant Pathology* 55(2):85–97.

Verslues, P.E., Agarwal, M., Katiyar-Agarwal, S., Zhu, J., Zhu, J.K. 2006. Methods and concepts in quantifying resistance to drought, salt and freezing, abiotic stresses that affect plant water status. *Plant Journal* 45(4):523–539.

Vimal, S.R., Singh, J.S., Arora, N.K., Singh, S. 2017. Soil–plant–microbe interactions in stressed agriculture management: A review. *Pedosphere* 27(2):177–192.

Weber, R.W.S., Stenger, E., Meffert, A., Hahn, M. 2004. Brefeldin A production by *Phoma medicaginis* in dead pre-colonized plant tissue: a strategy for habitat conquest? *Mycological Research* 108(6):662–671.

Wei, G., Kloepper, J.W., Tuzun, S. 1991. Induction of systemic resistance of cucumber to *Colletotrichum orbiculare* by select strains of plant growth-promoting rhizobacteria. *Phytopathology* 81(12):1508.

Wilcox, H.E. and Wang, C.J.K. 1987. Ectomycorrhizal and ectendomycorrhizal associations of *Phialophora finlandia* with *Pinus resinosa*, *Picea rubens*, and *Betula alleghaniensis*. *Canadian Journal of Forest Research* 17(8):976–990.

Yadav, A.N. 2018. Biodiversity and biotechnological applications of host-specific endophytic fungi for sustainable agriculture and allied sectors. *Acta Scientific Microbiology* 1:1–5.

Yang, L., Ding, W., Xu, Y., Wu, D., Li, S., Chen, J., Guo, B. 2016. New insights into the antibacterial activity of hydroxycoumarins against *Ralstonia solanacearum*. *Molecules* 21(4):468.

Yehuda, Z., Shenker, M., Hadar, Y., Chen, Y. 2000. Remedy of chlorosis induced by iron deficiency in plants with the fungal siderophore rhizoferrin. *Journal of Plant Nutrition* 23 (11–12):1991–2006.

Yi, S.Y., Shirasu, K., Moon, J.S., Lee, S.G., Kwon, S.Y. 2014. The activated SA and JA signaling pathways have an influence on flg22-triggered oxidative burst and callose deposition. *PloS One* 9(2): e88951.

Zhao, Q., Yang, X.-Y., Li, Y., Liu, F., Cao, X.-Y., Jia, Z.H., Song, S.S. 2020. N-3-oxo-hexanoyl-homoserine lactone, a bacterial quorum sensing signal, enhances salt tolerance in Arabidopsis and wheat. *Botanical Studies* 61(1):1–12.

3 Biochar-Microbe Interactions

An Integrated Mitigation Approach for Emerging Pollutant Management

Pratibha Tripathi and Puja Khare
CSIR, Lucknow, India

CONTENTS

3.1 INTRODUCTION

Exponential enhanced global urbanization and industrialization have increased the exploitation of natural resources, extravagant discharge of environmental contaminants such as heavy metals, by-products of disinfectants, polycyclic aromatic hydrocarbons (PAHs), greenhouse gases (GHGs), etc., lead to air, water, and soil pollution. Several remediation techniques and incineration methods have been applied to tackle the contaminants but, due to their inefficiency and non-feasibility, a sustainable

DOI: 10.1201/9781003147091-3

cost-effective method is still needed (Bucheli-Witschel and Egli 2001; Yousaf et al. 2016; Harindintwali et al. 2020). Bioremediation is a biological approach driven by the effective use of microbes, plants, and enzymes. For a few decades, bioremediation has emerged as potential cost-effective mitigation strategy for the mitigation of several environmental contaminants (Gavrilescu et al. 2015; Tripathi et al. 2020). But several biological systems-related problems such as environmental conditions, retarded growth of exogeneous microbe due to nutritional deficiency, and rhizosphere incompetence have led to the restricted performance of exogeneous microbe in contaminated soil (Chuaphasuk and Prapagdee 2019). Conventional carrier materials such as vermiculite and peat often have limitations such as unavailability and rapid decline, and the unearthing of these materials could have adverse environmental effects (Herrmann and Lesueur 2013; Wu et al. 2019). Hence, large-scale production and application of these exogeneous microorganisms outside the laboratory are the biggest obstacles to their successful implementation. Therefore, a suitable eco-friendly carrier material is required for better survival and distribution of inoculums with broad-spectrum usage.

3.2 BIOCHAR

Biochar is a carbon-rich, solid, porous material prepared through pyrolysis (temperature range 250–700°C) of biomass (agriculture waste, manure, wood, etc.) under limited oxygen conditions (Yadav et al. 2019; Harindintwali et al. 2020; Nigam et al. 2021). Employment of agro-waste for biochar preparation and its subsequent use as soil fertilizer/conditioner is found as a sustainable solution of disposal than direct burning in open land. Soil amendment with biochar is well known for increasing soil fertility through improved bulk density, cation exchange capacity (CEC), nutrient retention, water-holding capacity, carbon sequestration, and soil microbial activity, which ultimately leads to plant growth promotion (Yousaf et al. 2016; Palansooriya et al. 2019; Borgohain et al. 2020; Jain et al. 2020; Pankaj 2020). The alkaline nature of biochar helps in acid buffering and its highly porous structure is a major characteristic that could be utilized as a solution to many agriculture problems such as the presence of heavy metals, GHGs, nutrient depletion, etc. These characteristics of biochar provide a favorable environment for better plant growth under contaminated soil conditions. The role of biochar prepared at various pyrolysis temperatures in the removal of emerging contaminants has been shown in Table. 3.1.

Biochar has been reported to be rich in nutrient contents such as C, P, N, K, etc., and acts as a slow-release fertilizer; hence, it is beneficial for long-term microbial survival (Masiello et al. 2013; Ye et al. 2019). Moreover, biochar enhances soil microbial activities and act as a biostimulant to the indigenous bacterial population (Palansooriya et al. 2019; Ma et al. 2020a). Accordingly, biochar can be used as a suitable carrier material for microbial inoculums with enhanced bioremediation potential (Chen et al. 2014; Wu et al. 2019). As soil pH plays a major role in modification of the indigenous microbial community structure and due to the difference in pH sensitivities of microbiota (bacteria and fungi) variable responses of the native soil microbial community have been expected toward biochar-induced changes in soil pH (Zhu et al. 2017). The performance of biochar for microbial or chemical

TABLE 3.1
Role of Biochar in the Removal of Emerging Pollutants

Biochar Type	Pyrolysis Temperature (°C)	Pollutant	Effect	References
Bamboo and rice straw	≥ 500	Cd, Cu, Pb, and Zn	Minimized heavy metal Cd uptake in *Sedum plumbizincicola*	Lu et al. 2014
Sugarcane bagasse	700	Cd, Pb, and Zn	Reduced accumulation of Cd, Pb, and Zn in soil and plant tissue	Roy et al. 2014
Coconut shell	—	Cd and Zn	Reduced concentration of Cd and Zn in willow twigs	Břendova et al. 2015
Water hyacinth	450	Cd and Pd	Reduced exchangeable Cd and Pb with increasing dose	Yin et al. 2016
Eucalyptus saligna wood	450	Cd	Reduced concentration of bioavailable Cd in wheat	Yousaf et al. 2016
Tobacco stalk biochar	—	Cd and Pb	Limited mobility and bioavailability of Cd and Pb in calcareous soil	Cheng et al. 2018
Peanut shell and Wheat straw	350–500	Cd and Pb	Decreased heavy metal concentration in rice grains	Xu et al. 2018
Wood, sludge, manure, crop residue, biowaste	400–500	Nitrogen cycling	Ameliorated soil physicochemical properties and alteration of soil microbial functional genes associated with N cycling	Xiao et al. 2019
Rice straw	300, 500, 700, and 900	Cu, Zn, Cd, and Pb	Minimized uptake of heavy metal in plant tissue	Yang et al. 2020
Rice straw	600	PAHs	Effectively removed PAHs from coking plant soil	Zhang et al. 2020
Eucalyptus wood	400	GHG emission	Improved soil C sequestration with alleviated net greenhouse gases	Puga et al. 2020
Rice straw	—	Methane (CH_4)	Decreased emission of methane from paddy soil	Qi et al. 2021

transformation depends on certain environmental conditions like the type of feedstock, pyrolysis conditions, pH of soil and biochar, and texture and CEC of soil (Abit et al. 2012; Wei et al. 2019; Yadav et al. 2019). Studies have shown that biochar prepared at a low temperature retains low C:N ratio and its early mineralization mediates modifications in soil microbial community structure. Moreover, the recalcitrant nature of biochar is another potential factor for its long-lasting impact on agriculture soil. Aging is also an important factor that influence biochar-microbe interactions and their influence in soil remediation. These variable biochar properties largely affect biochar-microbe interactions and are a potential reason for the lack of understanding of its properties and uses. This chapter deals with the composite use of biochar and microbes for pollutant remediation and the benefits of this interaction on plants grown in contaminated soils. A comparative benefit of biochar over conventional carrier materials is shown in Figure 3.1.

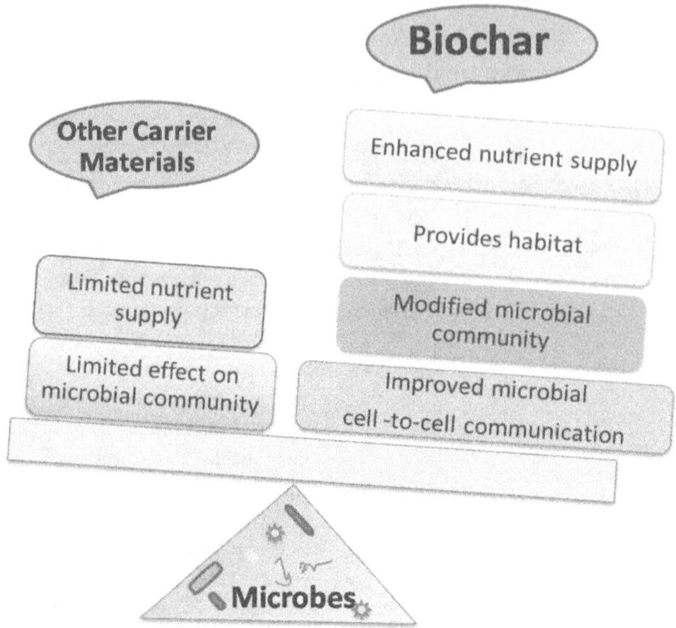

FIGURE 3.1 Advantages of biochar over conventional carrier materials of microbial inoculants.

3.3 EMERGING POLLUTANTS AND THEIR MITIGATION STRATEGIES ADOPTED

Ever-increasing industrial and agricultural growth has generated many environmental issues, such as the release of various toxic gases and chemicals that lead to water, air, and soil pollution. The presence of such contaminants in the environment negatively affects us in different ways and the lack of clean water and healthy soil has an adverse impact on crop production (Margenat et al. 2017; Jain et al. 2020). Researchers have developed many mitigation strategies to remediate pollutants for safe food crop production to meet global food demand without compromising human health (Gavrilescu et al. 2015; Yousaf et al. 2016; Tripathi et al. 2020). The utilization of biochar has been proved as an efficient remediation strategy for many environmental pollutants such as heavy metals, PAHs, and emission of GHGs, etc. (Zhu et al. 2017; Cheng et al. 2018; Yadav et al. 2019; Harindintwali et al. 2020; Nigam et al. 2021). Biochar reduces the mobility of many inorganic and organic contaminants through immobilization and adsorption methods. Biochar exhibits high surface area to volume ratio and the presence of active functional groups (–COOH and –OH) on its surface show positive affinity toward hazardous and persistent contaminants (heavy metals, dioxins, and PAHs) led to their dissipation (Jain et al. 2020; Zhao et al. 2020; Nigam et al. 2021). But efficiency of biochar largely depends on its preparative conditions (pyrolysis temperature, feed stock type, aging, etc.), and due to variable biochar characteristics (nutrient content, active functional groups, etc.) (Wei et al. 2019) its impact on soil remediation is inconsistent and unpredictable.

A brief discussion of emerging environmental pollutants and their remediation strategies adopted to date is given below.

3.3.1 Greenhouse Gases

Today, the reduction of GHGs emission from agricultural soil is a challenging issue and there is a need to check conventional farming methods (Buragienė et al. 2019). Agriculture activities have a major role in global GHG emissions, with CO_2 (5–20%), CH_4 (15–30%), and N_2O (80–90%) of the atmosphere derives annually from the agricultural soils. Soil properties like C and N content and C/N ratio have the most important role in CH_4 and N_2O emissions. Biochar application in soil posing elevated N, low C, and C/N ratio revealed higher N_2O emissions and generally decreased CH_4 emissions. Despite this, it can offset GHGs emissions as the biochar preparation procedure locks up the biogenic C content and subsequently delays the discharge of that C back to the atmosphere (Ashiq et al. 2020). Biochar has an efficient influence on the soil microbial functionality and diversity of microbial taxa that are responsible for CO_2 production. In soil, an intermediate short duration raised emission of CO_2 has been observed after biochar amendment, which could be due to enhanced mineralization of indigenous soil organic content or due to biotic or abiotic factors in the mediated release of biochar particles. In a recent study, biochar decreased the emission of methane in a flooded paddy soil for two seasons (Qi et al. 2021). The application of biochar improves soil NH_4^+ retention capacity probably due to sorption of nitrogen into its large pores, which facilitates reduced leaching of NO_3^- and volatilization of ammonia (Xiao et al. 2019; Ashiq et al. 2020). Recently, Puga et al. (2020) reported that biochar amendment ameliorates soil C sequestration and can also be instrumental in mitigation of net GHGs. Hence, biochar can be used as a soil amendment and has been found as a potential strategy to alleviate climate change by limiting the emission of GHGs and carbon sequestration.

3.3.2 Organic Contaminants

Environmental contamination with organic pollutants such as pesticides, pharmaceuticals, petrochemicals, PAHs, poly-chlorinated biphenyls, etc., is a global issue, and the search for efficient strategies for the remediation of such contaminant-polluted sites are therefore of the utmost importance. The conventional methods used to eliminate these organic xenobiotics include physical (such as evaporation and adsorption), chemical (such as chemical oxidation and photochemical degradation), and biological techniques (such as biodegradation) (Qiao et al. 2020; Tripathi et al. 2021). The high cost and risk of secondary pollution caused by physical and chemical methods has resulted in these techniques being not workable. Hence, there is a need for an efficient, realistic, and workable mitigating biological method. In a recent study of Zhang et al. (2020) removal efficiency of 12 types of biochars was evaluated for remediation of highly PAH-polluted soil from a coking plant. The study revealed that rice straw biochars prepared at relatively high pyrolysis temperature posed high mineral nutrients and thus were recommended for high PAH-contaminated soils (Zhang et al. 2020).

3.3.3 Heavy Metals

Heavy metals/metalloids such as As, B, Cd, Cr, Cu, Si, and Zn in abundance are catergorized as hazardous metals, and their high concentration may have detrimental effects on normal plant growth (Tripathi et al. 2020; Das et al. 2021). The toxicity and bioavailability of heavy metals largely depend on their chemical forms and soil characteristics, mainly soil pH. Numerous physical, chemical. and biological mitigation strategies have been adopted to date for heavy metal contaminated soil remediation and among them all chemical extraction is highly efficient but is costly and non-biodegradable, and has the threat of secondary pollutant generation (Bucheli-Witschel and Egli 2001; Harindintwali et al. 2020). Among biological methods phytoremediation is revealed to have multifarious advantages over other remediation strategies but it is generally time-consuming (Harindintwali et al. 2020). Thus, the quest for an efficient, rapid, cost-effective, and sustainable remediation technique is of importance. Several recent studies of the underlying mechanisms utilized by biochar for heavy metals stabilization or immobilization in the soil have been undertaken (Břendova et al. 2015; Andrey et al. 2019; Wu et al. 2019; Harindintwali et al. 2020). Biochar can either directly absorb heavy metals through various processes (as electrostatic interactions, reduction, precipitation, complexation, and cation exchange capacity (CEC)) or biochar application can indirectly enhance soil metal retention capacity by altering soil properties, microbial activities, and soil organic carbon (He et al. 2019). Biochar preparation properties (raw material, pyrolysis temperature, aging, etc.) are highly responsible for its efficiency in soil heavy metal stabilization (Abit et al. 2012; Borgohain et al. 2020). Grass-based biochars are more promising than wood-based biochars in decreasing heavy metal leaching and their bioaccumulation in plant tissues (O'Connor et al. 2018; Wang et al. 2020). Moreover, straw-based biochar has a relatively poor heavy metal adsorption potential than manure-based biochar, due to its lower alkaline nature and lesser ash content (Xiong et al. 2017; Wei et al. 2019). Pyrolysis temperature also plays a significant role in metal immobilization, as biochars produced at lower temperatures were revealed to be less efficient as compared to those at higher temperatures (Zhu et al. 2017; Wei et al. 2019). Nonetheless, other factors such as soil properties, native microbial population, application rate, mixing depth, and even environmental surroundings may also have significant impact on the immobilization or stabilization of heavy metal in the soil (O'Connor et al. 2018).

3.4 BIOCHAR-MICROBE INTERACTIONS

Biochar present properties that are required to be a suitable carrier for microbes such as huge surface area, nutrients, pore size, and high affinity to bacteria (Hale et al. 2014; Wu et al. 2019; Xiao et al. 2019). Owing to these characteristics, biochar has been reported as a suitable carrier material for many PGP microbes (Hale et al. 2014; Wu et al. 2019). Several recent studies showed that biochar microbe composite showed a positive synergy and efficiency in remediation of contaminated soil (Wu et al 2019; Qiao et al. 2020; Zhao et al. 2020). Wu et al. (2019) stated that active functional groups on PGP bacteria and biochar are responsible for successful immobilization and synergistic interaction. Recently, an efficient PAHs-degrading bacterial consortium was developed by immobilizing magnetic-floating biochar gel beads (Qiao et al. 2020). Moreover, biochar amendment alleviates soil pH levels and possibly acts as an additive to soil microbes resulting in improved biodegradation (Yousaf et al. 2016; Kätterer et al. 2019; Jain et al. 2020). Later on, free radical formation and electron transfer between microbial cell and biochar has been reported to catalyze microbial degradation and transformation of emerging contaminants (Zhu et al. 2017). However, biochar-microbe interaction and their efficiency for contaminants remediation are largely dependent on biochar properties. The potential reasons and mechanisms underlying beneficial biochar-microbe interactions as proposed by some researchers are listed below and shown in Figure 3.2.

3.4.1 Biochar Acts as a Microbial Shelter

Biochar is assumed to act as a shelter for microorganisms and provide more space per unit volume than soil (Quilliam et al. 2013). Live microbial cells adhere to the large surface areas of biochar and acquire shelter as a habitat (Abit et al. 2012). Differential colonization patterns in bacterial cells and fungal hyphae have been observed in biochar's external and internal pores (Quilliam et al. 2013). This heterogeneous trend of microbial colonization on the pores and surface of biochar could be due to limited nutrient availability as compared to the soil, the presence of natural organic matter, or toxic substances (heavy metals, PAHs, etc.) in biochar pores that create an obstacle for microbes to adhere (Quilliam et al. 2013; Zhu et al. 2017).

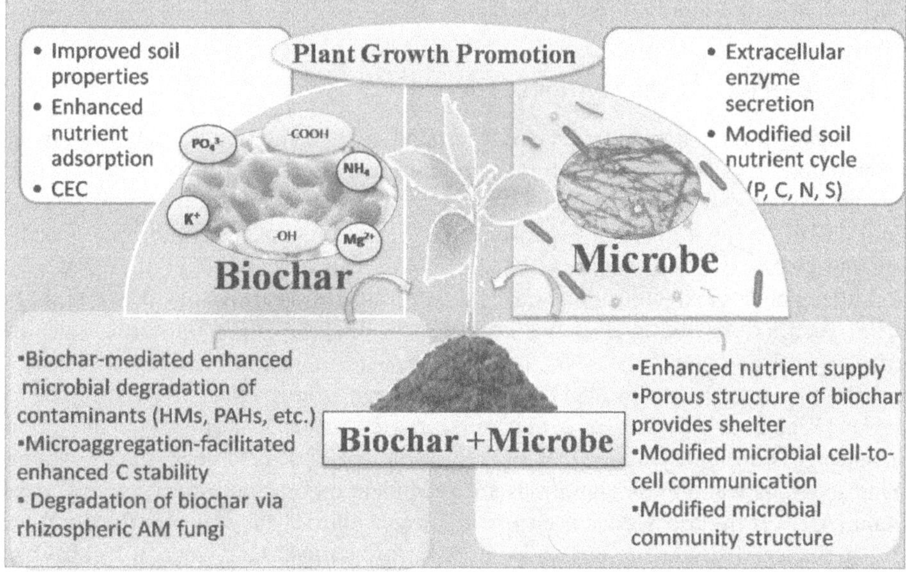

FIGURE 3.2 Biochar-microbe interaction mechanisms in soil.

Moreover, the biochar age is regarded as one of the crucial factors that affect the duration of microbial colonization (both for bacterial cells and fungal hyphae), and adjustment in the aging period enhanced the colonization on the pores and surface of biochar (Quilliam et al. 2013). For example, during the co-culture of *Geobacter metallireducens* and *Methano sarcinabarkeri*, bacterial cells were observed to colonize on biocharpores and surfaces in 20 days (Zhang et al. 2018). Similarly, cells of *Geobacter metallireducens* and *G. sulfurreducens* were found to colonize on biochar surface within 10 days (Chen et al. 2014). In a recent study, coconut shell biochar-*Bacillus* composite improved soil bacterial and fungal population as compared to their individual and control treatments, which might be due to the supply of nutrients to and habitat for microbes (Ma et al. 2020a).

3.4.2 Biochar Provides Nutrients to Microbes

Biochar retains many nutrients (C, P, K, Mg, Na, N, etc.), which release with variable rates into the soils so it can act as a slow-release fertilizer and provide prolonged benefits for microbial survival (Zhu et al. 2017; Ye et al. 2019). Biochar amendments improve soil CEC, which is an important parameter of soil capacity for cationic nutrients and provides higher nutrients to promote microbial functionality and decrease nutrient loss via leaching (Lehmann, 2011). Biochar absorbs and provides nutrition to soil microbes through the nutrient cations and inorganic anions sorption to active groups (–COOH and –OH) of its surface (Masiello et al. 2013; Zhu et al. 2017). The type of feedstock and pyrolysis temperature are the key parameters in determining the nutrient content of biochar (Masiello et al. 2013; Zhu et al. 2017; Borgohain et al. 2020; Gurtler et al. 2020). Biochars prepared from crop residues (husk, straw, etc.) and manure are commonly rich in ash content as compared to biochar of wood and thus can provides more nutrients (Akhter et al. 2015).

3.4.3 Biochar Alters Soil Properties

Soil physio-chemical properties play a crucial role in the determination of plant growth and productivity. Biochar amendment modulates soil properties such as pH, cation exchange capacity, nutrient retention, and water-holding capacity (Yousaf et al. 2016). Biochar also improves soil NH_4^+ retention capacity, limits NO_3^- leaching and metals mobility due to its strong binding capacity (Ashiq et al. 2020). The soil micro-ecological environment can be determined by various biochemical properties such as enzymatic activities because soil enzymes have important roles in nutrient cycling (Zhu et al. 2017). Biochar alleviates native microbial communities structure as revealed from altered gene abundance of microbially driven N and C cycles (Xiao et al. 2019). Xiao et al (2019) reported that biochar-mediated alteration in soil pH from highly acidic ($pH < 5$) to acidic (pH 5.5–6.5) improved N-cycling gene abundance and these alterations are dependent on biochar properties and cover plant. All soil microbes do not have similar sensitivities for soil pH (Rousk et al. 2010). Bacterial strains are known as more sensitive toward a change of soil pH but are more tolerant to a restricted pH range than fungal strains (Rousk et al. 2010). Consequently, bacteria and fungi could have variable responses to biochar-mediated differences in soil pH and could probably result in an overall modified microbial population due to differential reaction of the fungal and bacterial microbial population toward biochar-induced soil pH changes. In other report, enhanced microbial and enzymatic activities were observed after the combined application of *Bacillus* sp. and coconut shell biochar than control treatments (Ma et al. 2020a).

3.4.4 Biochar Alters Microbial Cell-to-Cell Signaling

The AHL (N-acyl-homoserine lactone) is an intercellular signaling molecule that is utilized by several gram-negative bacterial isolates in soil for intraspecific communication and regulation of gene expression (Masiello et al. 2013). Gao et al. (2016) reported that biochar alters the cell-to-cell communication of microorganisms through either sorption or enhanced hydrolysis of signaling molecules such as AHL (Gao et al. 2016). The mechanism of sorption is largely influenced by pyrolysis temperature and affects the adsorption potential of biochar (Masiello et al. 2013; Wei et al. 2019). Biochar prepared at elevated temperature (700°C) poses a comparative larger specific surface area than that prepared at low temperature (300°C) and thus could adsorb signaling molecules and interrupt the intercellular signal transfer to a greater extent than low-temperature prepared biochars (Masiello et al. 2013). Soil pH also affects biochar efficiency to increase or decrease the signal transfer that regulates specific soil microbial functionality (Zhu et al. 2017). Moreover, few bacterial communication signaling molecules (like AHL) are pH-sensitive; hence, hydrolysis of AHLs induced by biochar via enhanced soil pH decreased the bioavailable AHLs, which resulted in inactivation or reduced microbial intracellular communication (Gao et al. 2016). However, the less pH sensitivity of many fungal communication signals (such as farnesol, a fungal autoinducer) caused a shift in the fungi to bacteria ratio due to biochar application and probably leads to the alteration in the soil microbial population structure (Gao et al. 2016).

3.4.5 Biochar Alters the Toxic Effect of the Pollutant to Microbes

The biochar soil amendment can lower the toxic effects of soil pollutants to the microbial soil population (Koltowski et al. 2017). Toxicity of various potential soil pollutants such as heavy metals (As, Zn, Cu, Cd, etc.) and organic pollutants (phenols, PAHs, etc.) can be altered through biochar immobilization followed by their decreased bioavailability, which not only decreases soil pollutant toxicity to microbes but also enhances microbial population (Harindintwali et al. 2020). Biochar could also enhance the biotransformation and biodegradation of contaminants through sorption, electron transfer, and free radicals (Zhu et al. 2017; Wu et al. 2019; Harindintwali et al. 2020). The alkalinity of peanut biochar (pyrolyzed at 550°C and 750°C) can reduce the pH shock on *Pseudomonas citronellolis* by neutralizing the rapid acidic intermediates production of phenol (Zhao et al. 2020). Koltowski et al. (2017) reported that the use of willow biochar decreased the microbe mortality in soil contaminated with heavy metals and PAHs and also lowered the leachate toxicity to the gram- negative bacterial strain *Vibrio fischeri*.

3.5 BIOCHAR-MICROBE INTERACTIONS: IMPACT ON EMERGING CONTAMINANTS

The microbial adherence on surfaces and pores of biochar largely depends on its physiological characteristics and properties (Zhu et al. 2017). The combination of biochar-microbes could increase the contaminant's biodegradation as most of the co-inoculation studies revealed a synergistic interaction between them. Under contaminated conditions, biochar reduces the toxicity of pollutants on microorganisms via reducing their concentration and increasing bio-sorption, whereas tolerant microbes improve the pollutant adsorption potential of biochar (Talha et al. 2018). Hence, biochar-bacteria combination may remunerate for their particular flaws and thus attain efficient use of both materials in pollutant remediation. In a batch experiment enhanced biodegradation of congo red dye was reported by a biochar-bacteria composite prepared through immobilization of *Brevibacillus parabrevis*, a gram-positive bacteria, in biochar prepared from coconut shell (Talha et al. 2018). Zhao et al. (2020) reported stimulated phenol degradation by *Pseudomonas citronellolis* in the presence of biochar, as compared to the calcium alginate immobilization system. Ma et al. (2020a) showed that biochar and immobilized *Klebsiella* sp. formulation application in soil improves both bacterial richness and diversity. Moreover, biochar not only alleviates the bacterial diversity but also acts as a biostimulant for some indigenous polyacrylamide-reducing bacterial taxa. In another relevant study, significant removal of cadmium and arsenic in broth was observed through a composite of modified magnetic biochar (Fe_3O_4 loaded) and *Bacillus* sp. K1 immobilized on alginate beads (Wang et al. 2020). Because of the adverse chemical properties, adsorption or precipitation of As(III) with other cations (i.e. Cd(II)) is impossible for biochar alone at the same time, due to controlling pH or Eh. This combination of magnetic biochar and bacteria poses additional biosorption sites ($-NH_2$ and –OH groups) on the biochar surface and thus reveals enhanced removal of heavy metals than magnetic biochar alone. Reduced Cd and Pb mobility and bioavailability in calcareous soil after biochar amendment was recorded because of enhanced soil nutrients, alleviated bacterial diversity, and community structure (Cheng et al. 2018). The use of biochar with *Bacillus subtilis* strain as a co-sorbent revealed more efficiency than the biochar application alone (Wang et al. 2018). These studies revealed that during adverse environmental conditions biochar acts as a protectant for microbes. However, not much research has been conducted on the effect of biochar-microbe interactions on plants grown in contaminated soils; hence, the impact of biochar-microbe interactions on plants cultivated in contaminated soils conditions is not well understood. Recently, Andrey et al. (2019) have shown that the combined use of biochar and bacteria reduces the heavy metal content in all the *Hordeum vulgare* tissues to a greater extent than single-use of either biochar or bacteria, but the exact mechanism behind this advantageous synergistic biochar-bacteria interaction in soil remediation is still largely unanswered. Several biochar-microbe composites used for bioremediation in emerging contaminants are provided in Table 3.2.

Despite the benefits of biochar for microbes, several significant negative impacts of biochar on the remediation of organic pollutants should not be ignored. Biochar contains some microbial growth-suppressing compounds such as benzene (the main product of biochar preparation during charcoal burning), methoxyphenols and

TABLE 3.2
Role of Biochar-Microbe Interaction in Remediation of Emerging Pollutants

Microorganism Used	Biochar Feedstock Type	Pyrolysis Temperature (°C)	Effect	References
Pseudomonas putida and an unidentified indigenous bacterium	Wood chip, bamboo leaf, orange peel, and pine needle	100, 300, 400, and 700	Biochar produced at 400°C enhanced bio-dissipation of polyaromatic hydrocarbons (PAHs)	Chen et al. 2012
Enterobacter cloacae	Pinewood	300	Enhanced survival of *Enterobacter cloacae*	Hale et al. 2014
Mycobacterium gilvum	Rice straw, sewage sludge, and pig manure	500	Enhanced biodegradation of PAHs	Xiong et al. 2017
Brevibacillus parabrevis	Coconut shell	550	Improved bioremediation of Congo red dye	Talha et al. 2018
Bacillus subtilis	Pulverized cotton stalk	400 and 600	Enhanced survival of bacterial inoculants	Tao et al. 2018
UV-mutant *Bacillus subtilis*	Different stocks	—	Improved removal of heavy metal from aqueous solution	Wang et al. 2018
Bacillus sp.	Pinewood	—	Significant reduced heavy metal accumulation in plant tissue	Andrey et al. 2019
Serratia marcescens	Wheat straw	—	Improved Cd accumulation of *Chrysopogon. zizanioides*	Wu et al. 2019
Bacillus sp.	Coconut shell	800	Improved biomass and soil microbial activity with decreased Cd accumulation in ryegrass	Ma et al. 2020a
Klebsiella sp.	Coconut shell	500	Accelerated polyacrylamide biodegradation	Ma et al. 2020a
Pseudomonas frederiksbergensis	Maize straw	400	Decreased bioavailability of both Cd and Cu with enhanced soil microbial activities	Tu et al. (2020)
Consortia of marine microbes	Magnetic floating biochar	—	Efficient removal potential of high molecular weight PAHs	Qiao et al. 2020
Bacillus sp.	Modified rice straw (Fe_3O_4 loaded)	500	Efficient removal of Cd(II) and As(III) under high contaminated aqueous solution	Wang et al. 2020
Pseudomonas citronellolis	Peanut shells	350, 550, and 750	Accelerated phenol degradation	Zhao et al. 2020

phenols (pyrolysis products of lignin and hemicelluloses), carboxylic acids, ketones, furans (sorbed volatile organic compounds on biochar), and PAHs. Although in freshly prepared biochar VOCs act as a C source and supports the growth of certain microbes e.g. *Bacillus mucilaginosus* (Sun et al. 2015), at elevated concentrations VOCs exhibited substantial toxicity to microbes (Zhu et al. 2017). As discussed earlier, biochar-mediated biodegradation is highly dependent on its various physicochemical properties, which affect the microbial activities and soil properties (Masiello et al. 2013; Sun et al. 2015; Gao et al. 2016; Palansooriya et al. 2019). The strong affinity between biochar and organic contaminants could highly affect their bio-availability to microbes (Zhao et al. 2020). The addition of biochar revealed a significant decrease (90%) in 2,20,4,40-tetra bromodiphenyl ether degradation due to probable slow desorption of this organic pollutant (Xin et al. 2013). Gurtler et al. (2020) reported that biochar generated via fast pyrolysis at 600°C has more biocidal activity against *E. coli* than biochar generated at low temperatures. Also, the aging of biochar (for 2 years) was found to reduce its biocidal activity. Moreover, biochar with high pH levels, rich in ash content, and toxic residues (heavy metals, etc.) can have an adverse impact on soil microbial populations and plants used for phytoremediation (Harindintwali et al. 2020). A gradual increase in the pyrolysis temperature revealed low mobility with enhanced stability of heavy metals in biochar (Yang et al. 2020).

These findings revealed that feedstock type and pyrolysis program (temperature, hold time, heating rate) is the pioneer parameter that decides the functional active groups and potential of biochar to alleviate soil cation exchange capacity (Abit et al. 2012). Thus, biochar effectiveness is highly reliant on pyrolysis temperature, pH levels, and/or age. Moreover, as per biochar toxicity standards, content of total contaminants (As, Cd, Pb, etc.) in biochar should be below the permissible threshold limit as recommended by International Biochar Initiative (2013). Hence, considering the suitability of biochar determination of the above parameters is crucial to achieving a beneficial biochar-microbe composite with desired environmental benefits for practical application.

3.6 CONCLUSION AND FUTURE PROSPECTS

The present chapter focused on biochar-microbe interactions and their effects on environmental pollution remediation. Accordingly, we have listed the key physicochemical properties of biochar that influence microbes, interactions among microbes with various kinds of biochar, and their combined effect on contaminant degradation. Thus, a well-planned biochar-microbe biocomposite could be considered as a efficient low-cost strategy for the sustainable remediation of contaminated soil in practical application. However, more advanced metagenomic studies of soil-microbial genes will be crucial in order to discover the underlying mechanisms of biochar-microbe interactions that connect biochar characteristics with microbial activities.

ACKNOWLEDGMENTS

The authors are thankful to the Director, CSIR-CIMAP for his support and encouragement. PT acknowledges ICMR (Indian Council of Medical Research, New Delhi) for financial support in the form of a Research Associate Fellowship (File No. 45/8/2019/MP/BMS).

REFERENCES

Abit, S.M., Bolster, C.H., Cai, P. and Walker, S.L., 2012. Influence of feedstock and pyrolysis temperature of biochar amendments on transport of *Escherichia coli* in saturated and unsaturated soil. *Environmental Science & Technology*, *46*(15), pp.8097–8105.

Akhter, A., Hage-Ahmed, K., Soja, G. and Steinkellner, S., 2015. Compost and biochar alter mycorrhization, tomato root exudation, and development of *Fusarium oxysporum* f.sp. *lycopersici*. *Frontiers in Plant Science*, *6*, p.529.

Andrey, G., Rajput, V., Tatiana, M., Saglara, M., Svetlana, S., Igor, K., Grigoryeva, T.V., Vasily, C., Iraida, A., Vladislav, Z. and Elena, F., 2019. The role of biochar-microbe interaction in alleviating heavy metal toxicity in *Hordeum vulgare* L. grown in highly polluted soils. *Applied Geochemistry*, *104*, pp.93–101.

Borgohain, A., Konwar, K., Buragohain, D., Varghese, S., Dutta, A.K., Paul, R.K., Khare, P. and Karak, T., 2020. Temperature effect on biochar produced from tea (*Camellia sinensis* L.) pruning litters: A comprehensive treatise on physico-chemical and statistical approaches. *Bioresource Technology*, *318*, p.124023.

Břendová, K., Tlustoš, P. and Száková, J., 2015. Biochar immobilizes cadmium and zinc and improves phytoextraction potential of willow plants on extremely contaminated soil. *Plant, Soil and Environment*, *61*(7), pp.303–308.

Bucheli-Witschel, M. and Egli, T., 2001. Environmental fate and microbial degradation of aminopolycarboxylic acids. *FEMS Microbiology Reviews*, *25*(1), pp.69–106.

Buragienė, S., Šarauskis, E., Romaneckas, K., Adamavičienė, A., Kriaučiūnienė, Z., Avižienytė, D., Marozas, V. and Naujokienė, V., 2019. Relationship between CO_2 emissions and soil properties of differently tilled soils. *Science of the Total Environment*, *662*, pp.786–795.

Chen, S., Rotaru, A.E., Shrestha, P.M., Malvankar, N.S., Liu, F., Fan, W., Nevin, K.P. and Lovley, D.R., 2012. Promoting interspecies electron transfer with biochar. *Scientific Reports*, *4*, p.5019.

Cheng, J., Li, Y., Gao, W., Chen, Y., Pan, W., Lee, X. and Tang, Y., 2018. Effects of biochar on Cd and Pb mobility and microbial community composition in a calcareous soil planted with tobacco. *Biology and Fertility of Soils*, *54*(3), pp.373–383.

Chuaphasuk, C. and Prapagdee, B., 2019. Effects of biochar-immobilized bacteria on phytoremediation of cadmium-polluted soil. *Environmental Science and Pollution Research*, *26*(23), pp.23679–23688.

Das, P., Khare, P., Singh, R.P., Yadav, V., Tripathi, P., 2021. Arsenic-induced differential expression of oxidative stress and secondary metabolite content in two genotypes of *Andrographis paniculata*. *Journal of Hazardous Materials*, p.124302. doi:10.1016/j.jhazmat.2020.124302.

Gao, X., Cheng, H.Y., Del Valle, I., Liu, S., Masiello, C.A. and Silberg, J.J., 2016. Charcoal disrupts soil microbial communication through a combination of signal sorption and hydrolysis. *Acs Omega*, *1*(2), pp.226–233.

Gavrilescu, M., Demnerová, K., Aamand, J., Agathos, S. and Fava, F., 2015. Emerging pollutants in the environment: Present and future challenges in biomonitoring, ecological risks and bioremediation. *New Biotechnology*, *32*(1), pp.147–156.

Gurtler, J.B., Mullen, C.A., Boateng, A.A., Masek, O., Camp, M.J., 2020. Biocidal activity of fast pyrolysis biochar against *Escherichia coli* O157: H7 in soil varies based on production temperature or age of biochar. *Journal of Food Protection*, *83*(6), pp.1020–1029.

Hale, L., Luth, M., Kenney, R. and Crowley, D., 2014. Evaluation of pinewood biochar as a carrier of bacterial strain *Enterobacter cloacae* UW5 for soil inoculation. *Applied Soil Ecology*, *84*, pp.192–199.

Harindintwali, J.D., Zhou, J., Yang, W., Gu, Q. and Yu, X., 2020. Biochar-bacteria-plant partnerships: Eco-solutions for tackling heavy metal pollution. *Ecotoxicology and Environmental Safety*, *204*, p.111020.

He, M., Shen, H., Li, Z., Wang, L., Wang, F., Zhao, K., Liu, X., Wendroth, O. and Xu, J., 2019. Ten-year regional monitoring of soil-rice grain contamination by heavy metals with implications for target remediation and food safety. *Environmental Pollution, 244*, pp.431–439.

Herrmann, L. and Lesueur, D., 2013. Challenges of formulation and quality of biofertilizers for successful inoculation. *Applied Microbiology and Biotechnology, 97*(20), pp.8859–8873.

International Biochar Initiative, 2013. Standardized product definition and product testing guidelines for biochar that is used in soil. *IBI Biochar Stand.* https://www.biochar-international.org/wp-content/uploads/2018/04/IBI_Biochar_Standards_V1.1.pdf.

Jain, S., Khare, P., Mishra, D., Shanker, K., Singh, P., Singh, R.P., Das, P., Yadav, R., Saikia, B.K. and Baruah, B.P., 2020. Biochar aided aromatic grass [*Cymbopogon martini* (Roxb.) Wats.] vegetation: A sustainable method for stabilization of highly acidic mine waste. *Journal of Hazardous Materials, 390*, p.121799.

Kätterer, T., Roobroeck, D., Andrén, O., Kimutai, G., Karltun, E., Kirchmann, H., Nyberg, G., Vanlauwe, B. and de Nowina, K.R., 2019. Biochar addition persistently increased soil fertility and yields in maize-soybean rotations over 10 years in sub-humid regions of Kenya. *Field Crops Research, 235*, pp.18–26.

Kołtowski, M., Charmas, B., Skubiszewska-Zięba, J. and Oleszczuk, P., 2017. Effect of biochar activation by different methods on toxicity of soil contaminated by industrial activity. *Ecotoxicology and Environmental Safety, 136*, pp.119–125.

Lehmann, J., Rillig, M.C., Thies, J., Masiello, C.A., Hockaday, W.C. and Crowley, D., 2011. Biochar effects on soil biota–a review. *Soil Biology and Biochemistry, 43*(9), pp.1812–1836.

Lu, K., Yang, X., Shen, J., Robinson, B., Huang, H., Liu, D., Bolan, N., Pei, J. and Wang, H., 2014. Effect of bamboo and rice straw biochars on the bioavailability of Cd, Cu, Pb and Zn to *Sedum plumbizincicola. Agriculture, Ecosystems and Environment, 191*, pp.124–132.

Ma, H., Wei, M., Wang, Z., Hou, S., Li, X. and Xu, H., 2020a. Bioremediation of cadmium polluted soil using a novel cadmium immobilizing plant growth promotion strain *Bacillus* sp. TZ5 loaded on biochar. *Journal of Hazardous Materials, 388*, p.122065.

Ma, L., Hu, T., Liu, Y., Liu, J., Wang, Y., Wang, P., Zhou, J., Chen, M., Yang, B. and Li, L., 2020b. Combination of biochar and immobilized bacteria accelerates polyacrylamide biodegradation in soil by both bio-augmentation and bio-stimulation strategies. *Journal of Hazardous Materials*, p.124086.

Margenat, A., Matamoros, V., Díez, S., Cañameras, N., Comas, J. and Bayona, J.M., 2017. Occurrence of chemical contaminants in peri-urban agricultural irrigation waters and assessment of their phytotoxicity and crop productivity. *Science of The Total Environment, 599*, pp.1140–1148.

Masiello, C.A., Chen, Y., Gao, X., Liu, S., Cheng, H.Y., Bennett, M.R., Rudgers, J.A., Wagner, D.S., Zygourakis, K. and Silberg, J.J., 2013. Biochar and microbial signaling: Production conditions determine effects on microbial communication. *Environmental Science and Technology, 47*(20), pp.11496–11503.

Nigam, N., Khare, P., Ahsan, M., Yadav, V., Shanker, K., Singh, R.P., Pandey, V., Das, P., Yadav, R., Tripathi, P. and Govind Sinam, G., 2021. Biochar amendment reduced the risk associated with metal uptake and improved metabolite content in medicinal herbs. *Physiologia Plantarum.* 173(1), pp.321–339.

O'Connor, D., Peng, T., Zhang, J., Tsang, D.C., Alessi, D.S., Shen, Z., Bolan, N.S. and Hou, D., 2018. Biochar application for the remediation of heavy metal polluted land: A review of in situ field trials. *Science of The Total Environment, 619*, pp.815–826.

Palansooriya, K.N., Wong, J.T.F., Hashimoto, Y., Huang, L., Rinklebe, J., Chang, S.X., Bolan, N., Wang, H. and Ok, Y.S., 2019. Response of microbial communities to biochar-amended soils: A critical review. *Biochar, 1*(1), pp.3–22.

Pankaj, U. 2020. Multifarious benefits of biochar application in different soil types. In: (Eds.) Singh, J.S., & Singh, C., "Biochar Application in Agricultural and Environmental Management". Springer Nature, Switzerland. pp. 259–272.

Puga, A.P., Grutzmacher, P., Cerri, C.E.P., Ribeirinho, V.S. and de Andrade, C.A., 2020. Biochar-based nitrogen fertilizers: Greenhouse gas emissions, use efficiency, and maize yield in tropical soils. *Science of The Total Environment*, *704*, p.135375.

Qi, L., Ma, Z., Chang, S.X., Zhou, P., Huang, R., Wang, Y., Wang, Z. and Gao, M., 2021. Biochar decreases methanogenic archaea abundance and methane emissions in a flooded paddy soil. *Science of the Total Environment*, *752*, p.141958.

Qiao, K., Tian, W., Bai, J., Wang, L., Zhao, J., Song, T. and Chu, M., 2020. Removal of high-molecular-weight polycyclic aromatic hydrocarbons by a microbial consortium immobilized in magnetic floating biochar gel beads. *Marine Pollution Bulletin*, *159*, p.111489.

Quilliam, R.S., Glanville, H.C., Wade, S.C. and Jones, D.L., 2013. Life in the "charosphere"—Does biochar in agricultural soil provide a significant habitat for microorganisms? *Soil Biology and Biochemistry*, *65*, pp.287–293.

Rousk, J., Bååth, E., Brookes, P.C., Lauber, C.L., Lozupone, C., Caporaso, J.G., Knight, R. and Fierer, N., 2010. Soil bacterial and fungal communities across a pH gradient in an arable soil. *ISME Journal*, *4*(10), pp.1340–1351.

Roy, A., Dkhar, D.S., Tripathi, A.K., Singh, N.U., Kumar, D., Das, S.K. and Debnath, A., 2014. Growth performance of agriculture and allied sectors in the North East India. *Economic Affairs*, *59*, p.783.

Sun, D., Meng, J., Liang, H., Yang, E., Huang, Y., Chen, W., Jiang, L., Lan, Y., Zhang, W. and Gao, J., 2015. Effect of volatile organic compounds absorbed to fresh biochar on survival of *Bacillus mucilaginosus* and structure of soil microbial communities. *Journal of Soils and Sediments*, *15*(2), pp.271–281.

Talha, M.A., Goswami, M., Giri, B.S., Sharma, A., Rai, B.N. and Singh, R.S., 2018. Bioremediation of Congo red dye in immobilized batch and continuous packed bed bioreactor by *Brevibacillus parabrevis* using coconut shell bio-char. *Bioresource Technology*, *252*, pp.37–43.

Tao, S., Wu, Z., He, X., Ye, B.C. and Li, C., 2018. Characterization of biochar prepared from cotton stalks as efficient inoculum carriers for *Bacillus subtilis* SL-13. *Bioresources*, *13*(1), pp.1773–1786.

Tripathi, P., Khare, P., Barnawal, D., Shanker, K., Srivastava, P.K., Tripathi, R.D. and Kalra, A., 2020. Bioremediation of arsenic by soil methylating fungi: Role of *Humicola* sp. strain 2WS1 in amelioration of arsenic phytotoxicity in *Bacopa monnieri* L. *Science of The Total Environment*, *716*, p.136758.

Tripathi, P., Yadav, R., Das, P., Singh, A., Singh, R.P., Kandasamy, P., Kalra, A. and Khare, P., 2021. Endophytic bacterium CIMAP-A7 mediated amelioration of atrazine induced phyto-toxicity in *Andrographis paniculata*. *Environmental Pollution*, p.117635.

Tu, C., Wei, J., Guan, F., Liu, Y., Sun, Y. and Luo, Y., 2020. Biochar and bacteria inoculated biochar enhanced Cd and Cu immobilization and enzymatic activity in a polluted soil. *Environment International*, *137*, p.105576.

Wang, L., Li, Z., Wang, Y., Brookes, P.C., Wang, F., Zhang, Q., Xu, J. and Liu, X., 2020. Performance and mechanisms for remediation of Cd (II) and As (III) co-contamination by magnetic biochar-microbe biochemical composite: Competition and synergy effects. *Science of The Total Environment*, *750*, p.141672.

Wang, T., Sun, H., Ren, X., Li, B. and Mao, H., 2018. Adsorption of heavy metals from aqueous solution by UV-mutant *Bacillus subtilis* loaded on biochars derived from different stock materials. *Ecotoxicology and Environmental Safety*, *148*, pp.285–292.

Wei, S., Zhu, M., Fan, X., Song, J., Li, K., Jia, W. and Song, H., 2019. Influence of pyrolysis temperature and feedstock on carbon fractions of biochar produced from pyrolysis of rice straw, pine wood, pig manure and sewage sludge. *Chemosphere*, *218*, pp.624–631.

Wu, B., Wang, Z., Zhao, Y., Gu, Y., Wang, Y., Yu, J. and Xu, H., 2019. The performance of biochar-microbe multiple biochemical material on bioremediation and soil micro-ecology in the cadmium aged soil. *Science of the Total Environment*, *686*, pp.719–728.

Xiao, Z., Rasmann, S., Yue, L., Lian, F., Zou, H. and Wang, Z., 2019. The effect of biochar amendment on N-cycling genes in soils: A meta-analysis. *Science of The Total Environment*, *696*, p.133984.

Xin, J., Liu, R., Fan, H., Wang, M., Li, M. and Liu, X., 2013. Role of sorbent surface functionalities and microporosity in 2, 2′, 4, 4′-tetrabromodiphenyl ether sorption onto biochars. *Journal of Environmental Sciences*, *25*(7), pp.1368–1378.

Xiong, B., Zhang, Y., Hou, Y., Arp, H.P.H., Reid, B.J. and Cai, C., 2017. Enhanced biodegradation of PAHs in historically contaminated soil by *M. gilvum* inoculated biochar. *Chemosphere*, *182*, pp.316–324.

Xu, C., Chen, H.X., Xiang, Q., Zhu, H.H., Wang, S., Zhu, Q.H., Huang, D.Y. and Zhang, Y.Z., 2018. Effect of peanut shell and wheat straw biochar on the availability of Cd and Pb in a soil–rice (*Oryza sativa* L.) system. *Environmental Science and Pollution Research*, *25*(2), pp.1147–1156.

Yadav, V., Jain, S., Mishra, P., Khare, P., Shukla, A.K., Karak, T. and Singh, A.K., 2019. Amelioration in nutrient mineralization and microbial activities of sandy loam soil by short term field aged biochar. *Applied Soil Ecology*, *138*, pp.144–155.

Yang, T., Meng, J., Jeyakumar, P., Cao, T., Liu, Z., He, T., Cao, X., Chen, W. and Wang, H., 2020. Effect of pyrolysis temperature on the bioavailability of heavy metals in rice straw-derived biochar. *Environmental Science and Pollution Research*, pp.1–11.

Ye, S., Zeng, G., Wu, H., Liang, J., Zhang, C., Dai, J., Xiong, W., Song, B., Wu, S. and Yu, J., 2019. The effects of activated biochar addition on remediation efficiency of co-composting with contaminated wetland soil. *Resources, Conservation and Recycling*, *140*, pp.278–285.

Yin, D., Wang, X., Chen, C., Peng, B., Tan, C. and Li, H., 2016. Varying effect of biochar on Cd, Pb and As mobility in a multi-metal contaminated paddy soil. *Chemosphere*, *152*, pp.196–206.

Yousaf, B., Liu, G., Wang, R., Zia-ur-Rehman, M., Rizwan, M.S., Imtiaz, M., Murtaza, G. and Shakoor, A., 2016. Investigating the potential influence of biochar and traditional organic amendments on the bioavailability and transfer of Cd in the soil–plant system. *Environmental Earth Sciences*, *75*(5), p.374.

Zhang, G., He, L., Guo, X., Han, Z., Ji, L., He, Q., Han, L. and Sun, K., 2020. Mechanism of biochar as a biostimulation strategy to remove polycyclic aromatic hydrocarbons from heavily contaminated soil in a coking plant. *Geoderma*, *375*, p.114497.

Zhang, P., Zheng, S., Liu, J., Wang, B., Liu, F. and Feng, Y., 2018. Surface properties of activated sludge-derived biochar determine the facilitating effects on *Geobacter* co-cultures. *Water Research*, *142*, pp.441–451.

Zhao, L., Xiao, D., Liu, Y., Xu, H., Nan, H., Li, D., Kan, Y. and Cao, X., 2020. Biochar as simultaneous shelter, adsorbent, pH buffer, and substrate of *Pseudomonas citronellolis* to promote biodegradation of high concentrations of phenol in wastewater. *Water Research*, *172*, p.115494.

Zhu, X., Chen, B., Zhu, L. and Xing, B., 2017. Effects and mechanisms of biochar-microbe interactions in soil improvement and pollution remediation: A review. *Environmental Pollution*, *227*, pp.98–115.

4 Next-Gen Approaches to Understand Plant-Microbe Interactions Providing Abiotic Stress Tolerance in Plants

Romit Seth
NC State University, Kannapolis, USA
and
CSIR-IHBT and CSKHPKV, Palampur, India

Abhishek Bhandawat
CSIR-IHBT, Palampur, India
and
National Agri-Food Biotechnology Institute, Mohali, India

Tony Kipkoech Maritim
CSIR-IHBT, Palampur, India
and
AcSIR, Ghaziabad, India
and
KALRO–Tea Research Institute, Kericho, Kenya

Ram Kumar Sharma
CSIR-IHBT, Palampur, India
and
AcSIR, Ghaziabad, India

CONTENTS

DOI: 10.1201/9781003147091-4

4.1 INTRODUCTION

Global climate change in the form of abiotic factors such as drought, heat, and salinity has affected bioremediation and agriculture productivity. The use of synthetic fertilizers acts like a punishment of the soil due to the long-term effects on the edaphic microflora and fauna. Such agricultural abuse has depleted the natural treasures of essential nutrients by killing off the plant-beneficial fungi and bacteria. Moreover, the acquisitive behaviour to feed expanding human populations has extensively affected the natural soil health leading to future food insecurity (Battisti and Naylor 2009). Plants, being sessile in nature, often experience the rapid fluctuations of harsh environmental conditions. However, the intrinsic genomic complexities in plants play an important role in regulating the internal molecular programming to cope with such harsh environmental conditions. However, plants generally share a relationship with the diverse microbial community in the form of the "Rhizosphere" beneath the ground; or "Phyllosphere" above the ground (Bulgarelli et al. 2013). In the rhizosphere, the roots of the plants provide unique ecological niches for soil microbiota; while the phyllosphere extends up to some extent above ground (Hartmann et al. 2009). This is further alleviated by a beautiful relationship between plants and microorganisms (bacteria and fungi), which is often termed "symbiosis" (Ma, Vosátka, and Freitas 2019). During such beneficial mutualism, plants provide the prepared food (carbon assimilates in the form of starch and sucrose) to the associated microorganisms, while, in return, microorganisms assist plants to enhance water and nutrient absorption from the soil. Such interactions are key to the survival and adaptation of both partners in harsh environments, wherein the microorganisms alleviate abiotic stress response in plants by providing Induced Systemic Tolerance (IST) (Suarez et al. 2015). The intrinsic genetic and metabolic ability of microbes associated with the plants plays an important role in regulating ISTs.

4.2 EFFECTS OF PLANT-MICROBE INTERACTIONS OVER THEIR ABIOTIC STRESS TOLERANCE

4.2.1 Physiological Effects Regulating Molecular Insights Underlying Abiotic Stress in Plants

Plants possess an innate ability to smartly sense, manage or escape the harsh environmental conditions for their sustenance. The perception to abiotic stimuli involves an interactive cross-talk of metabolites regulating various biosynthetic pathways and networks, wherein the roots seems to be more sensitive to such conditions (Meena et al. 2017). Factors like prolonged heat or drought-like conditions result in a decrease in the water withholding capacity, root growth, delayed flowering and seed set, subsequently declining the yield and productivity of plants. However, the primary effect of adverse conditions was generally observed at the cellular level, which is followed by physiological symptoms. Some of the recent physiological, biochemical and molecular studies reported microbe-induced stress tolerance in plants activated *via* induced systemic tolerance (Farrar, Bryant, and Cope-Selby 2014). The close symbiotic relationship between the microbial community (microbiota) and the roots of the plants enhances the nutrients, water and salt absorption in a harsh environmental system (Figure 4.1).

Recently, a microbial consortia-based approach has been widely used, which comprises a combination of different microbial strains complementing each other to induce tolerance in plant genotypes during harsh environmental conditions (De Vrieze et al. 2018). This was well studied in most plants like grapevines, maize, tobacco and capsicum (Table 4.1), wherein multiple combinations are known to increase the plant growth-promoting (PGP) effect in contrast to single inoculants, bacterial combinations with low PGP effects (Compant et al. 2019; Pankaj et al. 2019; Pankaj 2020).

Such a mechanism is a complex phenomenon with dynamic and real-time changes at transcriptional, metabolic and physiological levels during various environmental cues in cells (Atkinson and Urwin 2012). Hence, to understand the mechanistic insights of such approaches, recently developed multi-omics tools can be used efficiently. Moreover, recent advances in sequencing technology have impressively opened doorways to augment the research in this area (Unamba, Nag, and Sharma 2015).

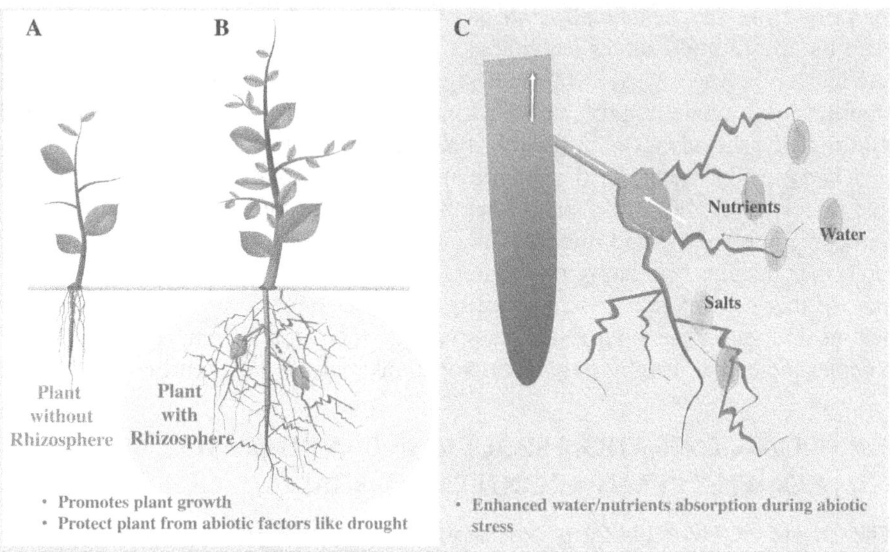

FIGURE 4.1 Plant-microbe interaction alleviating abiotic stress tolerance. (A) Plant without rhizosphere and (B) plant with rhizosphere. (C) Symbiosis between plant and the microbial community-enhancing nutrients and water absorption during harsh environmental conditions.

TABLE 4.1
Bacterial Consortia Inducing Abiotic Stress Tolerance in Plants

Plant	Consortia (Bacteria)	Stress	Consortia Effect	References
Blue maize	*Azospirillum brasilense* Sp7, *Pseudomonas putida* KT2440, *Sphingomonas* sp. OF178, and *Acinetobacter* sp. EMM02	Desiccation	Increase in shoot/root dry weight, along with plant height and diameter	(Molina-Romero et al. 2017)
Nicotiana attenuate	*Pseudomonas frederiksbergensis* A176, *Arthrobacter nitroguajacolicus* E46, *Pseudomonas azotoformans* A70, *Bacillus cereus* CN2, *Bacillus mojavensis* K1, and *Bacillus megaterium* B55	Natural wilt disease	Decrease in plant necrosis	(Santhanam et al. 2015)
Vitis vinifera and *Capsicum annuum*	*Bacillus* sp. S4, *Delftia* sp. S8, *Sphingobacterium* sp. S6, *Acinetobacter* sp. S2, and *Enterobacter* sp. S7,	Drought	Increased root, aerial fresh biomass, and enhanced photosynthesis efficiency	(Rolli et al. 2015)

4.2.2 Multi-omics Approaches to Study Microbe-Induced Systemic Tolerance in Plants

Deciphering the complex insights of plant-microbe interactions is a promising aspect for unraveling the beneficial/pathogenic microbes effect for crop improvement. Here the use of multi-omics technologies has recently become boon to explore the relationship between various molecules, constituting the organism, to interpret the plant response (genetic, proteomic and metabolic) to various biotic/abiotic factors. Additionally, the advances in bioinformatics to handle large data-set analysis have enriched our knowledge of the microbial community in a composite environment like rhizosphere, alleviating stress tolerance in plant system. Here, meta-transcriptomic, metaproteomic and metagenomic approaches can be used to understand the molecular insights regulating plant interaction with the diverse microbial community. In this chapter, we will specifically focus on some of the special features of next-generation sequencing (NGS) approaches, their application in understanding complex molecular insights of plant microbial interactions and their future prospects.

4.3 NEXT-GENERATION SEQUENCING: A BOON TO UNRAVEL COMPLEX PLANT-MICROBE CROSS-TALKS

The advent of NGS platforms has revolutionized the genomics/transcriptomics approaches in a cost-effective manner. Recently, multiple studies have been carried out to dissect the complex molecular insights regulating various physiological processes (Bhandawat et al. 2017; Jayaswall et al. 2016). The high-throughput data obtained with millions of reads from sequencing platforms can be efficiently used to understand the genomic/epigenomics variants and regulation of genes and their allelic isoforms expression (Bhandawat et al. 2019; Dhyani et al. 2020; Singh et al. 2019). Additionally, for identification of key transcription factors regulating genes/isoforms, the next-gen transcriptomics approaches currently available act as boon, which previously seemed to be akin to finding a needle in a haystack. Recently, various sequencing platforms have become available from Illumina, PacBio, Thermo Fisher Scientific, Oxford Nanopore, etc. (Table 4.2). The PacBio RSII sequencing platform is widely used for sequencing the long reads (~1400 bp), using the single molecule real time sequencing technology (SMRT) cells. These SMRT cells in sequel II are known to generate good reference-quality assemblies of about 48 microbial isolates (Rhoads and Au 2015). Despite providing long reads, the major drawback of PacBio lies in homopolymers errors, subsequently leading to a higher error rate in sequencing the DNA with high repeats. Another sequencer from Thermo Fisher Scientific, ion-torrent proton II is based on the principle of semi-conductor sequencing technology using emulsion PCR amplification. The ion torrent is known to provide approximately 25 giga bases of data with an average length of 175 base pairs (Marine et al. 2020). The step up advancement in Oxford Nanopore Technology provides the next cutting-edge features with easier interface and longer sequencing reads (~9000 bases) (Jain et al. 2016). However, the major drawback in the nanopore is its low-throughput data with a high error rate (van Dijk et al. 2018). Here, the Illumina NGS sequencing platforms have gained advantages over the other platforms, which provide both higher sequencing depth and coverage of data, reducing the sequencing error probability to study high-throughput transcriptomics and genomics profiling. Moreover, recently, several studies have successfully utilized Illumina platforms for dissecting complex traits in non-modal plant species (Seth et al. 2021; Nag et al. 2020; Parmar et al. 2019; Singh et al. 2017).

TABLE 4.2
Different Sequencing Platforms Available for High/Low-Throughput Next-Gen Sequencing

Criterion	Illumina	PacBio	Thermo Fisher Scientific	Oxford
Instrument	Nova Seq	RSII	Ion-torrent Proton II	Nanopore Minion
Technology	Sequencing by synthesis	Single-molecule real-time sequencing	Semi-conductor sequencing	Real-time sequencing
Amplification	Bridge PCR	None	emPCR	None
Data output	High throughput	Low throughput	Low throughput	Low throughput
Data generation	6 Tera bases	1 Giga base	25 giga bases	—
Average read length	2 × 150 bp	14,000 bp	175 bp	9000 bp
Sequencing reads	Short reads	Long reads	Short reads	Long reads
Run time	44 h	2 h	5 h	6 h
Drawbacks	High cost	High cost	Homopolymer errors	High error rate

4.3.1 Steps in NGS Sequencing

The NGS is carried out in three major steps, i.e. sample library preparation, paired end or single end sequencing, followed by bioinformatics data analysis. For cDNA library preparations, high quality RNA with a RNA integrity number (RIN) of more than 9 is required. The RIN value of the extracted RNA is generally assessed on BioAnalyzer using RNA 7500 series II Chip (Agilent Technologies, CA, USA). Moreover, genomic DNA should have a high molecular weight, be intact and un-sheared, with absorbance ratios 260/280 and 260/230 of ~1.8 and >2.0, respectively.

4.3.1.1 Library Preparation

Before heading toward the sequencing, the first step comprises library preparation from high quality genomic DNA or expressed RNA. The libraries preparation includes genomics and transcriptomics library preparation protocols for genomic DNA (Genomics) and complementary DNA (transcriptomics) synthesis. The genomics library preparation involves PCR-free libraries preparation protocols using Nextera DNA Library Prep Kit (Illumina), which requires the initial step of mechanical shearing or enzymatic fragmentation step of high-quality DNA. The mechanical shearing of DNA or RNA can be performed using a specific frequency of sound wave-producing instruments (focused ultrasonicators such as Covaris). However, automated DNA size selection/fragmentation from 100bp to 50 Kb can be efficiently carried out using the BluePippin system which works on the principal of pulse field electrophoresis (Sage Science). The fragmented DNA are purified using bead purification methods to discard the unwanted strands at every step of library preparation. Subsequently, the end repairs, followed by 3' polyadenylation is performed before the adaptor ligation step. The adapter sequences are 10 base pairs of unique sequences often referred as "adapter indices" in the Illumina library preparation kit. These are required for demultiplexing of libraries in bio-informatics data analysis, wherein each adapter index represents each specific sample. The adaptor ligated template sequences obtained are assessed for their quality using the fluorometer and BioAnalyzer (Figure 4.2A).

The complementary DNA (cDNA) libraries preparation generally follow Illumina Truseq RNA Sample prep v2 LS Protocol (Illumina Inc., CA, USA) to construct adaptor ligated cDNA libraries for the Illumina platform (Zhong et al. 2011). This requires high-quality RNA, which is also purified using the bead purification method (RNA purification beads: Oligo-dT) for mRNA enrichment and purification. The enriched mRNA undergoes fragmentation steps, followed by the reverse transcription method to synthesize the first strand of complementary DNA. For second strand synthesis of cDNA, DNA polymerase are used, which is generally followed by left-over single strand removal using RNAseH. The complementary DNA is further purified using the bead purification method, followed by the end repairs for 3' polyadenylation before proceeding towards the adaptor ligation step. The adaptor ligation step constitutes ligation of "adapter indices" provided in the Illumina library preparation kit, required for demultiplexing of libraries in bio-informatics data analysis. The fluorometer and BioAnalyzer are used to assess the quality of adaptor ligated libraries prepared (Figure 4.2B).

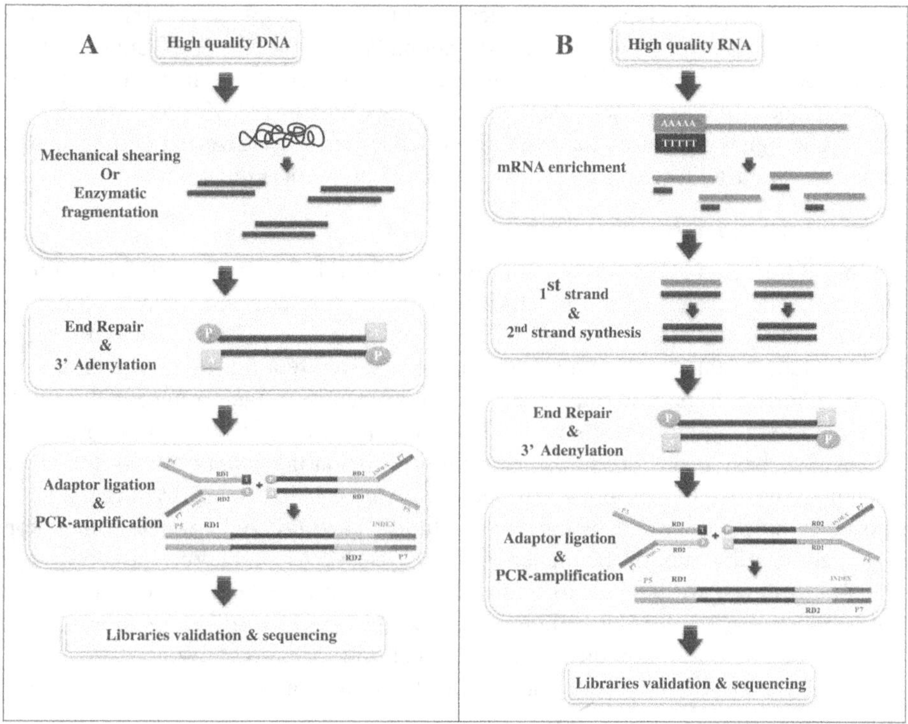

FIGURE 4.2 Libraries preparation protocols for sequencing: (A) Genomic DNA (gDNA) libraries preparation; (B) complementary DNA (cDNA) libraries preparation.

4.3.1.2 cDNA Sequencing

The cDNA sequencing is comprised of PCR-based amplification of prepared libraries to facilitate sequencing by enhancing the signal intensity. The amplification step is based on microbeads-mediated spatial library fragmentation with the help of emPCR (454, PGM) or bridge amplification on glass surface of flow cell (Illumina). In the case of the bead-based amplification method, the spatially separated water droplets in water-oil emulsion were enriched in a semi-conductor chip of Ion PGM, picotiterplate in 454 and glass surface hybridization in solid sequencing platforms (Mardis 2008).

The Illumina sequencing platforms for genomics/transcriptomics sequencing uses "sequencing by synthesis" chemistry for both paired-end and single-end high-throughput sequencing (Kartalov and Quake 2004). Here, single-end and paired-end represents sequencing reads, wherein each base pair is captured with specific signal intensity providing data-set in an image format. Prior to sequencing, the adaptor ligated template cDNA sequences were clonally amplified using the bridge amplification method (Grada and Weinbrecht 2013). Initially, the adaptor ligated sequences are hybridized into their complimentary sequences available in the flow cell. The hybridized sequences undergo extension followed by double stranded DNA denaturation, wherein the forward template is washed away, and remaining is the covalently bonded reverse template. The bridge amplification step of reverse templates is carried out, providing clusters of both forward and reverse templates in the flow cell. Subsequently, the reverse strands cleave off and wash away, with the remaining forward strands to initiate sequencing using "sequencing by synthesis" chemistry (Kartalov and Quake 2004). The signal intensities captured by the sequencers are converted into the nucleotide bases in fastQ format using the base-calling protocol (Figure 4.3).

4.3.1.3 NGS Data Analysis

The high-throughput data analysis protocol specifically depends on the question of the study. The prediction and identification of key transcription factors regulating expression of genes/isoforms requires condition-specific good experimental design. Initially, before starting the data analysis, the first step requires demultiplexing and quality filtering of the raw sequenced reads obtained from the sequencer. This will filter out the low-quality reads with adaptor contamination. In order to capture the active microbes in the microbial community alleviating abiotic stress tolerance in plants, meta-transcriptomics approaches can be efficiently use by targeting 16s ribosomal profiling. Some of the recent studies have utilized Simple Annotation of Metatranscriptomes by Sequence Analysis (SAMSA) standalone pipeline, specifically designed to handle large meta-transcriptomics data set (Westreich et al. 2018). Moreover, the DIAMOND toolkit is widely used for the annotation of read sequences against the reference database (Buchfink, Xie, and Huson 2015). In one recent study, the diamond tool was efficiently used to annotate large microbiome

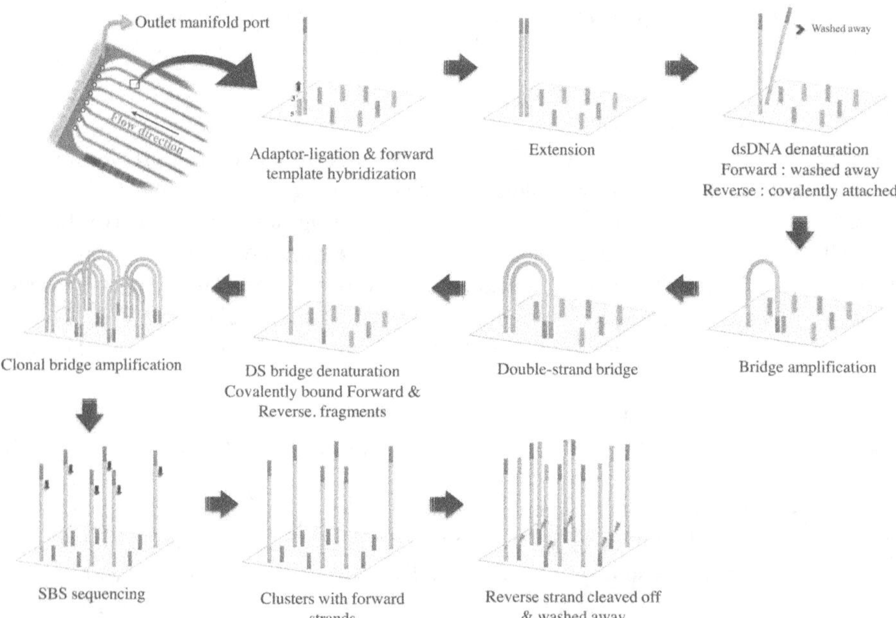

FIGURE 4.3 cDNA sequencing using "sequencing by synthesis" chemistry in Illumina platforms.

data collected from topsoil samples comprising of 7560 subsamples (Bahram et al. 2018). The DESeq2 differential expression (R Scripts) were generally used to identify differentially expressed genes obtained from sub system analysis.

The transcriptional profiling requires assembly of high quality sequenced reads, wherein in the presence of a reference-genome, reference-guided differential gene expression analysis can be carried out using the Tuxedo pipeline (Trapnell et al. 2012). This Linux-based pipeline has been widely used in various NGS data analyses, which helps in the identification of significant differentially expressed genes and isoforms, along with their genomic annotation. However, in the absence of a reference genome, the sequenced reads can be assembled *de novo* using various assembly pipelines. Some of the most widely available assembly tools include CLC workbench, TRINITY-RNASeq, Oases, SOAP*denovo*, etc. (Haas et al. 2013; Schulz et al. 2012; Xie et al. 2014). While performing *de novo* differential expression analysis, the *de novo* assembly should be of high quality to capture the novel transiently expressed genes. A recent study has utilized the concatenated assembly approach to obtain the high-quality *de novo* assembled sequences to target the transiently expressed genes during physiological processes (Seth et al. 2019). Furthermore, to identify the differentially expressed genes, the EdgeR or DESeq2 R-scripts has been widely used in most of the transcriptomics studies (Seth et al. 2020). Moreover, considering the limited available information of microbial reference genomes, the functional profiling of diverse microbial communities is quite challenging. Here, the Functional Mapping and Analysis Pipeline (FMAP) is widely used to predict differentially enriched gene families and pathways in meta-genomics/meta-transcriptomics data (Kim et al. 2016). Recently, an automated meta-genomics pipeline (SqueezeMeta) has also been developed, which supports co-assembly, merged assembly and a sequential assembly approach of related meta-genome with individual genome retrieval (Tamames and Puente-Sánchez 2019). Additionally, independent gene ontology and pathway enrichment analysis can be efficiently performed using various publicly available databases such as WEGO and Kyoto Encyclopaedia of Genes and Genomes (KEGG) pathway databases (Kanehisa et al. 2019; Ye et al. 2018). The methodologies and pipelines discussed in this chapter may assist researchers to study plant-associated microbial communities alleviating abiotic stress tolerance using NGS approaches.

4.3.2 Current Applications of NGS in Elucidating PMI

Advances in high-throughput sequencing technologies allow for deeper understanding of plant microbe interaction and identification of novel microbial communities associating with plants. Several studies to understand plant microbial interactions using meta-genomics/transcriptomics approaches have been carried out successfully in many plant species (Knief 2014). Among them, Chaparro et al. (2014) used meta-transcriptomics to uncover systematic changes in microbial genes triggered by *Arabidopsis thaliana* root exudates. Moreover, Knief et al. (2012) used a meta-proteogenomic approach to unravel key molecular insights of bacterial and archaeal interactions with rice above and below the ground surface. Similarly, Delmotte et al. (2009) integrated both meta-genomic and metaproteomic approaches to unravel the adaptation mechanism used by diverse microbial communities associated with different plants. Additionally, the analysis of multiple phyllospheric metagenomes identified diverse classes of phototrophic bacteria coexisting together on plant surfaces (Atamna-Ismaeel et al. 2012).

4.4 CONCLUSION AND FUTURE PROSPECTS

A beautiful symbiotic plant-microbe relationship is governed by a complex cross-talk between them, wherein the plant uses microbial assistance to endure the harsh environmental conditions. The advent of modern NGS technology has opened various doorways to comprehensively unravel the plant-microbe cross-talk. This has substantially fastened the preparation of meta-genomic/transcriptomic libraries with reduced sequencing costs per base. The deeper sequencing provided by new era sequencing platforms not only provides identity, but also the physiological properties to the plant-associated microbial communities. Recently established single cell genome sequencing has enabled sequencing of diverse large phyla of uncultured microorganism species. Hence, the meta-transcriptomics approaches available now can assist in identifying the active microbes regulating gene expression during stress tolerance alleviation in plants. This will not only contribute to enhancing our understanding of the rhizosphere and phyllosphere under changing environmental conditions but also can be used to develop healthier agriculture systems in the future.

REFERENCES

Atamna-Ismaeel, Nof, Omri Finkel, Fabian Glaser, Christian von Mering, Julia A Vorholt, Michal Koblížek, Shimshon Belkin, and Oded Béjà. 2012. Bacterial Anoxygenic Photosynthesis on Plant Leaf Surfaces. *Environmental Microbiology Reports* 4 (2): 209–216.

Atkinson, Nicky J, and Peter E Urwin. 2012. The Interaction of Plant Biotic and Abiotic Stresses: From Genes to the Field. *Journal of Experimental Botany* 63 (10): 3523–3543.

Bahram, Mohammad, Falk Hildebrand, Sofia K Forslund, Jennifer L Anderson, Nadejda A Soudzilovskaia, Peter M Bodegom, Johan Bengtsson-Palme, Sten Anslan, Luis Pedro Coelho, and Helery Harend. 2018. Structure and Function of the Global Topsoil Microbiome. *Nature* 560 (7717): 233–237.

Battisti, David S, and Rosamond L Naylor. 2009. Historical Warnings of Future Food Insecurity with Unprecedented Seasonal Heat. *Science* 323 (5911): 240–244.

Bhandawat, Abhishek, Gagandeep Singh, Romit Seth, Pradeep Singh, and R.K. Sharma. 2017. Genome-Wide Transcriptional Profiling to Elucidate Key Candidates Involved in Bud Burst and Rattling Growth in a Subtropical Bamboo (*Dendrocalamus hamiltonii*). *Frontiers in Plant Science* 7: 2038. doi:10.3389/fpls.2016.02038.

Bhandawat, Abhishek, Vikas Sharma, Pradeep Singh, Romit Seth, Akshay Nag, Jagdeep Kaur, and Ram Kumar Sharma. 2019. Discovery and Utilization of EST-SSR Marker Resource for Genetic Diversity and Population Structure Analyses of a Subtropical Bamboo. *Dendrocalamus hamiltonii. Biochemical Genetics* 57 (5): 652–672. doi:10.1007/s10528-019-09914-4.

Buchfink, Benjamin, Chao Xie, and Daniel H Huson. 2015. Fast and Sensitive Protein Alignment Using DIAMOND. *Nature Methods* 12 (1): 59–60.

Bulgarelli, Davide, Klaus Schlaeppi, Stijn Spaepen, Emiel Ver Loren Van Themaat, and Paul Schulze-Lefert. 2013. Structure and Functions of the Bacterial Microbiota of Plants. *Annual Review of Plant Biology* 64: 807–838.

Chaparro, Jacqueline M, Dayakar V Badri, and Jorge M Vivanco. 2014. Rhizosphere Microbiome Assemblage Is Affected by Plant Development. *ISME Journal* 8 (4): 790–803.

Compant, Stéphane, Abdul Samad, Hanna Faist, and Angela Sessitsch. 2019. A Review on the Plant Microbiome: Ecology, Functions, and Emerging Trends in Microbial Application. *Journal of Advanced Research* 19 (September): 29–37. doi:10.1016/j.jare.2019.03.004.

De Vrieze, Mout, Fanny Germanier, Nicolas Vuille, and Laure Weisskopf. 2018. Combining Different Potato-Associated *Pseudomonas* Strains for Improved Biocontrol of *Phytophthora infestans*. *Frontiers in Microbiology* 9: 2573.

Delmotte, Nathanaël, Claudia Knief, Samuel Chaffron, Gerd Innerebner, Bernd Roschitzki, Ralph Schlapbach, Christian von Mering, and Julia A Vorholt. 2009. Community Proteogenomics Reveals Insights into the Physiology of Phyllosphere Bacteria. *Proceedings of the National Academy of Sciences* 106 (38): 16428–16433.

Dhyani, Praveen, Balraj Sharma, Pradeep Singh, Mamta Masand, Romit Seth, and Ram Kumar. 2020. Industrial Crops & Products Genome-Wide Discovery of Microsatellite Markers and, Population Genetic Diversity Inferences Revealed High Anthropogenic Pressure on Endemic Populations of *Trillium govanianum*. *Industrial Crops & Products* 154 (June): 112698. doi:10.1016/j.indcrop.2020.112698.

Farrar, Kerrie, David Bryant, and Naomi Cope-Selby. 2014. Understanding and Engineering Beneficial Plant-Microbe Interactions: Plant Growth Promotion in Energy Crops. *Plant Biotechnology Journal* 12 (9): 1193–1206.

Grada, Ayman, and Kate Weinbrecht. 2013. Next-Generation Sequencing: Methodology and Application. *Journal of Investigative Dermatology* 133 (8): e11.

Haas, Brian J, Alexie Papanicolaou, Moran Yassour, Manfred Grabherr, Philip D Blood, Joshua Bowden, Matthew Brian Couger, David Eccles, Bo Li, and Matthias Lieber. 2013. De Novo Transcript Sequence Reconstruction from RNA-Seq Using the Trinity Platform for Reference Generation and Analysis. *Nature Protocols* 8 (8): 1494–1512.

Hartmann, Anton, Michael Schmid, Diederik Van Tuinen, and Gabriele Berg. 2009. Plant-Driven Selection of Microbes. *Plant and Soil* 321 (1–2): 235–257.

Jain, Miten, Hugh E Olsen, Benedict Paten, and Mark Akeson. 2016. The Oxford Nanopore MinION: Delivery of Nanopore Sequencing to the Genomics Community. *Genome Biology* 17 (1): 239.

Jayaswall, K, P Mahajan, G Singh, R Parmar, R Seth, A Raina, MK Swarnkar, AK Singh, R Shankar, and RK Sharma. 2016. Transcriptome Analysis Reveals Candidate Genes Involved in Blister Blight Defense in Tea (*Camellia sinensis* (L) Kuntze). *Scientific Reports* 6 (1): 30412. doi:10.1038/srep30412.

Kanehisa, Minoru, Yoko Sato, Miho Furumichi, Kanae Morishima, and Mao Tanabe. 2019. New Approach for Understanding Genome Variations in KEGG. *Nucleic Acids Research* 47 (D1): D590–D595.

Kartalov, Emil P, and Stephen R Quake. 2004. Microfluidic Device Reads up to Four Consecutive Base Pairs in DNA Sequencing-by-Synthesis. *Nucleic Acids Research* 32 (9): 2873–2879.

Kim, Jiwoong, Min Soo Kim, Andrew Y Koh, Yang Xie, and Xiaowei Zhan. 2016. FMAP: Functional Mapping and Analysis Pipeline for Metagenomics and Metatranscriptomics Studies. *BMC Bioinformatics* 17 (1): 1–8. doi:10.1186/s12859-016-1278-0.

Knief, Claudia, Nathanaël Delmotte, Samuel Chaffron, Manuel Stark, Gerd Innerebner, Reiner Wassmann, Christian Von Mering, and Julia A Vorholt. 2012. Metaproteogenomic Analysis of Microbial Communities in the Phyllosphere and Rhizosphere of Rice. *ISME Journal* 6 (7): 1378–1390.

Knief, Claudia. 2014. Analysis of Plant Microbe Interactions in the Era of Next Generation Sequencing Technologies. *Frontiers in Plant Science* 5: 216.

Ma, Ying, Miroslav Vosátka, and Helena Freitas. 2019. Editorial: Beneficial Microbes Alleviate Climatic Stresses in Plants. *Frontiers in Plant Science* 10 (May): 1–2. doi:10.3389/fpls.2019.00595.

Mardis, Elaine R. 2008. The Impact of Next-Generation Sequencing Technology on Genetics. *Trends in Genetics* 24 (3): 133–141.

Marine, Rachel L, Laura C Magaña, Christina J Castro, Kun Zhao, Anna M Montmayeur, Alexander Schmidt, Marta Diez-Valcarce, et al. 2020. Comparison of Illumina MiSeq and the Ion Torrent PGM and S5 Platforms for Whole-Genome Sequencing of Picornaviruses and Caliciviruses. *Journal of Virological Methods* 280: 113865. doi:https://doi.org/10.1016/j.jviromet.2020.113865.

Meena, Kamlesh K, Ajay M Sorty, Utkarsh M Bitla, Khushboo Choudhary, Priyanka Gupta, Ashwani Pareek, Dhananjaya P Singh, et al. 2017. Abiotic Stress Responses and Microbe-Mediated Mitigation in Plants: The Omics Strategies. *Frontiers in Plant Science* 8 (February): 1–25. doi:10.3389/fpls.2017.00172.

Molina-Romero, Dalia, Antonino Baez, Verónica Quintero-Hernández, Miguel Castañeda-Lucio, Luis Ernesto Fuentes-Ramírez, María del Rocío Bustillos-Cristales, Osvaldo Rodríguez-Andrade, Yolanda Elizabeth Morales-García, Antonio Munive, and Jesús Muñoz-Rojas. 2017. Compatible Bacterial Mixture, Tolerant to Desiccation, Improves Maize Plant Growth. *PloS One* 12 (11): e0187913.

Nag, Akshay, Shruti Choudhary, Mamta Masand, Rajni Parmar, Abhishek Bhandawat, Romit Seth, Gopal Singh, Praveen Dhyani, and Ram Kumar Sharma. 2020. Spatial Transcriptional Dynamics of Geographically Separated Genotypes Revealed Key Regulators of Podophyllotoxin Biosynthesis in *Podophyllum hexandrum. Industrial Crops and Products* 147 (October 2019): 112247. doi:10.1016/j.indcrop.2020.112247.

Pankaj U. 2020. Bio-fertilizers for Management of Soil, Crop, and Human Health. In: (Eds.) C Singh, S Tiwari, JS Singh & AN Yadav, "*Microbes in Agriculture and Environmental Development*". CRC Press/Taylor & Francis, pp. 71–85.

Pankaj, U, Singh, G, and Verma, RK. 2019. Microbial Approaches in Management and Restoration of Marginal Lands. In: (Eds.) JS Singh, "*New and Future Developments in Microbial Biotechnology and Bioengineering: Microbes in Soil, Crop and Environmental Sustainability*". Elsevier, NY. pp. 295–305.

Parmar, Rajni, Romit Seth, Pradeep Singh, Gopal Singh, Sanjay Kumar, and Ram Kumar Sharma. 2019. Transcriptional Profiling of Contrasting Genotypes Revealed Key Candidates and Nucleotide Variations for Drought Dissection in *Camellia sinensis* (L.) O. Kuntze. *Scientific Reports* 9 (1): 7487. doi:10.1038/s41598-019-43925-w.

Rhoads, Anthony, and Kin Fai Au. 2015. PacBio Sequencing and Its Applications. *Genomics,Proteomics & Bioinformatics* 13 (5): 278–289. doi:https://doi.org/10.1016/j.gpb.2015.08.002.

Rolli, Eleonora, Ramona Marasco, Gianpiero Vigani, Besma Ettoumi, Francesca Mapelli, Maria Laura Deangelis, Claudio Gandolfi, Enrico Casati, Franco Previtali, and Roberto Gerbino. 2015. Improved Plant Resistance to Drought Is Promoted by the Root-Associated Microbiome as a Water Stress-Dependent Trait. *Environmental Microbiology* 17 (2): 316–331.

Santhanam, Rakesh, Arne Weinhold, Jay Goldberg, Youngjoo Oh, and Ian T Baldwin. 2015. Native Root-Associated Bacteria Rescue a Plant from a Sudden-Wilt Disease that Emerged during Continuous Cropping. *Proceedings of the National Academy of Sciences* 112 (36): E5013–E5020.

Schulz, Marcel H, Daniel R Zerbino, Martin Vingron, and Ewan Birney. 2012. Oases: Robust de Novo RNA-Seq Assembly across the Dynamic Range of Expression Levels. *Bioinformatics* 28 (8): 1086–1092.

Seth, Romit, Abhishek Bhandawat, Rajni Parmar, Pradeep Singh, Sanjay Kumar, and Ram Kumar Sharma. 2019. Global Transcriptional Insights of Pollen-Pistil Interactions

Commencing Self-Incompatibility and Fertilization in Tea [*Camellia Sinensis* (L.) O. Kuntze]. *International Journal of Molecular Sciences* 20 (3): 539. doi:10.3390/ijms20030539.

Seth, Romit, Tony Kipkoech Maritim, Rajni Parmar, and Ram Kumar Sharma. 2021. Underpinning the Molecular Programming Attributing Heat Stress Associated Thermotolerance in Tea [*Camellia Sinensis* (L.) O. Kuntze]. *Horticulture Research* 8 (1): 99. doi:10.1038/s41438-021-00532-z.

Singh, Gagandeep, Gopal Singh, Romit Seth, Rajni Parmar, Pradeep Singh, Vikram Singh, Sanjay Kumar, and Ram Kumar Sharma. 2019. Functional Annotation and Characterization of Hypothetical Protein Involved in Blister Blight Tolerance in Tea [*Camellia Sinensis* (L) O. Kuntze]. *Journal of Plant Biochemistry and Biotechnology* 28 (4): 447–459. doi:10.1007/s13562-019-00492-5.

Singh, Pradeep, Gagandeep Singh, Abhishek Bhandawat, Gopal Singh, Rajni Parmar, Romit Seth, and Ram Kumar Sharma. 2017. Spatial Transcriptome Analysis Provides Insights of Key Gene (s) Involved in Steroidal Saponin Biosynthesis in Medicinally Important Herb *Trillium govanianum*. *Scientific Reports* 7 (1): 45295. doi: 10.1038/srep45295.

Suarez, Christian, Massimiliano Cardinale, Stefan Ratering, Diedrich Steffens, Stephan Jung, Ana M Zapata Montoya, Rita Geissler-Plaum, and Sylvia Schnell. 2015. Plant Growth-Promoting Effects of *Hartmannibacter diazotrophicus* on Summer Barley (*Hordeum vulgare* L.) under Salt Stress. *Applied Soil Ecology* 95: 23–30.

Tamames, Javier, and Fernando Puente-Sánchez. 2019. SqueezeMeta, a Highly Portable, Fully Automatic Metagenomic Analysis Pipeline. *Frontiers in Microbiology* 10 (JAN): 1–10. doi:10.3389/fmicb.2018.03349.

Trapnell, Cole, Adam Roberts, Loyal Goff, Geo Pertea, Daehwan Kim, David R Kelley, Harold Pimentel, Steven L Salzberg, John L Rinn, and Lior Pachter. 2012. Differential Gene and Transcript Expression Analysis of RNA-Seq Experiments with TopHat and Cufflinks. *Nature Protocols* 7 (3): 562–578.

Unamba, Chibuikem IN, Akshay Nag, and Ram K Sharma. 2015. Next Generation Sequencing Technologies: The Doorway to the Unexplored Genomics of Non-Model Plants. *Frontiers in Plant Science* 6: 1074. doi:10.3389/fpls.2015.01074.

van Dijk, Erwin L, Yan Jaszczyszyn, Delphine Naquin, and Claude Thermes. 2018. The Third Revolution in Sequencing Technology. *Trends in Genetics* 34 (9): 666–681.

Westreich, Samuel T, Michelle L Treiber, David A Mills, Ian Korf, and Danielle G Lemay. 2018. SAMSA2: A Standalone Metatranscriptome Analysis Pipeline. *BMC Bioinformatics* 19 (1): 1–11. doi:10.1186/s12859-018-2189-z.

Xie, Yinlong, Gengxiong Wu, Jingbo Tang, Ruibang Luo, Jordan Patterson, Shanlin Liu, Weihua Huang, Guangzhu He, Shengchang Gu, and Shengkang Li. 2014. SOAPdenovo-Trans: De Novo Transcriptome Assembly with Short RNA-Seq Reads. *Bioinformatics* 30 (12): 1660–1666.

Ye, Jia, Yong Zhang, Huihai Cui, Jiawei Liu, Yuqing Wu, Yun Cheng, Huixing Xu, Xingxin Huang, Shengting Li, and An Zhou. 2018. WEGO 2.0: A Web Tool for Analyzing and Plotting GO Annotations, 2018 Update. *Nucleic Acids Research* 46 (W1): W71–W75.

Zhong, Silin, Je-Gun Joung, Yi Zheng, Yun-ru Chen, Bao Liu, Ying Shao, Jenny Z Xiang, Zhangjun Fei, and James J Giovannoni. 2011. High-Throughput Illumina Strand-Specific RNA Sequencing Library Preparation. *Cold Spring Harbor Protocols* 8: 940–949. doi:10.1101/pdb.prot5652.

5 Bio-Nanotechnology in Plant Protection

A Novel Frontier for Risk Mitigation toward Sustainable Agriculture

Rakesh Yonzone and Bimal Das
Uttar Banga Krishi Viswavidyalaya, Majhian, India
M. Soniya Devi
Rani Lakshmi Bai Central Agricultural University
Jhansi, India

CONTENTS

DOI: 10.1201/9781003147091-5

5.1 INTRODUCTION

Nanotechnology also known generic technology is considered to be the next generation frontier in agricultural science and holds a very important position in transforming agricultural technologies and in sustainable production of food. It has the potentialities to revive the agriculture and food industry (Kuzma and VerHage 2006). It mostly deals with the matter of size 1–100 nanometers (Singh et al. 2015) which has proved to provide some better built facilitates, cost effective and lasting products that has wide applications and is synthesized by different physical, chemical and biological methods. Its important application includes household, medicine and food industries; in addition to these, some of the important agricultural applications include nanoherbicides, nanofertilizers, and importantly, nanopesticides. The application of nanotechnology in agriculture helps in increasing yields, reducing pesticide application, the early and rapid detection of pathogens and toxic chemicals in food, smart delivery of pesticides and fertilizers, and use in food processing and packaging, as well as in food security regulations (Moraru et al. 2003; Chau et al. 2007).

Due to the increasing demand of nanomaterials in agricultural industries, this technology has received major attention and has opened up novel applications giving innovative solutions to various challenges that humans are facing today. While this technology provides various opportunities for innovation, its uses has the potential to resolve the number of issues raised due to climate change and use of excessive inputs. With the advancement of nanotechnology and its applications, it uses have led to the fabrication of various nanomaterials having unique properties and that are target specific. The use of nanomaterials in the agriculture sector as a fertilizer, pesticide and sensor for plants and in the detection of soil health has resulted in increased crop production and in improving soil health. The opportunities in the application of nanotechnology in plants, animals and insect control for developing countries has been highlighted by Chen and Yada (2011) and Rai and Ingle (2012). Its application in a country like India includes the areas of nanoinputs, nanofood systems, nanobiotechnology and nano-remediation (Subramanian and Tarafdar 2011).

Bio-nanotechnology has emerged as the integration of biotechnology and nanotechnology for producing eco-friendly nanoparticles for various agricultural appliances. It combines the principle of physical, biological and chemical approaches for producing nanoparticles having specific functions. This technology also involves multidisciplinary approaches with the utilization of biosensors and, mostly importantly, the convergence of chemistry, biology, biophysics, engineering photonics and nanomedicine. The particle produced has some unusual properties that are completely different from what they exhibit on a microscale, giving unique systematic applications. The applications of bio-nanotechnology have provided an opportunity to revolutionize different fields in producing various products that were hardly possible to produce through the application of conventional systems. Their application has the potential to change agriculture plant protection sectors by allowing sustainable management of crops and conservation of animal and plants production. However, research endeavors have so far have mostly focused on two groups, viz., the post-harvest food arena and next-generation pesticide formulations (Shrivastava and Dash 2009; Sozer and Kokini 2009; Sekhon 2010; Mousavi and Rezaei 2011; Khot et al. 2012).

Different strategies have been used to produce nanoparticles, among which include chemical and physical methods such as solvothermal synthesis, chemical reduction, sol-gel techniques, inert gas condensation, and laser ablation, but there are some limitations with these methods, such as feeble reproducibility, polydispersity, involvement of costly instruments as well as being energy intensive and eco-hazardous. To overcome these challenges, biological systems for synthesizing nanoparticles have served as prominent sources due to their remarkable variations in complex structural and functional activities. In addition, a number of single-celled organisms are currently reported to produce different types of particles, which have a nanoscale range from intracellular and/or extracellular. Realizing the importance of producing nanoparticles from natural biological systems, researchers are now able to develop a new alternative technology for synthesizing nanoparticles with the help of microorganisms. These biological systems have now proved to offer promising management products for sustainable plant protection, which have some unique properties in controlling plant pathogenic diseases and pests. Among the various biological methods for the synthesis of nanomaterial most research relies upon plant, bacteria, fungi and viruses for fabrication of inorganic nanomaterial. Different types of encapsulated pesticides, nanocapsules and nanoemulsions are produced from nanoparticles for better efficiency of pest and disease control in plants (Anjali et al. 2012). However, the most dependent and reliable source is from fungal-mediated synthesis, which is also known as mycosynthesis of nanoparticles. The nanomaterial biosynthesis form of fungi is mostly produced by targeting the extracellular region, which includes binding and transforming metals by metabolites, from polymers to specific polypeptides, as well as the metabolism-dependent activity reaction. One of the most important reasons for preferring fungi to produce nanoparticles over prokaryotes is the secretion of higher amounts of metabolites, resulting in a significantly higher production of nanoparticles through the biosynthetic method.

The application of different biologically based nanomaterials is an emerging area of research as these materials have the potential to enhance sustainable plant protection against various pests and diseases thereby increasing agricultural production, which is indeed necessary to meet the future demand of a growing population. Various research studies conducted globally indicate that the nanomaterials produced from biological sources at certain concentration ranges have the potential to overcome the challenges of invading pathogens to protect plants and thereby increase the yields of crops and reduce environmental pollution. Their application in crop production has been widely recognized through the production of crop protection products through nanofertilizers (Anjali et al. 2012) and also through precision farming using nanosensors (Kalpana-Sastry et al. 2009). The development of innovative nanomaterials is an emerging priority of the research from the point of view of both agricultural and environmental protection to ascertain food security for future generations and promotes sustainable agriculture.

5.2 NANOTECHNOLOGY IN AGRICULTURE

With the growing importance of bio-nanotechnology in agriculture, researchers are now actively involved in gaining knowledge about the application of nanoparticles in various fields for sustainable plant production. There are various applications of bio-nanotechnology in agriculture, some of the important aspects are discussed below.

5.2.1 Application as Nanofertilizers

Fertilizers play an important role in increasing food grain production. It has been reported that fertilizer contributes about 35–40% in production of various crops. The realization and utilization of nanofertilizer to increase production has resulted a drastic increase in the use of chemicals. Due to these overdoses and imbalanced fertilizer applications, various chemicals residues have caused pollution in groundwater leaving a major portion of fertilizer in the soil and causing eutrophication in aquatic systems. To overcome the challenges of chemical fertilizers such as imbalanced fertilization, deficiency of nutrients, organic matter degradation or low fertilizer use efficiency, nano-based fertilizer has proved to be quite promising tools. It has been found in pearl millet and cluster bean that foliar application of 640 mg/ha at 40 ppm concentration of nanophosphorous could give a yield that is equivalent to the application given by 80 kg phosphorous per ha (Raliya 2012; Tarafdar 2012). Even though nanofertilizer can be an effective approach for plant nutrient supply, only limited studies have been carried out and more studies are needed to know its impact on biochemical, nutritional, morphological and physiological changes along with fate in plant systems and soil.

5.2.2 Application as Nanoherbicides

Among various limitations in sustainable production of crops, weeds play a vital role in the production system and can even jeopardize the total harvest. Most of the herbicides that are available on the market today are able to manage only the aboveground plant parts of weeds and none are able to inhibit activities of the belowground parts. This causes hindrance in crop growth in new seasons and lowers productivity. The nanoherbicides can be a promising tool to overcome the negative challenges of weeds and can result in increasing yield thereby lowering damage to soil. Nanoparticles can be used to produce encapsulated nanoherbicides that can be utilized when it is desired such as in a rain system, where if there is spell of rain it will be available and remain protected under the natural environment. With this technique there is every possibility of developing herbicides that are target specific and encapsulated. For example, the use of nanoherbicides targeted to specific receptors in the roots of weeds helps in killing weeds by entering into the system and inhibiting glycoses of reserved food in roots which causes starvation for food in weeds. Use of this technology can help in overcoming the negative effect of chemicals in protecting beneficial microorganisms present in the soil and reducing synthetic herbicides applications. Also, with the combination of active ingredients with smart delivery systems the management of weeds will be effective and need-based application can be carried out based on field conditions.

5.2.3 Application as Nanopesticides

Pesticides play a vital role in the management of various invading pests and diseases and increasing crop production. They help in minimizing the pest population below the economic threshold level and have an effect on the control of various diseases. However, in order to effectively control pests and diseases in the long term, protection of active ingredients from environmental degradation is essential. Therefore, in order to promote persistence, nanotechnologies with nanoencapsulation approaches play a vital role in improving the pesticidal values. These are mostly comprised of nanoparticles that have active ingredient that are encapsulated within thin-wall sacs or shells that act as a protective coat. This coating helps in providing better control of pests and diseases, leaving very little damage to the non-targeted microorganisms and residues in the soil. Another important approach that can be used in protecting active ingredients and promoting persistence is the "controlled release of the active ingredient." The nano-based pesticides are specific and the rate of application is also very low as they are effective at 8–15 times smaller quantities than the classical formulation applied.

5.2.4 Application as Biosensors to Detect Nutrients and Contaminants

Nanotechnology application as a biosensor to detect soil health and environment protection is an emerging approach. It is also a rapid way to detect the sensitivity of pathogens and pollutants with molecular precision. Since the evaluation of soil fertility has been performed with same protocol over the decades, the sensitivity may not correspond with present-day farming systems. Therefore, to overcome these challenges, accurate sensors are needed and the application of nanotechnologies proves to be the best option. For example, ZnO nanowires are used to make gas sensors, i.e., Electronic nose (E-nose), which is widely used to identify different odor types. Foul odors are the characteristic volatile compounds produced by the fermenting microorganisms or by rotting microorganisms. The detection of these odors through application of nanotechnologies can prevent food loss through the rapid detection of the degrading products.

5.2.5 Application as Water Management

Utilization of nanotechnologies offers a potential novel approaches for management of surface water, wastewater and groundwater, which are contaminated by toxic chemicals, metal ions, different organic and inorganic solutes and microorganisms. The applications are effective in managing waterborne pathogens that are harmful not only to plants but also to human health. They are used to identify pathogens and filter the contaminated water with the utilization of filters made up of nanofiber membranes and also with the use of nanobiocides. They are used to produce biofilms, where mats of bacteria are wrapped in natural polymers to manage various contaminants in water.

5.2.6 Application in Animal Science

Nanotechnology has the potential and future orientation to provide appropriate solutions in detection and treatment of domesticated animals and for providing food items, veterinary care, medicines and vaccinations. Application of medications such as antibiotics and vaccines and probiotics for treating various diseases in animals will be much more effective when applied in nanoscale. Then nanolevel use of medicine has multilateral properties and increasing medicinal efficiencies. Apart from these, time of drug release, capacity planned and self-regulator capabilities provide an added advantage in the use of nanotechnology in the drug industry.

5.2.7 Application in Fisheries and Aquaculture

The application of nanotechnologies to revolutionize fisheries and aquaculture has increased tremendously. Various technologies such as nanosensors, gene delivery, drug delivery and DNA nanovaccines have been potential tools to solve various problems related to the nutrition and health of fish. This approach has also been a potential tool in the production of quality products by detecting bacterial contamination during packaging, producing strong flavors and monitoring the quality of color and safety.

5.2.8 Application in the Nanofood Industry

The importance of nanotechnology in the food industry has increased tremendously over the past as it is used for a number of food and food packaging products. The impact of nanotechnology in food processing ranges from basic processing of food to nutrition delivery to packaging. Nanotechnology can be used to develop nanofood through various engineering of food ingredients and to improve texture, flavor, taste, color and the ingredients of food without losing its nutritional value. Other applications include the waste utilization for nanoproduct developments and nanofilms for extending the shelf life of perishables fruits, flowers and vegetables.

5.2.9 Application in Seed Science

Quality seeds are the basic inputs that decide the fate of productivity of any crop. The application of nanotechnologies in seed science helps in the production of quality disease and pest-free seeds thereby enhancing the germination percentage. Carbon nanotubes have been found to be effective in improving seed germination in tomato. The mechanism shows that the carbon nanotubes (CNTs) served as new pores that allowed water permeation through penetration of the seed coat and acted as a gate to channelize the water from the substrate into the seeds which facilitated tomato seed germination (Khodakovskaya et al. 2009).

5.3 BIO-NANOTECHNOLOGY IN PLANT PROTECTION

5.3.1 Bioagents as a Source of Nanoparticles

Microorganisms, viz., fungi, bacteria, actinomycetes and yeast, are used for synthesis of various metal nanoparticles with different shapes and sizes. After bacteria and fungi, many other types of organisms such as actinomycetes and algae were discovered which could help in nanomaterial synthesis (Ahmad et al. 2003; Lengke et al. 2006). The mechanism of synthesis of nanoparticles from multicellular and unicellular microorganisms varies; however, the actual mechanism used to synthesize the nanoparticle is based on hypothesis only. The nanomaterials from these microorganisms are synthesized using the regular chemical media and the apparatus and do not require a heavy instrument. The selection of the microorganism in synthesizing nanoparticles is important because not every fungus or bacteria is capable of producing nanoparticles, and only those isolates that are resistant or tolerant to metal can synthesize nanoparticles. These microbes can be screened using different specific enrichment and isolation methods (Figure 5.1).

Microbes produce nanomaterials in two ways: Intracellularly, i.e., within the cytoplasm or the cell wall or periplasmic space and extracellularly, by secretion of capping or reducing agents in the external surrounding media. As compared to extracellular production of nanoparticles, harvesting of intracellularly produced nanomaterials is a tedious job as it requires the separation of nanoparticles from the organic matrix and components of cells, whereas extracellularly produced nanoparticles are easier to harvest as they are free from the cell components. Therefore, from a large-scale production point of view, extracellular nanoparticle production is less cumbersome; hence, more research has been done on this type of synthesis. However, there are various factors affecting nanoparticles synthesis; for example, slight changes in pH, reducing temperature, and reducing agent concentration lead to different shapes and sizes of the nanoparticles for nanospheres, nanotriangles, nanorods, etc., and different types of nanoparticles serve different purposes.

5.3.2 Types of Nanoparticles

There are mainly two types of nanoparticles synthesized, i.e., Natural and Artificial that are mostly fabricated. Natural nanoparticles can be obtained from volcanic ashes, viruses, sea spray, forest fires, mineral composites, and from human origins such as diesel exhaust,

Flowchart of isolation of nanoparticles from microorganism

1. Isolation of microorganism

2. Identification of microorganism

3. Culturing of microorganism

4. Centrifugation of broth, washing and suspension

5. Challenging with metals

6. Examination of the supernatant

7. Confirmation

8. Characterization of the nanoparticles

FIGURE 5.1 Isolation of nanoparticles from a microorganism.

welding fumes, sand blasting, industrial effluents, cooking, smoke, etc. Some of the other naturally occurring nanomaterials exist in biological systems, viz., viruses (Capsids), bon matrix substances, etc.; however, artificial nanomaterials are manufactured from metal-based, quantum dots, composites, carbon-based and dendrimers. They are mostly produced by fabrication in different experiments.

5.3.2.1 Nanomaterials Can Be Classified Based on Their Sources in Four Classes

5.3.2.1.1 Carbon-Based

These nanomaterials are mostly composed of carbon and are mostly ellipsoids, tubes, hollow or spheres in shape. Those carbon nanomaterials which are ellipsoidal or spherical in shape are called fullerenes and those which are cylindrical in shape are referred to as carbon nanotubes (CNTs).

5.3.2.1.2 Metal-Based

These nanoparticles mostly include nano-gold, nano-silver, metal oxides or quantum dots, which are produced as a result of reduction of metal compounds.

5.3.2.1.3 Dendrimers

They are mostly nano-sized polymers which are built from branched units and their surface contains many chain ends that can be tailored to perform specific chemical functions. These branched properties are useful for catalysis and are also useful in delivering drugs due to their three-dimensional structure, which contains interior cavities where other molecules can be placed.

5.3.2.1.4 Composites

Composites mostly combine nanoparticles with other nanoparticles or help to combine nanoparticles to larger, bulk-type materials. They may be any combination of carbon-, metal- or polymer-based nanomaterials with any form of ceramic, metal or polymer bulk materials.

5.3.3 Synthesis of Nanoparticles

Two different approaches are followed in synthesis of nanomaterials as given below.

5.3.3.1 Top-Down Approach

The top-down method involves the breakdown of bulk-sized material into nanoscale structures or particles. In this process, micron-sized particles are broken down to nano-sized objects by giving high energy into the system. This method is much simpler and is based on either removal or division or miniaturization of bulk material to produce the required structure with the appropriate size and properties. The main disadvantage with the top-down approach is the surface structure imperfection, as pointed out by Petit et al. (1993). In this method, the top-down approach involves various instruments—electron beam lithography, high-energy wet ball milling, gas-phase condensation, aerosol spray—atomic force manipulation is being used. Processes like homogenization, ball-milling and laser ablation fall under this category. Although this is being followed commercially for various applications, the size distribution of resultant particles is very wide. Recently, work has been done to use enzyme-degradation systems for the production of nanomaterials with low-energy input.

5.3.3.2 Bottom-Up Approach

The bottom-up (self-assembly) approach refers to the buildup of a material from the bottom as atom-by-atom, molecule-by-molecule or cluster-by-cluster. Here, atoms/ions/molecules are used as precursors and they are assembled to form nano-size particles by chemical and/or self-assembly processes. Many of these techniques used in the bottom-up approach are still under development for use in commercial production of nanopowders. This approach mainly focuses on the chemical synthesis of nanoparticles that are luminescent and involves reverse-micelle, organometallic route, hydrothermal synthesis, colloidal precipitation, sol-gel synthesis, electrodeposition and template assisted sol-gel. The advantage here is the low energy requirement and narrow size distribution.

5.3.3.3 Synthetic Approaches to Metallic Nanoparticles

Recent research on nanoparticles and nanoscale materials has generated a great deal of interest among scientists. This interest was due to the different characteristics of nanoparticles, such as surface reactivity, optical and magnetic properties, specific heats and melting point. The surface-dependent properties were due to the results of a high ratio of surface to bulk atoms since these two states represent a relation between atomic and bulk materials. In nanoscale, both metal and metal oxide materials have well-defined molecular orbitals as standard bulk material, which is represented by electronic band structure. Metallic particles were of great industrial importance, so an understanding of small cluster to bulk materials properties was needed. The method of preparation of nanoscale colloidal metals is of great interest for the scientist for its diverse applications.

5.3.3.3.1 Wet Chemical Preparations

These methods used "Top-down" and "Bottom-up" approaches in preparation of nanostructure metal colloids. It has been reported that the former method involves mechanical grinding of bulk metals and subsequent stabilization, resulting in nano-sized materials by the addition of colloidal protecting agents (Amulyavichus et al. 1998). Metal vapor techniques provide versatile routes for the production of a wide variety of nanostructured metal colloids in the laboratory (Furstner 2007). However, this technique has some limitations since it was difficult to get narrow size distribution. The "Bottom-up" method of wet chemical nanoparticle preparations mainly depends on chemical reduction method of metal salts, electrochemical pathways or controlled deposition of meta-stable organometallic compounds. It was reported that a variety of stabilizers were used during preparation of nanoparticles, e.g., donor ligands, surfactants and polymers. These were mainly utilized for controlling the growth of nanoclusters that are primarily formed and for preventing them from agglomeration. It was reported that the chemical reduction method of transition metal salts by the presence of stabilizing agents lead to the formation of zerovalent metal colloids in organic or aqueous media. Metal salts reduced to give zerovalent metal atoms at an embryonic stage of nucleation were observed (Leisner et al. 1996). These will further collide with solutions of metal atoms, metal ions or clusters to form an irreversible seed of stable nuclei. Protectants are being required in nanostructured colloidal metals in stabilizing and preventing agglomeration. The two basic modes responsible for stabilization were electrostatic and steric force. It was observed that electrostatic stabilization mainly involves columbic repulsion between the particles due to electrical double layers formed by ions adsorbed at the particle surface and the corresponding counter ions. Steric stabilization was achieved by coordinating organic molecules that act as protective shield on the metallic surfaces (Bradley and Lang 1994).

5.3.3.3.2 Reducing Agents

Reducing agent types greatly affect the resulting nanoparticles and their structure. It was observed that silver when reacting with strong reducing agents produced small nuclei called "seeds" (Leisner et al. 2007). During the process of ripening, these small nuclei grow to produce colloidal metal particles of a size of 1–50 nm with narrow size distributions. The mechanism behind the particle formation was the agglomeration of zerovalent nuclei in the "seed," or alternatively due to the collision of already formed nuclei with reduced metal atoms. Henglein's group followed stepwise the reductive formation of Ag^{3+} and Ag^{4+} clusters by the spectroscopic method. The results of the spectroscopic method suggested involvement of an autocatalytic pathway in which metal ions were adsorbed and reduced at the zerovalent cluster surface. The size of resulting metal colloids can be determined by the relative rate of particle growth and nucleation. However, the process that takes place between or during particle growth and nucleation cannot be analyzed separately. The advantage of the salt reduction method in liquid phase was that it can be easily reproducible and allows colloidal nanoparticles with a narrow size distribution.

5.3.3.3.3 *Electrochemical Synthesis*

Many versatile preparations methods for nanostructured monometallic and bimetallic colloids were developed between 1994 and the present (Furstner 2007). One among them was electrochemical synthesis of colloid nanoparticles. There were certain advantages to the electrochemical pathway compared to other method of preparation. Firstly, it prevents contamination with by-products which are produced by chemical reducing agents and the products can be easily separated from the precipitate. Another advantage was that it helps in the formation of size-selective particles. It was found in an experiment where palladium was used as an anode in electrochemical cells to give (C8H17) $N^{+}Br^{-}$ stabilized (Pd0) particles, indicating that particle size mainly depends on current density. The high current densities results in small palladium particles, whereas low current densities gave larger particles. TEM image and small angle X-ray scattering (SAXS) observations found that particle size was not controlled by single parameters but depended on the following parameters: reaction time and temperature, distance between the electrodes and polarity of the solvent (Reetz and Helbig 1994).

5.3.4 Characterization of Nanoparticles

Confirmation of the synthesized nanoparticle is accomplished by using various approaches for their application. It can be done as a preliminary means by observing the change in color, physical look, structure, composition and properties. Further, the specific peaks given by particle within the range 250 to 800 nm also ascertain the formulations (Sunkar et al. 2012). The myco-synthesized nanoparticles can be characterized by employing various techniques, viz., UV-visible spectroscopy, Fourier transform infrared spectroscopy, atomic force microscopy, scanning electron microscopy, transmission electron microscopy, dynamic light scattering, energy dispersive X-ray spectroscopy, and eta potential measurements and X-ray diffraction which are discussed below.

5.3.4.1 UV-Visible Spectroscopy

This is one of the vital tools for the identification, characterization and study of nanoparticles. This works on the principle of quantification of light which is absorbed and scattered by a solution or a compound, i.e., measuring the surface plasmon resonance (SPR) frequency of the particles in the solution. The plasmonic nanoparticles are sensitive to shape, size, concentration, agglomerate state of the nanoparticle and exhibit optical properties. Whereas the non-plasmonic nanoparticles are not sensitive to properties of dispersion, they have concentration and size-dependent optical properties as plasmonic nanoparticles. The intensity of the light is measured when a sample is kept in between the light source and photodetector. It has to be recorded before and after passing through the sample. At a desired range of wavelength, a measurement scan is performed for generating the wavelength dependent extinction spectrum. The number of nanoparticle present is depicted by the intensity of the peak. There is a positive relation of the particle concentration to the breadth and height of the peak. This technique has certain advantages as SPR can be utilized in various applications such as optical sensing, biosensors, bioimaging, etc. Moreover the sample preparation is very simple and doesn't require skilled labor. However, the technique is not applicable for solid samples and is restricted only to homogenous solution.

5.3.4.2 Fourier Transform Infrared Spectroscopy

This technique helps in the determination of the mycosynthesized nanoparticle functional group by measuring the wavelength with infrared of light. The principle of the Fourier transform infrared spectroscopy is that, according to the type of bonds and the elements, the chemical bonds of a molecule vibrate at particular frequencies. For measurement, the specimen spot is placed to a modulated infra-red beam. The measured transmittance and reflectance at various frequencies is again translated to an IR absorption plot. The spectral pattern is analyzed and matched with the known identified materials signature in the FTIR library. It is also used for studying the presence or absence of a protein molecule in a solution. Moreover, it helps in identifying the presence of metal and their oxides.

5.3.4.3 Atomic Force Microscopy

Atomic force microscopy (AFM) is an instrument used for imaging a small probe (< 1 μm) by scanning over its surface. The physical property of the images and its surface topography can be measured by AFM through visualizing the wave of probe. It is a suitable tool for determining a sample's angstrom scale's roughness. Besides this, it also gives quantitative measurement of the characteristics of the sample (e.g., capacitance measurements), as well as feature sizes measurement (e.g., step height). The instrument has various applications, as it has the provision of quantifying surface roughness and cleanliness; tench shape on patterned wafers can be assessed and can analyze the whole wafer (300, 200, 100 mm); it has high-spatial resolution; and it also has the ability to image the insulating and conducting samples. However, there is the possibility of creating problems with oddly shaped and very rough samples. The range limit for scanning is only of 100 μ laterally and 5 μ in z direction. Also, it has a potential of tip-induced errors.

5.3.4.4 Transmission Electron Microscopy

Transmission electron microscope (TEM) is useful for capturing the images of nanoparticle samples by transmission of beams of electrons through the sample. This high magnification microscopy technique basically assists in measuring the morphology, particle size and size distribution of the nanoparticle. The sample is usually prepared on carbon-coated copper grids by adding a single drop of the nanoparticle solution. The excess sample can be removed by absorbent paper. The grid is placed in incubation overnight at room temperature in a vacuum desiccator. Researchers observed various shapes of diamond, rod-like, spherical, and cubic with various sizes (1–100 nm in diameter). Even though the technique is very useful, it requires expertise and needs more time in sample preparation. It also requires proper maintenance since it involves highly sophisticated instrumentation.

5.3.4.5 Scanning Electron Microscopy

Scanning electron microscope (SEM) is an instrument which is very useful for determining the topology and visualizing the surface of the sample. The high electrical conductivity of metal nanoparticles makes them easy to scan them by SEM. It has also a high magnifying power offering better resolution and depth of field than an optical microscope. SEM can be carried out by placing samples on black surfaces so as to prevent unlikely scattering of the incident beam. However, it is a very expensive and sophisticated technique. Also, for analysis of the sample it requires a well skilled technician.

5.3.4.6 Energy Dispersive X-ray Spectroscopy

The technique, also referred as energy dispersive X-ray (EDX) analysis, is a micro-analysis technique, which is usually carried out to determine the composition of the sample. The technique is utilized in conjunction with the electron microscope. During electron bombardment of the sample, the emitted X-rays from the sample are used for determining the said composition. After bombarding the sample by electron beam of SEM from the atoms the electrons are ejected. The vacancies of the electron are filled by another electron from a higher shell, so as to balance the difference of energy between the two electrons, as an X-ray is being emitted. The emitted X-ray number is measured by EDS detector and plots their value against the energy. The X-ray which was emitted from the element is characterized by the energy of the X-ray. EDS can also determine the chemical composition in a semi quantitative mode by peak height ratio in relation to a standard. However, it cannot detect those elements which are very light.

5.3.4.7 Dynamic Light Scattering and Zeta Potential Measurements

Dynamic light scattering (DLS) is an efficient tool for observing the characteristics of bio-nanoparticles and other colloidal solutions. This technique can check the particle surface charge and its size distribution. DLS is based on the interaction of light with the particle. The diameter of the particle in solution can be obtained by observing the scattered light intensity in the colloidal solution. The determination can be done by observing the diffusion motion of its particle where smaller particle move rapidly and scatter less light as compared to larger particles. The technique is utilized for measuring the particle range from 1–500 nm; however, it is cumbersome for recording the size of agglomerated particles (Akbari et al. 2011). Moreover, the instrument (at 25°C with 90° scattering angle) also analyzed the charge and mean size of the particles for the dried sample of pure filtered nanoparticles after dissolving in solvent (Sharma et al. 2016). Besides these, the time-dependent nanoparticle formation can be sorted out by applying the technique.

The effective charge on the surface of the nanoparticle is measured by zeta potential and quantifies its colloidal nanoparticle charge stability. The charged nanoparticle can be screened by an inclination in the ion concentration of opposite charges around the particle surface. The peak area and peak number involving the charge on the particles measures zeta potential. Negative charge shows that the particle size is smaller than 100 nm and vice versa. The smaller particles indicate more stability (Sunkar et al.2012; Birla et al. 2013). The pH of the solution is one of the factors which determines the magnitude of the nanoparticles charge. At a specific pH, the charge of the particles surface may be reduced to zero, which is referred as isoelectric point.

Nature of Nanoparticle	Zeta Potential
Highly stable	>40 mV
Moderately stable	20–40 mV
Minimally stable	5–20 mV
Tend to aggregate	0–5 mV

5.3.4.8 X-ray Diffraction

X-ray diffraction is the most widely used technique for studying simple or complicated crystalline structure, size, and their diffraction patterns. Due to a variation in nanometer scale structure, the material property can be affected. Still, the structure of the crystal can be determined by separating the atom from one another by 0.1 nm, approximately. There is a unique diffraction pattern of X-ray for each crystalline substance. The scattering intensities of X-ray is directly proportional to the electron number in the atom. Therefore, X-rays are scattered efficiently by heavy atoms and weakly by light atoms. For every crystalline material, a unique fingerprint X-ray powder pattern is characterized by the following three features of pattern of diffraction—the peaks number, the peak position and its intensities. The Joint Committee on Powder Diffraction Standards–International Center for Diffraction Data (JCPDS-ICDD) is the widely used database for identifying crystal structure. The powder XRD is based on the principle of Bragg's law (ʎ = $2d$ sinθ). The technique can also be applied for sample purity measurement, unit cell dimensions determination, determination of nanostructure thickness, roughness and density, etc.

5.4 NANOPESTICIDE USAGE: EMERGING THREATS OR OPPORTUNITIES FOR THE GROWERS?

Nanopesticide usage has gained tremendous importance over the past among the researcher and has been able to create many new opportunities for management of the various pests and diseases through is effective application. Some of them are innovative nanomaterials that cause less pollution and/or contamination to the environment. Some of the nanoparticles such as carbon nanotube and metal nanoparticles have proved to be effective for management of pollution in water and soil.

Even though their importance is increasing day by day on the one hand, on the other, many questions remain unanswered. For example, to what extent the nanopesticides may pose a risk to human health, beneficial microbes and the environment is not yet fully understood. Therefore, as suggested by precautionary principles the application of nanopesticides is to be kept minimal until their toxicity and fate are clearly understood.

Researchers these days believe that the application of nanopesticides has the potential to revolutionize pest and disease management practices, thereby reducing chemical hazards caused by synthetic pesticides. This technology can give sustainable protection measures, thereby increasing plant production and reducing residue toxicity in food and the soil. Scientists around the globe have now been actively involved in the development of a new generation of pesticides based on nanotechnology. Research is emerging at high speed in various agrochemical industry laboratories, but due to the as of yet lack of awareness of this technology by the public or government authorities, no product has so far been produced or become available in the market. Since nanopesticides have enhanced properties and new mechanisms involved for the management of pests and diseases, the realization of their importance will not take much time and their utilization will change in the near future. This may result inevitably in both advantages or benefits to the human and environmental health or will create a new risk to both.

Since nanopesticides consist of a large variety of products, their application can only be the intentional diffused input of large-scale engineered nanomaterials in the environment. Since innovation in any new material has both drawbacks and benefits to human health, environmental application of nanopesticides may reduce the contamination of chemical pesticides in soil, food and the environment through the reduction in synthetic pesticide usage but may also create new kinds of risks due to contamination of nanopesticides to soil and water that may be toxic to humans and the environment due to their enhanced transport, higher toxicity and longer persistence.

Our current level of knowledge does not allow for a fair assessment of the advantages and disadvantages of nanopesticide usage. Therefore, such an assessment requires more research and a greater understanding of both the benefits and risks of nanopesticides after their application on our overall health. More research on nanopesticide production and applications is therefore needed for maintaining the quality of the food chain and environment and thereby sustainable production and protection of agricultural crops.

5.5 IMPACT OF BIO-NANOPARTICLE APPLICATION

5.5.1 Response of Plants to Nanoparticles

Nanoparticles have unique properties and have proved their ability in modifying the genetic constitution of various crop plants, thereby helping in crop improvement (DeRosa et al. 2010; Jones 2006). There are various positive and negative impacts when nanoparticles are applied to plants. Some of the positive effects include increased percentage of germination, root and shoot growth, and an increase in vegetative biomass in many crops such as lettuce, spinach, pumpkin, cucumber, tomato, etc. Enhancement of many physiological parameters related to plant growth and development were also reported that include increased photosynthetic activity and nitrogen metabolism by MBNMs in a few crops, including spinach, and tomato by multiwalled carbon nanotubes (MWCNTs). Also they have countless positive effects on numerous plants during different growth stages from germination to harvesting. Besides the improvement in morphological characters at seedling level, enhancement of many physiological parameters had also been reported. These included nitrogen metabolism, chlorophyll content, activities of a number of enzymes related to photosynthesis, promoted levels of protein and mRNA and accelerated expression of gene particularly those related to stress (Rico et al. 2011). A group of researchers is currently working on metal oxide nanoparticles and carbon nanotubes to improve the germination of rain-fed crops. There was an enhancement on the yield of plant biomass, fruit characters and phytomedicine content, increase in germination of tomato seeds with the use of carbon nanotubes (Kole et al. 2013; Khodakovskaya et al. 2009). Both positive (Table 5.1) and negative effects (Table 5.2) of nanomaterials on higher plants have been reported (Rico et al. 2011). However, their effects depend on the crop species and method of treatment. The effect of NPs on plants varies from plant to plant and species to species and also the age of the targeted population.

5.5.2 Response of Soil to Nanoparticles

The impact of nanoparticle applications in the soil is one of the most important factors as the soil is the largest receptor of nanoparticles application. The soil is a naturally occurring matrix which consists of a large number of natural nanoparticles as primary particles or as an aggregates/agglomerates. The impact of the application of nanoparticles on soil health has lately been very topical in both arable and natural ecosystems. Their artificial application may have positive and negative impacts on the soil health as they may be less degradable and may have the potentialities to accumulate and persist in the soil for a longer time.

Many of the studies carried out in the past by different researchers have reported both the beneficial as well as harmful effects to the soil. There are many effects on the properties of the soil on the microscopic level (Ben-Moshe et al. 2012). The effect on soil owing to nanoparticle application depends upon various issues such as soil type, its concentration of application and enzymatic activities of soil, and its protection and diversity is one of the important factors in sustainable use and in preventing harmful effects to beneficial microorganism (Table 5.3). The negative effect on the various enzymatic activities and mostly on dehydrogenase enzyme activity was found to be largely affected when nanoparticles were applied in larger concentrations (Josko et al. 2014). In addition, a negative impact has also been reported on the rate of soil self-cleaning and nutrient balance which has hampered the regulation of plant nutrition and soil fertility improvement (Suresh et al. 2013). Therefore, more studies are needed in the future focusing on the significant impact of nanoparticles, their persistence and their impact on microbial biodiversity.

As well, the effect of nanoparticles on soil physical properties such as texture, structure, pH, organic matter content, etc., needs to be studied as they may influence natural soil's ability to detoxify the pollutants and may cause harmful effects on beneficial microorganism communities. It is also observed that the nanoparticles can

TABLE 5.1
Positive Response of Nanoparticles in Crop Plants

Nanoparticle	Particle Size (nm)	Plant	Effective Concentration	Observed Toxicity	Reference
Al	1–100	Radish, rape, lettuce Corn, cucumber Red kidney beans, ryegrass, radish, rape	2000 mg L^{-1} 10, 100, 1000 10,000 mg L^{-1} 2000 mg L^{-1}	No effect on germination No observed toxicity Improved root growth	Rico et al. 2011
Ag	2	Cucumber, lettuce	62, 100, 116 mg L^{-1}	Low to zero toxicity	Rico et al. 2011
Au	10	Cucumber, lettuce	62, 100, 116 mg L^{-1}	Positive effect on germination index	Rico et al. 2011
Si		Zucchini	1000 mg L^{-1}	No effect on the germination	Rico et al. 2011
Cu		Lettuce	0.013% (w/w)	Improved shoot/root	Rico et al. 2011
Fe_3O4	20	Pumpkin cucumber, lettuce	500 mg L^{-1} 62, 100, 116 mg L^{-1}	No toxic effect low to zero toxicity	Rico et al. 2011
Nanoanatase (TiO_2)	4–6	Spinach	0.25%	Enhanced mRNA expressions (51%), Protein levels (42%), activity of Rubisco Activase, Rubisco carboxylation, improved light absorbance	Rico et al. 2011
Rutile (TiO_2)		Spinach (naturally aged)	0.25–4%	Increased germination and vigor indices	Rico et al. 2011
Mixture of Au/ Cu		Lettuce	0.013% (w/w)	No effect on the germination Improved shoot/root ratio	Rico et al. 2011
Multiwalled carbon		Tomato	10–40 mg L^{-1}	Significant increase in germination nanotube rate, fresh biomass, and length of stem Significantly enhanced moisture content inside	Rico et al. 2011
Single-walled carbon nanotube	8	Onion, cucumber	104, 315, 1750 mg L^{-1}	Tomato seeds significantly increased root length	Rico et al. 2011

TABLE 5.2
Negative Effects of Nanoparticles in Crop Plants

Nanoparticle	Particle Size (nm)	Plant	Effective Concentration (n mg L^{-1})	Observed Toxicity	References
Ag	100	Zucchini	100, 500, 1000	Reduced transpiration (41–79%), reduced biomass (57–71%)	Rico et al. 2011
Ag	<100	Onion	100	Decreased mitosis Disturbed metaphase Sticky chromosome Cell wall disintegration and breaks	Rico et al. 2011
Si	10	Zucchini	1000	Completely inhibited germination	Rico et al. 2011
Al		Corn, lettuce	2000	Reduced root length	Rico et al. 2011
Zn	Ryegrass, lettuce, Corn, cucumber		2000	Reduced root length	Rico et al. 2011
CeO_2	7	Tomato Tomato, cucumber	2000 2000	Significantly reduced shoot growth (30%) Reduced germination	Rico et al. 2011
Multiwalled carbon nanotube (MWCNTs)	10–30	Zucchini Lettuce	1000 2000	Reduced biomass (38%) Reduced root length	Rico et al. 2011
Single-walled carbon nanotube (SWCNTs)	8	Tomato		Most sensitive in root reduction	Rico et al. 2011

TABLE 5.3
Effect of Nanoparticles on Properties of Soil

Nanoparticles	Effect	References
ZnO	Organic matter and pH	Waalewin-Koot et al. 2014
Zn	Bacterial communities	Garcia-Gomez et al. 2018
CuO	Change pH in paddy soil	Shi et al. 2018
Ag	Soilborne insect feeding and change in pH	Javed et al. 2019
Zn	Higher retention in calcareous soil causing less availability to the microorganism and influence on pH	Javed et al. 2019
AgNPs	Lethal effect toward earthworm and inhibiting locomotion of earthworm	Kwak et al. 2016

affect the mobility of pollutants; therefore, thorough studies are needed to gather the information about the toxicity level of NPs on different types of soil before a recommendation can be made. The nanoparticles may affect the particle size distribution and organic matter composition resulting in a change in microbial population and communities (Calvarro et al. 2014). Also, the soil properties and composition may change due to the use of high doses of nanoparticles.

5.5.2.1 Effect on Soil pH

Soil pH is an important factor that is directly related to soil health and fertility. It mostly determines the acidity or alkalinity and the availability of the nutrients to the plants. The soil pH depends upon various factors, and one of the important factors is the soil composition. Generally, 5–7 pH is considered as optimum range is favored by most of the plants for nutrient uptake and growth. The nanoparticles such as Zn, Ag, Au and Cu have been found to mainly influence the pH of soil due to their accumulation and create toxicity causing negative effects on many microorganisms. Some of the impacts of Nps application on soil pH are listed in Table 5.3.

5.5.3 Response of the Environment to Nanoparticles

The application of metal-based nanoparticles has raised a lot of concern for human health, as well as for environment. In the environment, there are various uses of nanoparticles, such as for removing contaminants from air, water and sewage. Also, they are being used as an environmental instrument, viz., green nanotechnology, sensors and in the reduction of greenhouse gases. In addition to these utilities of nanoparticles, they also cause hazards to the environment starting from their production to their disposal (Taghavi et al. 2013). The environment most affected has been expected to be the terrestrial one, especially the soil, where the greatest effect will be seen, as it is the largest repository of environmentally released nanoparticles such as metal-oxide nanoparticles. Therefore, more studies and an accurate determination of the effect of nanoparticles on the environment is needed to access real-time scenarios of harmful effects on soil health and also the threat to humans through the food chain. Nanoparticles based on carbon like fullerenes, nanotubes, natural inorganic compounds like asbestos and quartz, and metal oxides like iron and titanium, may have biological impacts on their surroundings, including the environment and human health (Taghavi et al. 2013). A series of safety regulations and evaluations of toxicological risks must be formulated before the application of nanoparticles as many nanoproducts are now commercially available in the form of fertilizers and pesticides in a market where the dose and type of nanoparticle is not properly documented, creating a threat to the environment and the inhabiting population.

To date, more than 1600 nanoproducts are available and more than 3000 products are on online data base lists, which are of various categories containing nanoparticles. This production is estimated to reach more than 1.6 million tons by the year 2020 where many products have not disclosed the actual nanoparticles used. Many more are expected to be made commercially available in the coming year with novel application and characteristics (Anonymous 2020). Because of high specificity and various chemical and physical properties and high surface volume ratio, their application and importance is gaining rapidly. Application of these nanoparticle in food packaging, agricultural products, pesticides and fertilizers has raised health concerns due to their slow degradation, which can cause major ecological, ethical, safety, policy making and regulation problems.

Most of the nanoproducts end up in the soil either through dumping directly or through discarded products or wastes. This will affect the properties of the soil and could also form a new kind of toxic substance by interacting with the pollutants causing risk and hindrance to the functioning of the soil microbial population, growth of plants and even human health. To date, there is a major knowledge gap regarding the behavior and potential toxic risk of Mo-NPs due to a lack of knowledge of their mechanisms and can only be overcome by the development of more studies on detection methods, quantification and identification. Therefore, it is essential to explore systematically and examine in depth the level of toxicity and mechanisms and accumulation in the soil and plants.

Silver particles have been found at water treatment plants in sewage sludge. The particles in the water can thus be ingested or absorbed by aquatic organisms. The microbial populations that are present in water are vulnerable to silver nanoparticles contamination, which may inhibit their growth and other forms of metabolism which are very essential for the wastewater treatments process. The nanoparticles also threaten both the aquatic and terrestrial microbial populations of many ecosystems. They may also enter through the bodies of various aquatic edible plants and organisms which indirectly are exposed to humans and other mammals. Skin diseases such as argyria or argyrosis rise due to exposure to silver salt or colloidal silver deposits of metallic silver (Eisler 1996).

The interaction of Mo-NPs in environmental conditions varies with plant to plants, species, microbes and even the type of medium. To date, most of the research on MO-NPs has been carried out on artificial media in controlled conditions which do not reflect the actual interaction mechanism of natural ecosystems. Therefore, quantitative research with joint action is needed to determine the effect on the environment, taking consideration of changes in climatic conditions and soil types so that the threshold level can be determined for sustainable application. This can help in making decisions about the correct dose of application at the correct time, such as at different crop growth stages, to get maximum benefits.

5.6 NANOPESTICIDES AS RISK MITIGATION TOWARD CHEMICAL HAZARDS

The application of nanotechnology has proven to be a promising tool in formulating many new products with wide application range. It has been successfully applied in several fields of agriculture such as food, fertilizers, biofuels, agrochemical industries, etc. Among all these fields, one of the important innovative based applications of nanotechnology includes the management of insect pests and diseases through nanopesticides. Nanopesticides based on SiO_2, Ag, ZnO and Cu formulation have a broad spectrum efficiency and are specific to the targets. Also, they help to reduce water requirements and pollution in the soil and environment during the application period. The Zn-based nanopesticides have proved to be excellent for dealing with pests and are also beneficial for plants as an essential nutrient. Besides this, they are also cost effective and safe as compared to chemical pesticides. Apart from Zn, silver (Ag) also has the potentialities of microbial, antibacterial, larvicidal, fungal, pesticidal and anti-viral activity and are effective and target specific. Various Ag-based nanoparticles which have been found effective against fungal pathogen *Fusarium graminearum* have been reported. Ag-based nanopesticides have the advantage of slow kinetic release, are stable and can help in the production of new generations of eco-friendly pesticides. The Ag nanopesticides synthesized from aloe vera are effective for the management of *H. armegera* (Rajakumar and Rahuman 2011). Besides these, larvicidal activities for controlling the mosquito hold true when silver nanoparticles (AgNPs) are extracted using form *Eclipta prostrate.* AgNP of 35–60 nm size showed the maximum efficacy in crude aqueous against *C. quinquefasciatus* and *A. subpictus.* The pesticidal effect of $AgNO_3$ can be incorporated effectively as an eco-friendly approach in management of *S. oryzae.* The extracts form *Catharanthus roseus* possess a good larvicidal property against the *Anopheles stephensi* (Panneerselvam et al. 2013). AgNPs was extracted from *Ficus religiosa*, Peepal tree, banyan tree and *Ficus benghalensis* successfully controlled the pest *H. armigera*, modulating the function of gut proteases (Kantrao et al. 2017). It was also confirmed that the AgNPs from *F. religiosa* and *F. benghalensis* with a concentration ranging between 50–70% was capable for reducing the larval weight and survival rate of *H. armigera.* But more effective results were found when larvicidal compounds were integrated with silver and showed more prominent results, as compared to other mixture of secondary metabolites (Devi et al. 2014).

At present, the use of nano-based pesticides are the cheapest and fastest methods for controlling pests and diseases. Many adverse effects and problems have been caused due to indiscriminate use of synthetic pesticides on beneficial insects, microorganisms, domestic animals, human health and the environment through direct or indirect exposure. The use of nanoscale pesticides, on the other hand, can be an appropriate solution to this problem and can help in overcoming all the suffering caused by pathogens in the crop plants. Many of the nanomaterials, such as iron oxide nanoparticles and polymeric nanoparticles, as well as gold nanoparticle can be synthesized easily and used as pesticides for managing different diseases and pests. Some sulfur-based nanoparticles (SNPs) have been found to be effective against fungal pathogens such as *Fusarium solani* and *Venturia inaequalis* (Rao and Paria 2013). The nanoparticles of size 35 nm are reported to be very effective in preventing fungal growth and can be used in protecting important crops such as potato, tomato, grapes and apple from diseases of various origins. They can also be used in organic management of pests and diseases due to the antimicrobial properties of the silver particle. Several nano-based diagnostic kits have been developed for easy and rapid detection of viruses.

5.7 CONCLUSION AND FUTURE PROSPECTS

Among various agricultural sectors plant protection is one of the busiest areas for the researcher in order to formulate active solutions to overcome the hazards of synthetic pesticides. Nanopesticide technologies are now combined with traditional methods of managing pests and diseases and have resulted in many effective strategies with better solubility, slow-releasing properties and avoidance of premature degradation. The use of biological compounds in capped and reducing agents to various nanoparticles has resulted in much more stability. It has now become possible to apply nanopesticides directly or indirectly as a vector to many inherent characteristics such as permeability, crystallinity, biodegradability, stiffness and thermal stability. Such properties have proven to be advantageous over synthetic pesticides that are generally used. There are various effect caused by these nano-carries to the insect pest such as indigestion, causing desiccation and death of insects. Their nano-size makes them effective to enter and act within the insect pest even in small doses. They help to protect the secondary metabolites and help to release bioactive compounds in a controlled way to act upon the insect pests. These are many possible ways through which nano-biopesticides can be synthesized using plant pesticidal secondary metabolites and can be applied in the form of sprays. This can effectively remove pest and disease infestations.

The important factors that need to be taken into account before the synthesis of any pesticides are:

- They should be easy to prepare.
- They should be effective at a very economical rate.
- They must be specific toward targeted pests.
- They must be effective toward a wide range of pests and diseases.
- They must be safe to all living beings.
- They must be eco-friendly.
- They must be non-toxic to farmers.
- They must not contain any kind of toxic chemicals and must be pre-approved by health authorities before use.
- They must not accumulate in food produce.
- They must not leave or produce unacceptable residues.
- The legal lethal dose concentration must be framed before its use.
- They must not affect the quality, flavor, texture and fragrance of the food.
- They must not be flammable, corrosive or explosive and should be accepted in the international market.
- They must be easy in application.

Considering the above factors, numerous studies have been carried out on the toxicity effect of nanoparticles on various pests and diseases. However, more studies are still needed to know the further effects upon the environment and health. Less is known about the effect of various nanoparticles on important pests such as *Brachytrypes portentosus*, *Helicoverpa armigera*, *Laphygma exigua*, *Episomus lacerta*, termites, etc. Therefore, in order to fill the gaps in knowledge extensive funding is needed to come up with more solutions and findings in the field of plant protection. The identification of various plant species containing secondary metabolites that are effective in repelling insects and controlling plant disease must done in the near future. Only then will the biologically synthesized nanoparticles that are effective against various pests and diseases be clarified and conclusions made about the specific resistance of a plant species for a particular insect pest and disease.

The importance of nano-biotechnology is accelerating day-by-day to every sector of agricultural science. The most important of these are in the field of plant protection. Scientists around the globe are actively involved in gathering information about the positive and negative effects of nanopesticides. There is no doubt that nanopesticides are more effective in managing pests and diseases over conventional control measures; however, the toxicity to every living being is the major emerging issue. The problem of pests is not only restricted to geographical locations but the methods we currently opt to manage this problem are very toxic to human health and the environment. Therefore, the vital challenges that we face today are the selection and production of eco-friendly methods of managing pests and diseases that are safe, effective and cheap considering all the negative issues of the synthetic pesticides. Bio-nanoparticles could be the alternative for sustainable management of insect pests and diseases even though much information is still required before bringing them into commercial application. This may assist in minimizing the harmful effects of dangerous synthetic pesticides that are in use today and help in the sustainable management of pests and diseases.

REFERENCES

Ahmad, A., P. Mukherjee., S. Senapati., D. Mandal., M. I. Khan., R. Kumar, and M Sastry. 2003. Extracellular biosynthesis of silver nanoparticle using fungus *Fusarium oxysporium. Surfaces B: Biointerfaces* 24 (4): 313–318.

Ahmad, A., P. Mukherjee., S. Senapati., D. Mandal., M.I. Khan., R. Kumar and M. Sastry. 2003. Extracellular biosynthesis of silver nanoparticles using the fungus *Fusarium oxysporum. Colloids and Surfaces Biointerface* 28: 313–318.

Akbari B., M. P. Tavandashti, and M. Zandrahimi. 2011. Particle size characterization of nanoparticles—a practical approach *Iranian Journal of Materials Science & Engineering.* 8 (2): 48–56.

Albright L. J. and E. M. Wilson. 1974. Sub lethal effects of several metallic salts—organic compounds combinations upon the heterotrophic microflora of a natural water. *Water Research* 8: 101–105.

Amulyavichus, A., A. Daugvila, R. Davidonis, and C. Sipavichus. 1998. Study of chemical composition of nanostructural materials prepared by laser cutting of metals. *Fizika of Metals and Metallography* 85: 111–117.

Anjali, C. H., Y. Sharma, A. Mukherjee, and N. Chandrasekaran. (2012) *A. indica* oil (*Azadirachta indic*a) nanoemulsion—a potent larvicide agent against *Culex quinquefasciatus. Pest Management Science*, 68: 158–163.

Anonymous 2020. Nanodata: nanotechnology knowledge base. https://euon.echa.europa.eu/nanodata.

Ben-Moshe, Tal., S. Frenk, S. D. Ishai., D. Minz., B. Berkowitz. 2012. Effects of metal oxide nanoparticles on soil properties. *Chemosphere.* 90 (2): 640–646.

Birla, S. S., C. G. Swapnil., K. G. Aniket and M. K. Rai. 2013. Rapid synthesis of silver nanoparticles from *Fusarium oxysporum* by optimizing physicocultural conditions *Scientific World Journal*, 1–12. Article id: 796018 https://doi.org/10.1155/2013/796018

Bradley, M. M., and P. J. Lang. 1994. Measuring emotion: the self-assessment manikin and the semantic differential. *Journal of Behavior Therapy and Experimental Psychiatry* 25: 49–59.

Calvarro, L. M., A. Santiago-Martin, J. Q. Gomez, C. González-Huecas, and J. R. Quintana. 2014. Biological activity in metal-contaminated calcareous agricultural soils: the role of the organic matter composition and the particle size distribution. *Environmental Science and Pollution Research* 21: 6176–6187.

Chau, Chi-Fai, Wu, Shiuan-Huei, Yen and Gow-Chin. 2007. The development of regulations for food nanotechnology. *Trends in Food Science & Technology* 18: 269–280.

Chen, H. and R. Yada. 2011. Nanotechnologies in agriculture: New tools for sustainable development. *Trends in Food Science and Technology* 22: 585–594.

Derosa, M, C., C. Monreal, M. Schnitzer, and Y. Sultan. 2010. Nanotechnology in fertilizers. *Nature Nanotechnology* 5: 91.

Devi, D. G., K. Murugan, and P. C. Selvam. 2014. Green synthesis of silver nanoparticles using *Euphorbia hirta* (Euphorbiaceae) leaf extract against crop pest of cotton bollworm, *Helicoverpa armegera* (Lepidoptera: Noctuidae). *Journal of Biopesticides* 7: 54–66.

Eisler, R. 1996. "A review of silver hazards to plants and animals," in Proceedings of the 4th International Conference on Transport Fate and Effects of Silver in the Environment, A. W. Andren and T. W. Bober, Eds., 143–44, Madison, WI, USA.

Furstner, A and P. W. Davies. 2007. Catalytic carbophilic activation: catalysis by platinum and gold π acids. *Chemie* 46: 3410–3449.

Garcia-Gomez, C., M. D. Fernandez, S. Garcia, et al. 2018. Soil pH effects on the toxicity of zinc oxide nanoparticles to soil microbial community. *Environmental Science and Pollution Research* 25, 28140–28152.

Javed, Z., K. Dashora, M. Mishra, V. D. Fasake, and A. Srivastva. 2019. Effect of accumulation of nanoparticles in soil health—a concern on future [sic]. *Frontiers in Nanoscience and Nanotechnology* 5: 1–9.

Jones, P.B.C. 2006. A nanotech revolution in agriculture and the food industry. Blacksburg, VA: Information Systems for Biotechnology; 2006. Available from: http://www.isb.vt.edu/articles/jun0605.htm. Accessed April 19, 2014.

Josko, I., P. Oleszczuk, and B. Futa. 2014. The effect of inorganic nanoparticles (ZnO, Cr2O3, CuO and Ni) and their bulk counterparts on enzyme activities in different soils. *Geoderma* 232: 528–537.

Kalpana-Sastry, R., H. B. Rashmi., N. H. Rao, and S. M. Ilyas. 2009. Nanotechnology and agriculture in India: the second green revolution? "Potential environmental benefits of nanotechnology: fostering safe innovation-led growth. Session 7. Agricultural nanotechnology, 7: 15–17.

Kantrao, S., M. A. Ravindra, S. M. D. Akbar, P. D. K. Jayanthi, and A. Venkataraman. 2017. Effect of biosynthesized silver nanoparticles on growth and development of *Helicoverpa armigera* (Lepidoptera: Noctuidae): interaction with midgut protease. *Journal of Asia Pacific Entomology* 20: 583–589.

Khodakovskaya, M., E. Dervishi, M. Mahmood, Y. Xu, Z. Li, and A. S. Biris. 2009. Carbon nanotubes are able to penetrate plant seed coat and dramatically affect seed germination and plant growth. *ACS Nano* 3: 3221–3227.

Khot, L. R., S. Sankaran, J. M. Maja., R. Ehsani, and E. W. Schuster. 2012. Applications of nanomaterials in agricultural production and crop protection: a review. *Crop Protection* 35: 64–70.

Kole, C., P. Kole, K. M. Randunu, P. Choudhary., R. Podila., P.C. Ke., A. M. Rao and R. K. Marcus. 2013. Nanobiotechnology can boost crop production and quality: first evidence from increased plant biomass, fruit yield and phytomedicine content in bitter melon (*Momordica charantia*). *BMC Biotechnology* 13: 37–46.

Kuzma, J. and P. VerHage. 2006. Nanotechnology in agriculture and food production: anticipated applications. Project on Emerging Nanotechnologies, Washington, DC. Available from: http://www. nanotechproject.org/process/assets/files/2706/94_pen4_agfood.pdf.

Kwak, H., J.Y.H. Kim., H.M. Woo., E. Jin., B.K. Min and S.J. Sim. 2016. Synergistic effect of multiple stress conditions for improving microalgal lipid production. *Algal Research* 19: 215–224.

Kwak, J. I. and Y. J. An. 2016 Trophic transfer of silver nanoparticles from earthworms disrupts the locomotion of springtails (Collembola). *Journal of Hazardous Materials* 315: 110–116.

Leisner, J. J., B. G. Laursen, H. Prevost, D. Drider, and P. Dalgaard. 2007. Carnobacterium: positive and negative effects in the environment and in foods. *FEMS Microbiological Review* 31: 592–613.

Leisner, T., C. Rosche, S. Wolf, F. Granzer, L. Woste. 1996. *Surface Review and Letters*. 3: 1105–1108.

Lengke, M. F., M. E. Fleet and G. Southam. 2006. Morphology of gold nanoparticles synthesized by filamentous cyanobacteria from gold (I)-thiosulfate and gold (III)-chloride complexes. *Langmuir* 22: 2780–2787.

Moraru, C. I., C. P. Panchapakesan, Q. Huang, P. Takhistov, S. X. Liu, and J. L. Kokini. 2003. Nanotechnology: a new frontier in food science. *Food Technology* 57: 24–29.

Mousavi, S. R. and M. Rezaei. 2011. Nanotechnology in agriculture and food production. *Journal of Applied Environmental and Biological Sciences* 1: 414–419.

Panneerselvam, C., K. Murugan, K. Kovendan, P. M. Kumar, and S. Ponarulselvam. 2013. Larvicidal efficacy of *Catharanthus roseus* Linn. (Family: Apocynaceae) leaf extract and bacterial insecticide *Bacillus thuringiensis* against *Anopheles stephensi* Liston. *Asian Pacific Journal of Tropical Medicine* 6: 847–853.

Petit, C., P. Lixon and M. P. Pileni. 1993. In situ synthesis of silver nanocluster in AOT reverse micelles. *Journal of Chemical Physics* 97: 12974–12983.

Rai, M. and A. Ingle. 2012. Role of nanotechnology in agriculture with special reference to management of insect pests. *Applied Microbiology and Biotechnology* 94: 287–293.

Rajakumar, G and A. A. Rahuman. 2011. Larvicidal activity of synthesized silver nanoparticles using *Eclipta prostrata* leaf extract against filariasis and malaria vectors. *Acta Tropica* 118: 196–203.

Rajakumar, G and A.A. Rahuman. 2011. Larvaidal activity of synthesized silver nanoparticles using *Eclipta prostrata* leaf extract against filariasis and malaria vectors. *Acta Tropica* 118: 196–203.

Raliya, R, and J.C. Tarafdar. 2012. Novel approach for silver nanoparticle synthesis using *Aspergillus terreus* CZR-1: mechanism perspective. *Journal Bionanoscience* 6: 12–16.

Rao, K. J. and S. Paria. 2013. Use of sulfur nanoparticles as a green pesticide on *Fusarium solani* and *Venturia inaequalis* phytopathogens. *RSC Advances* 3: 10471–10478.

Reetz, M. T. and W. Helbig. 1994. Size-selective synthesis of nanostructured transition metal clusters. *Journal of the American Chemical Society* 116: 7401–7402.

Rico, C. M., S. Majumdar, M. Duarte-Gardea, J. R. Peralta-Videa, and J. L. Gardea-Torresdey. 2011. Interaction of nanoparticles with edible plants and their possible implications in the food chain. *Journal of Agricultural Food Chemistry* 59: 3485–3498.

Sekhon, B. S. 2010. Food nanotechnology—an overview. *Nanotechnology Science and Application* 3: 1–15. 11.

Sharma, N., P. Madan and S. Lin. 2016. Effect of process and formulation variables on the preparation of parenteral paclitaxel-loaded biodegradable polymeric nanoparticles: a co-surfactant study. *Asian Journal of Pharmaceutical Science.* **11**(3):404–416.

Shi, J., J. Ye, H. Fang, S. Zhang, and C. Xu. 2018. Effects of copper oxide nanoparticles on paddy soil properties and components. *Nanomaterials (Basel, Switzerland)* 8: 839 DOI: 10.3390/nano8100839. PMID: 30332772; PMCID: PMC6215298.

Shrivastava, S. and D. Dash. 2009. Agrifood nanotechnology: a tiny revolution in food and agriculture. *Journal of Nanoparticle Research* 6: 1–14.

Singh, S., B. K. Singh., S. M. Yadav, and A. K. Gupta. 2015. Applications of nanotechnology in agricultural and their role in disease management. *Research Journal of Nanoscience and Nanotechnology* 5: 1–5.

Sozer, N. and J. L. Kokini. 2009. Nanotechnology and its applications in the food sector. *Trends in Biotechnology* 27: 82–89.
Subramanian, K. S. and J. C. Tarafdar. 2011. Prospects of nanotechnology in Indian farming. *Indian Journal of Agricultural Science* 8: 887–893.
Sudharsanam, A., R. S. Suresh., P. Logeshwaran., K. Venkateswarlu., M.M. Mallavarapu and R. Subashchandrabose. 2019. Potential of acid-tolerant microalgae, *Desmodesmus* sp. MAS1 and *Heterochlorella* sp. MAS3, in heavy metal removal and biodiesel production at acidic pH. *Bioresource Technology* 278: 9–16.
Sunkar, S., V. Nachivar, and S. K. R. Namasivayam. 2012. Biogenesis of antibacterial silver nanoparticles using the endophytic bacterium *Bacillus cereus* isolated from *Garcinia xanthochymu*. *Asian Pacific Journal of Tropical Biomedicine* 2: 953–959.
Suresh, Y., S. Annapurna, G. Bhikshamaiah, and A. K. Singh. 2013. Characterization of green synthesized copper nanoparticles: A novel approach. In *International Conference on Advanced Nanomaterials Emerging Engineering Technologies*, 63–67.
Taghavi, S. M., M. Momenpour, M. Azarian, et al. 2013. Effects of nanoparticles on the environment and outdoor workplaces. Electron Physician 5: 706–712.
Tarafdar, J. 2012. Perspectives of nanotechnological applications for crop production. *NAAS News* 12: 8–11.
Waalewijn-Kool, P. L., S. Rupp, S. Lofts, C. Svendsen and C. A. M. Gestel. 2014. Effect of soil organic matter content and pH on the toxicity of ZnO nanoparticles to *Folsomia candida*. *Ecotoxicology and Environmental Safety* 108: 9–15.

6 Metagenomics of Sustainably Bioremediating Microbial Diversity

Vyoma Mistry
Uka Tarsadia University, Surat, India
Abhishek Sharma
Uka Tarsadia University, Surat, India
and
Institute of Advanced Research, University for Innovation
Gandhinagar, India

CONTENTS

DOI: 10.1201/9781003147091-6

6.1 INTRODUCTION

The worldwide explosion of the human population has led to a rapid increase in anthropogenic activities like industrialization, urbanization, agriculture, mining, etc. As a result, a wide range of xenobiotics has been introduced into the environment (soil). This hazardous chemical can be organic or inorganic in nature. These include an array of organic compounds, aromatic and petroleum hydrocarbons, pesticides, volatile organic and phenolic compounds, and inorganic compounds like nitrates, salts, phosphates, and heavy metals. This hazardous compound contaminates the ecosystem, which adversely affects the structure, microbiome, fertility, plants, and the biogeochemical cycles of soil (Kumar et al. 2018).

Biogeochemical cycles that facilitate the movement of conserved matter through the biotic and abiotic portion of the ecosystem in soil play a pivotal role in delivering diverse ecosystem services, where the flow of energy dissolutes as heat with the recycling of chemicals. Biogeochemical cycles are important to living organisms including carbon, nitrogen, water, phosphorous, and sulfur. Having a large diversity in soil delivers fundamental ecosystem services (Smith et al. 2015). To maintain the ecological balance under stressed soil, several remediation techniques have been applied. These include physio-chemical treatment to remove contaminants from the soil; however, these methods lead to changes in the physio-chemical nature of the soil and the chemical method results in the accumulation of some chemicals into the soil, which will be hard to process further. Nature has its way to resolve the imbalance in the ecosystem, where it utilizes microorganisms as the main tool to overcome this problem. Therefore, while the world is moving toward bioproducts and biosolutions to the problem, "Bioremediation" of contaminated soil has gained the interest of researchers for several decades (Sardrood et al. 2013).

The term Bioremediation is made up of two parts: "Bio" means related to living organisms or life, and "To remediate" means redress the problem. Thus, bioremediation means solving the problems using microbial organisms (Sardrood et al. 2013) (Figure 6.1). Microorganisms have colonized most of the environment, and many of them are efficient degraders of pollution (Jonsson and Hallar 2014). By utilizing this property of organisms to remove the pollution is known as bioremediation, which is a sustainable approach. Many organisms can utilize toxic pollution by converting and modifying it to obtain energy and biomass production in the process (Tang 2007). Therefore, researchers are exploring microbiome diversity in dirtied areas to explore and discover organisms with the capacity to degrade a wide range of contaminated chemical substances (Shah 2017; Damodaran et al. 2019). Biodegradation and bioremediation are different terms. However, biodegradation can be a part of the bioremediation process. Microorganisms that can degrade the contamination will increase in number in the presence of contaminants, and when it is degraded the microbial population declines, leaving the residue which consists of harmless products, including CO_2, water, and cell biomass. However, the outcome and process of microbial bioremediation not only depends on microbial diversity. Apart from microbes, numerous biotic and abiotic factors affect the bioremediation process (Figure 6.2).

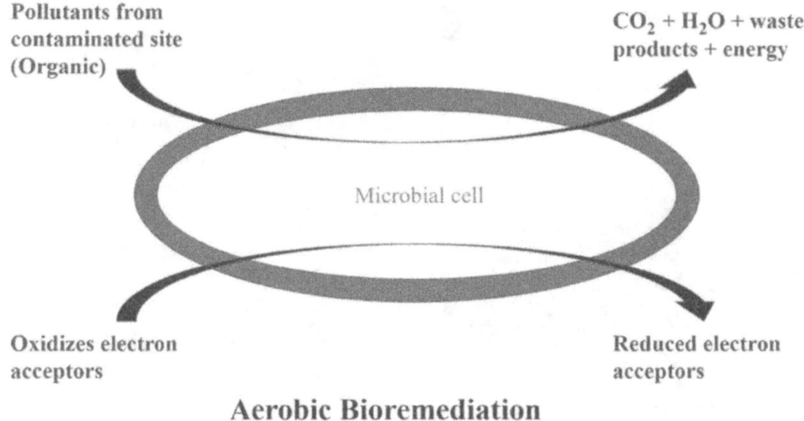

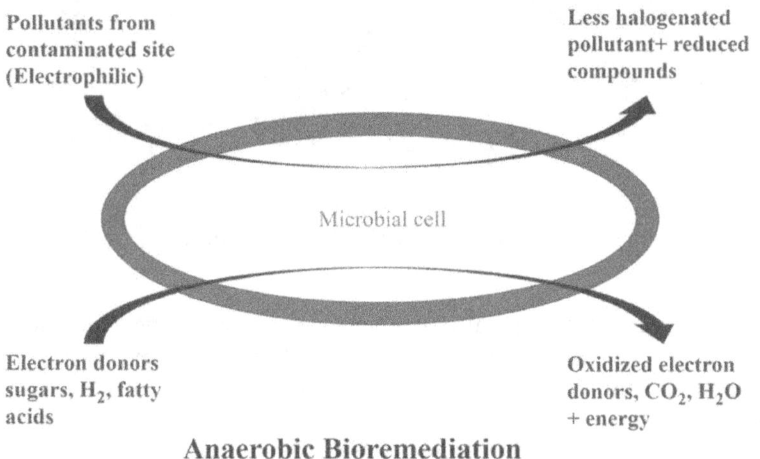

FIGURE 6.1 Microbial bioremediation mechanisms in aerobic and anaerobic conditions.

Despite the abiotic and biotic factors that affect the process, bioremediation is more advantageous over the chemical and physical processes in the following ways:

- It is a natural and harmless process, as the residue of treatment is generally CO_2, water, and cell biomass.
- There is no product risk of application of hazardous chemicals and no toxic waste products.
- It is a simple, lesser regress work method due to the natural participation of environment.
- It is sustainable, eco-friendly, and easy to implement.

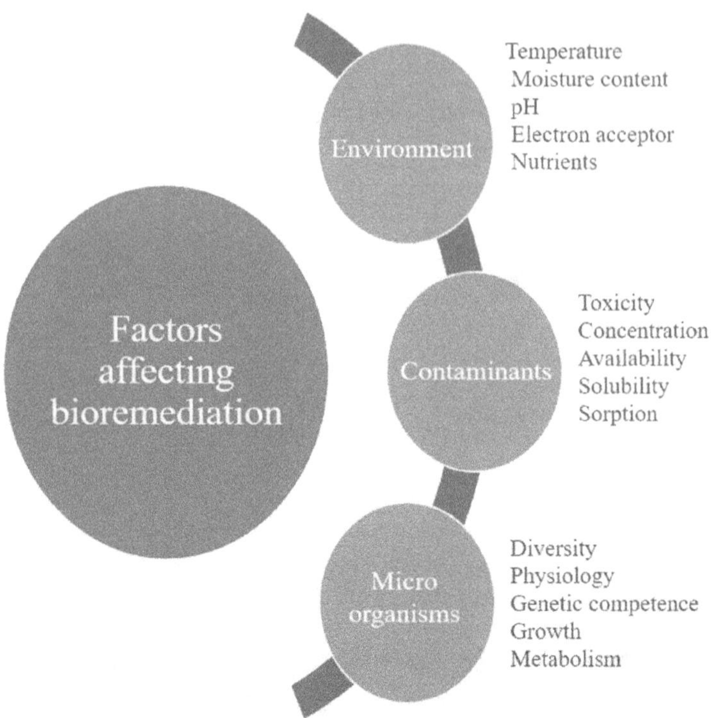

FIGURE 6.2 Biotic and abiotic factors affecting bioremediation.

Bioremediation technology also has some drawbacks. There are only a few organisms that act on a wide range of organic compounds. This limitation can be overcome by screening and detecting new potential microorganisms. Also, bioremediation takes a long time to act. However, to overcome these issues some genetic engineering techniques are in place (Sardrood et al. 2013). Additionally, genetic engineering can be performed to increase the proficiency of organisms with suitable properties (Sharma et al. 2022). In this chapter, efforts were made to shed light on the metagenomic approach of bioremediation of contaminated soil, which is a molecular approach in biogeochemical cycling for Plant-Soil-Microbes under stressed soil.

6.2 MICROBIAL BIOREMEDIATION

Microbial bioremediation denotes the use of microorganisms in the degradation of environmental contaminants. It is a highly sustainable and eco-friendly system that utilizes different microorganisms that work in a series of sequences to detoxify the contaminant (Kumar et al. 2018). Worldwide environmental contamination due to anthropogenic activities is a major issue nowadays. However, scientists around the globe are trying different physical and chemical methods of decontamination, but those methods are labor intensive, quite expensive, and often change the physiochemical nature of the soil, which ultimately generated interest and a growing awareness of the importance in "Bioremediation" (Das and Adholeya 2012). Bioremediation is classified into two parts: (A) Culture-dependent approach and (B) Culture-independent approach.

6.2.1 Culture-Dependent Approach

The culture-dependent method for bioremediation is also known as the traditional method, based on the cultivation of the microorganism. Culture-dependent methods are based on differential cellular, metabolic, morphological, and physiological traits of microorganisms (Varjani et al. 2018). Several culture-based techniques of soil bioremediation include cultivation of organisms using solid culture media, Most Probable Number (MPN) using liquid media, Colony hybridization, BIOLOG, etc. All these methods are based on enrichment, isolation, and characterization of microbial spices to perform their metabolic process. Bioremediation of contaminated soil can be done on-site and off-site. Based on this *in situ* and *ex situ* bioremediation methods are carried out accordingly. Table 6.1 summarizes the methods of *in situ* and *ex situ* bioremediation. In initial studies of *in situ* bioremediation, estimation of bacterial cell survival is observed using conventional "culture-dependent" methods.

6.2.1.1 Plating/CFU Count

Plating and CFU Count are conventional microbial methods. For this, 1 gm of a soil sample is collected from the contaminated site and is subjected to Serial dilution followed by plating. Plating is done on solid (sterilized) media, according to the dilution. The choice of media depends on microbial diversity present at the contaminated site. In general, a nutrient agar medium is used, which may be supplemented with the contaminant as the only energy source to enhance the growth and activity of particular organisms. The use of minimal culture media, culture dilution, and time of incubation support the retrieval of some slow-growing organisms.

The use of a minimal selected few nutrient media may not provide a natural condition for the growth of those organisms. A specific technique known as an encapsulation of cells in gel microdroplets was developed to culture the difficult to culture microbes for their cultivation at a large scale under the lower nutrient environment (Wu et al. 2015).

Plating followed by a Colony-Forming Unit (CFU) count is performed using both selective and non-selective mediums. CFU count generally uses the growth characteristics of organisms on a solid medium. CFU count is still the quickest method for the ecological survival monitoring of microbial diversity (Labana et al. 2005; George et al. 2008). CFU count simply counts the organism's colonies on a solid medium plate. However, CFU count may become difficult if organisms do not show distinct morphological characteristics. This drawback can be overcome by colony hybridization.

6.2.1.2 Colony Hybridization

Colony Hybridization is a culture-based approach that discriminates against organisms having morphological similarities. In this method, selective DNA/RNA probes are used, which bind to bacterial DNA in cells and distinguish organisms

TABLE 6.1
***In Situ* and *Ex Situ* Methods of Microbial Bioremediation**

Sr. No.	Technique	Details	References
***In Situ* Bioremediation**			
1.	Bioventing	Air and nutrients are applied to contaminated soil through wells that stimulate the organisms to aerobically degrade pollutants like semi-volatile compounds (SVCs) and non-chlorinated volatile organic compounds (VOCs) This method works best where contamination is deep—up to 60 cm and greater	Vidali 2001; Atlas and Philp 2005
2.	Biosparging	Sparging of sir under pressure through boreholes is done to enhance microbial activity below the water level This method applies to sites saturated with volatile compounds like gasoline and petroleum products	Sharma 2012
3.	Biostimulation	This is a natural attenuation process in which the growth and the metabolic activity of the indigenous microbial population are stimulated by supplementing the growth-limiting nutrients and electron acceptors or donors to degrade the contaminants	Crivelaro et al. 2010; Sardrood et al. 2013; Tyagi et al. 2011
4.	Bioaugmentation	This method improves the biodegradation ability through the addition of specialized and genetically engineered microbes to target specific contaminants	El Fantroussi and Agathos 2005; Thapa et al. 2012
***Ex Situ* Bioremediation**			
1.	Land farming	This is a land-based solid waste management system for contaminated soil, sludge, and sediment, which is dug and spread over a prepared bed To stimulate the degradation process, aeration is given until the contaminants become degraded	Vidali 2001; USEPA 2006
2.	Composting	Contaminated soil piles are generated, and composting is performed by aeration or continuously fed reactors The organic wastes are transformed by organisms into the hummus-like matter, which is known as "compost" Primarily three classes of organisms are used: Psychrophiles, Mesophiles, and Thermophiles	Vidali 2001; Sharma et al. 2022
3.	Biopiles	Biopiling is a mixed approach to land farming and composting The contaminated soil is stacked into piles and compost is added, which increases bioremediation due to the ideal conditions and pile structure Biopiles are 2–3 meters in height Diesel, Crude oil, and Lubricated oil-contaminated soil are treated with this method	Sardrood et al. 2013; Vidali 2001
4.	Bioreactors	The biological process is done in reactors to treat a relatively small amount of pollutants Slurry reactors or Aqueous reactors are vessels or tanks where microorganisms perform the biological reactions	Chikere et al. 2011; Vidali 2001; Sardrood et al. 2013

on a molecular basis. The definite DNA probes to target phylogenetically conserved genes; i.e., 16s rRNA are developed for microbial profiling (Richard et al. 1994; Wagner et al. 1994; Schuppler et al. 1995; Kowalchuk et al. 1999). Colony hybridization can be used to increase the selectivity of the culture-dependent methods; however, it is limited with the Viable But Not Culturable (VNBC) microorganisms. VNBC are viable organisms with active metabolisms but are temporarily nonculturable. Colony hybridization is also not a suitable method for these VNBC organisms.

6.2.1.3 Most Probable Number Technique

The drawback associated with solid media techniques forced the development of a new liquid culture-based approach utilizing the Most Probable Number (MPN) technique. MPN is a dilution-based method in which multiple sample replicates are analyzed and compared with a statistical table. Microtiter plate-based MPN is a miniature method of MPN having 96-well microtiter plates. Brown and Braddock in 1990 represented the method for oil degraders and crude oil as substrate. An inherent limitation of this method is the nature of the well and the small size, which results in a high risk of the well drying out (Chikere et al. 2011).

6.2.1.4 BIOLOG

BIOLOG is a culture-dependent approach of microbial activity analysis based on integration and hydrolysis of different carbon substrates. BIOLOG is also a Microtiter-based method in which 96 microtiter wells are used. Among them, 95 are filled with colorless tetrazolium dye and different carbon sources are placed in all 95 wells (*viz.* Carbohydrates, Amino acid, Amides, Amines, Carboxylic acid, etc) with the 96th well as control, devoid of any substrate. Soil samples are then added and inoculated at a constant temperature. Microbes present in soil oxidized the substrate and reduced the dye resulting in violet color formation, which is spectrophotometrically measured (Sabale et al. 2019). This allows rapid identification of over 2900 bacteria, yeast, and fungi. This method is used to study the impact of stress on soil microbial diversity.

6.2.2 Culture-Independent Approach

The limitations associated with culture-dependent approaches and advances in molecular techniques have enhanced the utilization of culture-independent methods. The soil contains the highest amount of prokaryotic diversity, as 1gm of soil may have 10 billion microbes and among them <1% are culturable (Torsvik et al. 2002). The soil microbes are a goldmine for the genes to perform several important functions and, therefore, the metagenomic is the best approach to study these types of organisms. Metagenomic approaches have enabled the understanding of the genetic potential of microbes by analyzing their DNA directly extracted from the environmental samples from contaminated sites, followed by PCR amplification (Figure 6.3). Therefore, the microbial diversity assessment is mostly based on the different tools and techniques that are associated with PCR amplification of metagenomic DNA, whereas, few are non-PCR associated as described below.

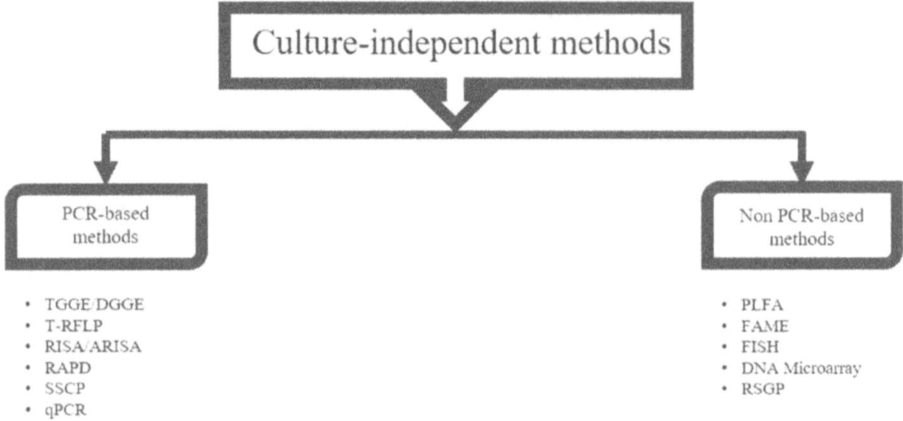

FIGURE 6.3 Cultivation of independent approach-based methods to study microorganisms in the contaminated environment.

6.2.2.1 PCR-Based Methods

6.2.2.1.1 Temperature/Denaturing Gradient Gel Electrophoresis (TGGE/DGGE)

Temperature Gradient Gel Electrophoresis (TGGE) and Denaturing Gradient Gel Electrophoresis (DGGE) techniques facilitate the separation of amplified PCR products of ribosomal DNA. They work on the principle of varying the concentration of the denaturing agent in DGGE and Temperature in TGGE. In the process, a GC clamp (a GC rich sequence) is attached to 5′- ends of the PCR primers and utilized as a special primer to anneal with PCR fragments.

However, DGGE and TGGE have the limitation of detection only the most populated organisms in the soil. As well, the single band in the gel can mislead as it may denote multiple microbial species and the same species can be represented by multiple bands (Dowd et al. 2008; Nübel 1997).

6.2.2.1.2 Terminal Restriction Fragment Length Polymorphism (T-RFLP)

Terminal Restriction Fragment Length Polymorphism uses a 5′ primer mounted with a fluorescent dye. The fluorescent-labeled primers facilitate the identification of amplified products, followed by their restriction digestion with one or more restriction enzymes. This restriction digestion results in the formation of fluorescent-labeled terminal restriction fragments, and its size determination was carried out using capillary methods or on a sequencing gel (Osborn et al. 2000). As different soil microbes contain diverse restriction sites, variations arise in the sequences of the amplified gene that provide unique "fingerprints" to a particular population of microbes. T-RFLP approaches provide a good understanding of changes in the structure and composition of soil communities (Rincon-Florez et al. 2013).

6.2.2.1.3 *Ribosomal Intergenic Spacer Analysis (RISA)/Automated Ribosomal Intergenic Spacer Analysis (ARISA)*

Ribosomal Intergenic Spacer Analysis (RISA) is described as a PCR-based approach applied for microbial diversity profiling by detecting the size variation of intergenic transcribed spacer (ITS) region present amid the bacterial 16S and 23S rRNA genes (Borneman and Triplett 1997). The ITS region present between the 16S and 23S rRNA genes is likely to carry the genes for tRNA and helps to differentiate the different bacterial strains due to the heterogeneity in the length and sequence of ITS. The ITS region of 16S and 23S rRNA genes is subjected to PCR amplification and separated under denaturing conditions over the polyacrylamide gel for sequence polymorphism detection with silver stain. Thus, RISA becomes an ideal approach for microbial diversity profiling in different environmental niches. However, the Automated Ribosomal Intergenic Spacer Analysis (ARISA) approach uses a fluorescent-labeled forward primer for the automated PCR amplification detection using the laser. Compared to RISA, ARISA provides fast and effective microbial diversity profiling and can be added with sophisticated techniques for fine-scale spatial and temporal resolution (Sabale et al. 2019).

6.2.2.1.4 *Random Amplified Polymorphic DNA*

Random Amplified Polymorphic DNA (RAPD) is a PCR-based approach used to identify the genetic relation and diversity among the organisms (Hadrys et al. 1992). RAPD utilizes the short primers of arbitrary nucleotide sequences to anneal with different genomic multiple locations at low temperatures, to allow dissolute pairing of primers; however, this often affects the reproducibility of results (Rincon-Florez et al. 2013). RAPD remains one of the most utilized techniques to develop DNA markers for genetic profiling and is less time-consuming and cheaper than T-RFLP.

6.2.2.1.5 *Single-Strand Conformation Polymorphism (SSCP)*

Developed by Peters et al. (2000), Single-Strand Conformation Polymorphism (SSCP) is a simple and sensitive screening technique for the detection of different genetic variants and genotyping in a wide range of organisms, ranging from microbes to humans. SSCP technique detects the genetic variations up to single-point mutation, single nucleotide polymorphism (SNPs), and other small changes in DNA through their differential electrophoretic mobility. The principle behind the SSCP working approach is based on the confirmation of the single-stranded DNA. The single-stranded DNA as a defined conformation and even a single nucleotide base alteration in sequence can cause a change in the conformation that consecutively changes the migration of DNA under non-denaturing electrophoresis. This change in electrophoretic movement due to base change will result in the display of different band patterns of wild-type and mutant DNA samples. SSCP protocols involve the PCR amplification of DNA samples of organisms to study, followed by heat-based denaturation of the PCR amplified double-stranded DNA segments to obtained single-stranded DNA. Denaturation is immediately followed by cooling of the denatured single-stranded DNA molecules to allow the self-annealing and finally, the DNA molecules are subjected to electrophoresis under denaturing conditions for the detection of a difference in single-stranded DNA electrophoretic mobility (Dong and Zhu 2005). The mobility shifts of DNA samples are visualized by several methods developed to visualize it using fluorescent-labeled PCR primers, silver staining, radioisotope labeling, and the latest, capillary electrophoresis.

6.2.2.1.6 Real-Time PCR (RT-PCR)

Real-Time PCR or Quantitative Real-Time PCR (RT-PCR/RT-qPCR) is a complementary DNA amplification technique that collects the data during the process of cDNA PCR amplification at the real time it occurs; therefore, it combines the amplification and detection at real time in a single step. Concerning microbial diversity profiling, RT-PCR provides quantification of the active microbial population from environmental samples, by targeting species-specific genes such as ribosomal and taxonomic and functional marker genes (Zhang 2010). Contrasting with the end-point detection of genes amplification through conventional PCR, the real-time amplification detection in RT-PCR is achieved by the chemistry of specific fluorescent molecules that correlates the amplification with fluorescent intensity (Higuchi et al. 1993). Therefore, the reaction is characterized at the scale of point in time of PCR cycle at which targeted cDNA amplificated detected for the first time. This point in time value of amplification, when the intensity of amplification fluorescent is higher than the background fluorescence, is termed as cycle threshold (Ct) value. Consequently, the lower the quantity of targeted cDNA in the reaction, the lower the fluorescent signal, which will yield high Ct, whereas higher the targeted cDNA, higher the fluorescent signal and lower the Ct value (Heid et al. 1996).

6.2.2.2 Non-PCR-Based Methods

6.2.2.2.1 Phospholipid Fatty Acid Analysis (PLFA) and Fatty Acid Methyl Ester (FAME) Analysis

Phospholipid Fatty Acid Analysis (PLFA) is a unique microbial lipid-based technique used for the quantification of the total viable microbial biomass in the environmental sample, thus helping to profile the microbial diversity. As phospholipid is one of the major components of the cell membrane of all microbial cells, PLFA is used to quantify a set of biomarkers to primarily detect microbial biomass under *in situ* conditions, thus preventing the *in vitro* culturing of microbial populations. As phospholipids denote the phenotypic characteristics of the microbial cells, PLFA provides phenotypic profiling of microbial populations (Joergensen and Wichern 2008). Compared to other microbial profiling techniques viz. 16S rRNA gene metabarcoding, PLFA provides better identification of microbial communities. Similar to PLFA, the Fatty Acid Methyl Easter (FAME) is another technique that uses fatty acid extraction and profiling for the characterization of microbial population and estimation of microbial biomass in soil ecosystems (Sabale et al. 2019).

6.2.2.2.2 Fluorescence In Situ Hybridization (FISH)

As the name suggests, Fluorescent *In Situ* Hybridization (FISH) is a technique that uses fluorescent molecules commonly known as fluorochrome conjugated with oligonucleotide probes, to identify and quantify the naturally occurring specific complementary sequence to it, situated on a chromosome in a biological sample. The hybridization of the fluorescent bonded probe with a chromosomal target is visualized under fluorescent microscopy. Primarily, the FISH is the weapon of cytogeneticists to hunt the DNA sequences over chromosomes for disease diagnosis, however, the sensitivity of this technique allows its utilization on microbial ecology and phylogenetic analysis (Moter and Göbel. 2000).

6.2.2.2.3 DNA Microarray and Reverse Sample Genome Probing (RSGP)

DNA-DNA hybridization is the main principle behind the development of identification techniques like FISH. Similarly, one more advanced technique known as DNA microarray was developed to assess the microbial species with great specificity. DNA microarray is a chip-based technology that utilizes oligonucleotides probes that bind to its naturally occurring complementary sequence viz. 16S rRNA and other microbe-specific genes from environmental samples. Compared to FISH, DNA microarray allows simultaneous evaluation of several genes (Cho and Tiedje 2001). DNA microarray is a more competent method for parallel analysis of a large number of specific sequences. Technically, in DNA microarray the target DNA molecule is labeled with a fluorescent molecule and then allowed to hybridize with the gene-specific complementary probe. As the DNA sequences on the microarray chip are from the microbial population, and probes are species and gene specific, the fluorescent detection through hybridization provides microbial identification. The role of DNA microarray in environmental microbiology is very important, specifically for the identification and profiling of microbial populations associated with biodegradation, etc. DNA microarray is further advanced in the form of Reverse Sample Genome Probing (RSGP), a hybridization-based approach to understanding the microbial community. The RSGP uses the genome microarray for the identification and presence of microbial network systems and dominant culturable microbial species in a given environment (Greene et al. 2000; Shekhar et al. 2020).

6.3 GENOME ENRICHMENT: METAGENOMIC LIBRARY DEVELOPMENT

Handelsman (1998) coined the term "Metagenomics." As >1% of soil microbes are unidentified and uncultivated, metagenomics provides a better approach to study the function of those microbes without culturing. Metagenomics provides direct extraction of genomic material from the environmental samples. However, the main challenge is extracting the genomic material of insufficient quality and quantity for further analysis (Pushpanathan et al. 2014). The active population of microbes within a contaminated environment can be assessed by genome enrichment. To selectively enrich DNA, RNA, phospholipids—Stable Isotope Probing (SIP) technique is used which is followed by metagenomics (Chen and Murell 2011). Figure 6.4 summarizes the flow pattern of metagenomic research in the contaminated soil environment, including the identification and screening of bioremediating genes constructed on functional and sequence-based screening methods.

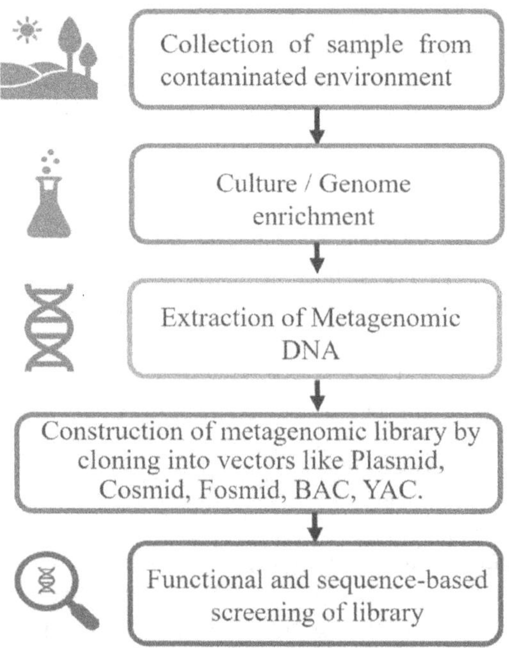

FIGURE 6.4 Work flow patterns of metagenomic research in the contaminated soil environment.

6.3.1 Extraction of Metagenome and Library Construction from the Contaminated Site

From the site of contamination, metagenomic DNA is directly extracted. Several methods are available to extract DNA from the environment. A commercial kit is available too, which is easy to use. However, the method used to extract the DNA should be chosen carefully, and the choice of method depends on the purpose of analysis (Pushpanathan et al. 2014). Total metagenomic DNA extracted from the contaminated site is necessary to represent all microbial genomes. Two different strategies are there for the isolation of metagenomic DNA: (1) Cell recovery (indirect) and (2) Direct lysis (direct).

The indirect method of cell recovery consists of the collection and purification of the cell from the matrix, followed by lysis and purification of genetic material (Gabor et al. 2003). The cell recovery method has a lower yield than the direct lysis method. The direct lysis method is the main strategy to perform and it provides greater DNA concentration. But when it comes to microbial diversity the indirect method is preferred, as some researchers suggested that a high quantity of DNA may compose eukaryotic DNA. Metagenome extraction can also be performed by physical, chemical, and enzymatic methods. To recover higher molecular weight DNA, the chemical method proved to be moderate and bead-beating by the physical method recovered more diversity in genomic material. However, low frequency in a positive hit is a major concern in metagenomics; thus, to increase the proportion of positive hit,

Genome enrichment is done, Genome or gene enrichment is the addition of a culture medium as the sole source of carbon and energy to obtain a greater proportion of the population (Eyers et al. 2004). This can also be done by labeling C^{13}, N^{15}, O^{18} of the substrate in Stable Isotope Probing (SIP) (Radajewski et al. 2000). SIP-DNA having different densities (heavy) and unlabeled DNA (light) can be differentiated by density gradient ultracentrifugation using cesium chloride or cesium fluoride. The heavy DNA fragment can be used for constructing a metagenomic library (Sar and Islam 2012). Selective media can also be used for enrichment purposes. Selective media containing pollutants as the sole source of energy will selectively increase that pollutant-degrading microbe (Entcheva et al. 2001). This way, genome enrichment increases positive hits during the screening procedure. However, the achievement of a high amount of intact, pure DNA is challenging in metagenomics. Yet, the knowledge of different methodologies eventually elevates the chances of obtaining pure and intact DNA.

After isolation and purification of the genetic material, metagenomic libraries are constructed. These consist of cloning of DNA fragment with a specific vector. The large DNA fragment of 25 to 200 kb used to construct the library. The choice of vector depends on the size of the DNA fragment. Construction of a large-insert metagenomic library makes use of vectors such as Bacterial Artificial Chromosome (BAC), Cosmid, and Fosmid. However, if the gene encoding compounds are smaller, then small inserts like plasmid are required. Plasmid supports DNA inserts which are less than 15 kb in size, whereas Cosmid supports DNA fragments from 25–35 kb, Fosmid from 24–40 kb, BAC from 100–200 kb, and Yeast Artificial Chromosome (YAC) over 40 kb (Riesenfeld et al. 2004). Therefore, vector choice and type of construction are highly dependent on the size of the gene that encodes a molecule that has to be screened.

6.3.1.1 Identification of Bioremediating Genes

The bioremediation outcome is greatly dependent on a variety of physicochemical and biological factors that need to be assessed continuously. Several identification and screening strategies are available to assess the bioremediation process. After generation of the metagenomic library, screening and identification of genes can be done on functional-based analysis and sequenced-based analysis. Sequence-based metagenomic analysis can be used for gene identification, metabolic pathway clarification, genome assemblages, etc. (Malla et al. 2018). Bioremediating genes are identified based on the genetic sequence. Identification of bioremediating genes includes three technologies:

1. PCR-based identification
2. Microarray metagenome profiling
3. Metagenome shotgun sequencing

6.3.1.1.1 PCR-Degenerated Primer-Based Identification

In this approach, the desired gene is identified based on conserved nucleotide sequences of known catabolic enzymes. By using a hybridization probe or degenerated primers amplification of gene is done. Primers having a consensus sequence of catabolic enzymes will amplify the gene that encodes for an enzyme (Hildebrand et al. 2004). The PCR-based microarray approach makes use of "Geochip". Geochip provides a linkage between the biogeochemical process and functional microbial activities. The chip has around 24,243 oligonucleotides probes and > 10,000 genes involved in bioremediation (He et al. 2007).

6.3.1.1.2 Microarray Metagenome Profiling

Microarray Metagenome Profiling is the profiling of the clones in the microbial community via DNA probe hybridization with the genomes of bacterial isolates, and environmental DNA with reference strains. DNA microarray has become one of the important approaches for gene content identification in microbial populations and their gene expression for their characterization and population profiling. The metagenomic libraries, i.e., genomic libraries developed from the environmental DNA, are the sole microarray metagenome profiling used for the characterization of uncultured microorganisms. This metagenomic library-based microarray approach is different from the conventional library screening approaches that allow the characterization of only a very few environmental microbial clones; the direct cloning of DNA from environmental samples also helps to avoid biases of cultivation and PCR. Thus, microarray metagenomic libraries are useful for the identification of novel genes from unculturable microbial species involved in the significant ecosystem processes (Sebat et al. 2003).

Technically, cosmid libraries prepared from the environmental soil samples are used to prepare the microarray plate that contains PCR amplified products from inserts of cosmid and 16S rRNA controls. Thereafter, this microarray plate is hybridized with DNA of reference strains obtained from the total soil and groundwater microbial community. Clones specifically hybridizing with the community genomic DNA but not with reference strains will be supposed to be originated from microbes that failed to isolate and culture. These DNA inserts from hybridized clones were then sequenced, and could be assigned to functions with their potential ecological importance viz. novel biodegradable genes.

6.3.1.1.3 Metagenome Shotgun Sequencing

Metagenome sequencing allows microbiologists to comprehensively sample all the genes from all microbes present in the given environmental samples for evaluation of bacterial diversity. The involvement of shotgun sequencing with metagenomics also helps to study and characterize the unculturable microbes. Thus, metagenome shotgun sequencing approach provides a clear picture of metabolism and genomics of unculturable microbial systems. The technical approach involves the isolation of total metagenomic DNA from the sample site, followed by metagenomic library preparation and, finally, it is subjected to random end-sequencing with Next-Generation Sequencing Systems. The sequence of every DNA fragment obtained is assembled to several contigs and scaffolds followed by assembly and annotation to assign their functions. For example, Deep shotgun sequencing of environmental samples has resulted in the identification of a 79-kb catabolic gene cluster responsible for the anaerobic degradation of benzoate and hydroxybenzoate from the seafloor or Black Sea (Kube et al. 2005).

The continuous use of these sequencing techniques has established a large set of metagenomic databases from polluted environments. Along with the databases, metagenome analysis tools like Integrated Microbial Genome/Microbiome (IMG/M) are also developed to access the genomic and metabolic functional capabilities of microbial populations, whose metagenome sequence is available at the Joint Genome Institute (JGI), MetaBioME (Markowitz et al. 2008; Sharma et al. 2010). Similarly, Orphelia is a tool used to discover the open reading frames (ORFs) in the short metagenomic sequences of unknown phylogeny (Hoff et al. 2009). However, the unavailability of reference genome sequences remains the major drawback of this technique.

6.3.2 Screening of Bioremediating Genes

After the identification of the specific microbes from environmental populations using the above-described techniques, major work is moving in the direction to discover the genes associated with bioremediation processes. The screening of genes associated with bioremediation is generally done with the help of approaches like functional screening, i.e., screening of genes with known or unknown function (Dias et al. 2014). The functional screening of genes can be done with the help of the following screening approaches:

1. Phenotype-based screening
2. Substrate-induced gene expression profiling (SIGEX)
3. Metabolite expression profiling (METREX)

6.3.2.1 Phenotype-Based Screening

The phenotype-based screening method is utilized for the screening of bioremediating genes. In this process, from the contamination site, the genomic DNA of the microbial population is directly isolated and then screened with metagenomic libraries for characterization of a particular gene encoding catabolic enzyme, molecule, and compound playing role in the process of bioremediation (Henne et al. 2000). The process of phenotype-based screening is performed by several strategies viz. (i) Enzymatic assays; (ii) Induced gene expression; (iii) Use of fluorescent catabolic probes for the direct detection of specific phenotype, and (iv) Host strain heterologous complementation (Simon and Daniel 2011; Felczykowska et al. 2012). These approaches have helped in the screening of genes associated with bioremediating enzymes like nitrilase, naphthalene dioxygenase, extra diol deoxygenase, and styrene monooxygenase (Robertson et al. 2004; Ono et al. 2007; Suenaga et al. 2007; Van Hellemond et al. 2007).

6.3.2.2 Metabolite-Regulated Expression Profiling

The Metabolite-Regulated Expression Profiling approach utilizes highly sensitive screening methods like Metrex to screen the poorly expressed metabolites through the fluorescence-activated cell sorting (FACS) technique. In this approach, the inducible promoters and reporter plasmids responsible for controlling the expression of the gene of interest are linked with fluorescent proteins and incorporated into cells. Now the cell has reporter plasmid and metagenomic DNA, that helps to identify the cells with poor gene expression due to the presence of a fluorescent-linked reported gene, followed by sorting with FACS and gene sequencing and annotation (Pushpanathan et al. 2014).

6.3.2.3 Substrate-Induced Gene Expression Profiling (SIGEX)

The Substrate-Induced Gene Expression Profiling (SIGEX) approach allows the identification of genes associated with xenobiotic degradation in a culture-independent manner; however, it lacks genomic integration. The SIGEX approach for the identification of genes and enzymes associated with bioremediation works similar to metabolite-regulated gene expression profile with the only difference that instead of metabolites, the expression of the genetic reporter is targeted. Hence, SIGEX is a phenotypic profiling approach utilizing substrate-induced gene expression and whole metagenome shotgun sequencing. SIGEX is a high-throughput promoter-trap technique that works on the principle that the bioremediating genes are associated with transcriptional regulators, which are activated in the presence of targeted substrates (Bouhajja et al. 2017). Hence, the green fluorescent protein (GFP) is added upstream to metagenomic DNA. So, when in the presence of a specific substrate, the genes responsible for bioremediation enzymes will induce and express with GFP, which will allow the isolation of clones subsequently using FACS (Uchiyama et al. 2005).

6.4 CONCLUSION AND FUTURE PROSPECTS

With the increasing understanding of the hazardous effect of pollutants, there has been an upsurge in research for the development of strategies to decontaminate the environment. The application of various characterization approaches of microbial diversity is a safe and efficient tool for the removal of harmful contaminants. Traditional methods of bioremediation involve culturing of microbes on different media. 1 gm of soil contains around 10 billion microorganisms but among them > 1% is cultivable. Metagenomics provides a broad spectrum to identify and screen these non-culturable microbes directly from a contaminated environment. Metagenomics with high through output methods provide deep insight into the interaction of an organism with its metabolites, which could be explored in the bioremediation of various harmful contaminants. Bioremediation is an eco-friendly and adaptable approach that is growing rapidly in the field of the environment, as the world is moving toward bioproducts and biological solutions. For better bioremediation results, an insight into the microbial capacity of degrading contaminants, their metabolic process, and a resistant mechanism will be more crucial in the construction of suitable microbial formulas, particularly for a contaminated site. An understanding of the contaminated site, its microbial diversity, and knowledge about the physiological condition of the soil are required to achieve better results. The gap between laboratory research and field application must be reduced. The use of a biocatalyst can be another alternative approach that will reduce difficulties such as low survival rate of the cell and generation of toxins during the remediation process. However, the large-scale application of this enzymatic approach will need more in-depth work. Metagenomic bioremediation is considered one of the potential tools to remove contaminants from the environment. A combination of various high through output methods will provide a comprehensive study of links between gene density, species diversity, protein production, gene expression, etc. Characterization and identification of bioindicators that may help to determine the use of a particular bioremediation process are required. Integration of data generated from various "OMICS" approaches is essential as the integration of Functional metagenomics, Proteomics, Transcriptomics, and Metabolomic data will provide a clear perspective on microbial consortia required for the bioremediation process. Overall, a more integrated and multidisciplinary approach is required for the proper assessment of bioremediation techniques.

REFERENCES

Atlas, R.M., Philp, J. 2005. Bioremediation. *Applied Microbial Solutions for Real-World Environmental Cleanup.* American Society of Microbiology Press. Washington DC. DOI:10.1128/9781555817596

Borneman, J., Triplett, E.W. 1997. Molecular Microbial Diversity in Soils from Eastern Amazonia: Evidence for Unusual Microorganisms and Microbial Population Shifts Associated with Deforestation. *Applied and Environmental Microbiology* 63 (7): 2647–2653.

Bouhajja, E., McGuire, M., Liles, M.R. Bataille, G., Agathos, S.N., George, I.F. 2017. Identification of Novel Toluene Monooxygenase Genes in a Hydrocarbon-Polluted Sediment Using Sequence- and Function-Based Screening of Metagenomic Libraries. *Appl Microbiol Biotechnol.* 101, 797–808.

Chen, Y., Murrell, J.C. 2010. When Metagenomics Meets Stable-Isotope Probing: Progress and Perspectives. *Trends in Microbiology* 18 (4): 157–163.

Chikere, C.B., Okpokwasili, G.C., Chikere, B.O. 2011. Monitoring of Microbial Hydrocarbon Remediation in the Soil. *Biotech* 1 (3): 117–138.

Crivelaro, S.H.R., Mariano, A.P., Furlan, L.T., Gonçalves, R.A., Seabra, P.N., Angelis, D.F. 2010. Evaluation of the Use of Vinasse as a Biostimulation Agent for the Biodegradation of Oily Sludge in Soil. *Brazilian Archives of Biology and Technology* 53 (5): 1217–1224.

Damodaran, T., Mishra, V.K., Jha, S.K., Pankaj, U., Gupta, G., Gopal, R. 2019. Identification of Rhizosphere Bacterial Diversity with Promising Salt Tolerance, PGP Traits and Their Exploitation for Seed Germination Enhancement in Sodic Soil. *Agricultural Research* 8(1):36–43.

Das, M., Adholeya, A. 2012. Role of Microorganisms in Remediation of Contaminated Soil. Satyanarayana, T., Johri, B. (eds.) *Microorganisms in Environmental Management. Dordrecht*: Springer. https://doi.org/10.1007/978-94-007-2229-3_4 pp. 81–111

Dias, R., Silva, L.C., Eller, M.R., Oliveira, V.M., Paula, S.O.D., Silva, C.C. 2014. Metagenomics: Library Construction and Screening Methods. *Metagenomics: Methods, Applications and Perspectives* 5: 28–34.

Dong, Y., Zhu, H. 2005. Single-Strand Conformational Polymorphism Analysis: Basic Principles and Routine Practice. In *Hypertension* (Eds. Fennell, J. P., Baker, A. H.) 149–158. New Jersey: Humana Press.

Dowd, S.E., Sun, Y., Secor, P.R., Rhoads, D.D. et al. 2008. Survey of Bacterial Diversity in Chronic Wounds Using Pyrosequencing, DGGE, and Full Ribosome Shotgun Sequencing. *BMC Microbiology* 8 (1): 43.

El Fantroussi, S., Agathos, S.N. 2005. Is Bioaugmentation a Feasible Strategy for Pollutant Removal and Site Remediation? *Current Opinion in Microbiology* 8 (3): 268–275.

Entcheva, P., Liebl, W., Johann, A., Hartsch, T., Streit, W.R. 2001. Direct Cloning from Enrichment Cultures, a Reliable Strategy for Isolation of Complete Operons and Genes from Microbial Consortia. *Applied and Environmental Microbiology* 67 (1): 89–99.

Eyers, L., George, I., Schuler, L., Stenuit, B., Agathos, S.N., Fantroussi, S. El. 2004. Environmental Genomics: Exploring the Unmined Richness of Microbes to Degrade Xenobiotics. *Applied Microbiology and Biotechnology* 66 (2): 123–130.

Felczykowska, A., Bloch, S.K., Nejman-Faleńczyk, B., Barańska, S. 2012. Metagenomic Approach in the Investigation of New Bioactive Compounds in the Marine Environment. *Acta Biochimica Polonica* 59 (4).

Gabor, E.M., Vries E.J., Janssen D.B. 2003. Efficient Recovery of Environmental DNA for Expression Cloning by Indirect Extraction Methods. *FEMS Microbiology Ecology* 44: 153–163.

George, I., Eyers L., Stenuit B., Agathos S.N. 2008. Effect of 2, 4, 6-Trinitrotoluene on Soil Bacterial Communities. *Journal of Industrial Microbiology & Biotechnology* 35 (4): 225–236.

Greene, E.A., Kay, J.G., Jaber, K., Stehmeier, L.G., Voordouw, G. 2000. Composition of Soil Microbial Communities Enriched on a Mixture of Aromatic Hydrocarbons. *Applied and Environmental Microbiology* 66 (12): 5282–5289.

Hadrys, H., Balick M., Schierwater B. 1992. Applications of Random Amplified Polymorphic DNA (RAPD) in Molecular Ecology. *Molecular Ecology* 1 (1): 55–63.

Handelsman, J., Rondon, M. R., Brady, S. F., Clardy, J., Goodman, R. M. 1998. Molecular Biological Access to the Chemistry of Unknown Soil Microbes: A New Frontier for Natural Products. *Chem. Biol.* 5: R245–R249.

He, Z., Gentry, T.J., Schadt, C.W., Wu, L. et al. 2007. GeoChip: A Comprehensive Microarray for Investigating Biogeochemical, Ecological and Environmental Processes. *ISME Journal* 1 (1): 67–77.

Heid, C.A., Stevens, J., Livak, K.J., Williams, P.M. 1996. Real Time Quantitative PCR. *Genome Research* 6: 986–994.

Henne, A., Schmitz, R.A., Bömeke, M., Gottschalk, G., Daniel, R. 2000. Screening of Environmental DNA Libraries for the Presence of Genes Conferring Lipolytic Activity on *Escherichia coli. Applied and Environmental Microbiology* 66 (7): 3113–3116.

Higuchi, R., Fockler, C., Dollinger, G., Watson, R. 1993. Kinetic PCR Analysis: Real Time Monitoring of DNA Amplification Reactions. *Biotechnology* 11: 1026–1030.

Hildebrand, M., Waggoner, L.E., Lim, G.E., Sharp, K.H., Ridley, C.P., Haygood, M.G. 2004. Approaches to Identify, Clone, and Express Symbiont Bioactive Metabolite Genes. *Natural Product Reports* 21 (1): 122–142.

Hoff, K.J., Lingner, T., Meinicke, P., Tech, M. 2009. Orphelia: Predicting Genes in Metagenomic Sequencing Reads. *Nucleic Acids Research* 37 (suppl_2): 101–105.

Joergensen, R. G., Wichern, F. 2008. Quantitative Assessment of the Fungal Contribution to Microbial Tissue in Soil. *Soil Biology and Biochemistry* 40 (12): 2977–2991.

Jonsson, A., Haller, H. 2014. *Sustainability Aspects of In Situ Bioremediation of Polluted Soil in Developing Countries and Remote Regions.* In Hernandez-Soriano M.C. (Eds), *Environmental Risk Assessment of Soil Contamination*, pp 57–86. London, United Kingdom, IntechOpen.

Kowalchuk, G.A., Naoumenko, Z.S., Derikx, P.J., Felske, A. et al. 1999. Molecular Analysis of Ammonia-Oxidizing Bacteria of the β Subdivision of the Class Proteobacteria in Compost and Composted Materials. *Applied and Environmental Microbiology* 65 (2): 396–403.

Kube, M., Beck, A., Meyerdierks, A., Amann, R., Reinhardt, R., Rabus, R. 2005. A Catabolic Gene Cluster for Anaerobic Benzoate Degradation in Methanotrophic Microbial Black Sea Mats. *Systematic and Applied Microbiology* 28 (4): 287–294.

Kumar, V., Shahi, S.K., Singh, S. 2018. Bioremediation: An Eco-Sustainable Approach for Restoration of Contaminated Sites. In Shah. M. P. (Eds) *Microbial Bioremediation & Biodegradation*, pp 115–136. Springer: Singapore.

Labana, S., Pandey, G., Paul, D., Sharma, N.K., Basu, A., Jain, R.K. 2005. Pot and Field Studies on Bioremediation of P-Nitrophenol Contaminated Soil Using *Arthrobacter protophormiae* RKJ100. *Environmental Science & Technology* 39 (9): 3330–3337.

Malla, M.A., Dubey, A., Yadav, S., Kumar, A., Hashem, A., Abd_Allah, E.F. 2018. Understanding and Designing the Strategies for the Microbe-Mediated Remediation of Environmental Contaminants Using Omics Approaches. *Frontiers in Microbiology* 9: 1132.

Markowitz, V.M., Ivanova, N.N., Szeto, E., Palaniappan, K. et al. 2008. IMG/M: A Data Management and Analysis System for Metagenomes. *Nucleic Acids Research* 36: D534–D538.

Moter, A., Göbel, U.B. 2000. Fluorescence In Situ Hybridization (FISH) for Direct Visualization of Microorganisms. *Journal of Microbiology Methods* 41 (2): 85–112. doi: 10.1016/s0167-7012(00)00152-4.

Nübel, U., Garcia-Pichel, F., Muyzer, G. 1997. PCR Primers to Amplify 16S rRNA Genes from Cyanobacteria. *Applied and Environmental Microbiology* 63 (8): 3327–3332.

Ono, A., Miyazaki, R., Sota, M., Ohtsubo, Y., Nagata, Y., Tsuda, M. 2007. Isolation and Characterization of Naphthalene-Catabolic Genes and Plasmids from Oil-Contaminated Soil by Using Two Cultivation-Independent Approaches. *Applied Microbiology and Biotechnology* 74 (2): 501–510.

Osborn, A.M., Moore, E.R., Timmis, K.N. 2000. An Evaluation of Terminal-restriction Fragment Length Polymorphism (T-RFLP) Analysis for the Study of Microbial Community Structure and Dynamics. *Environmental Microbiology* 2 (1): 39–50.

Peters, S., Koschinsky, S., Schwieger, F., Tebbe, C.C. 2000. Succession of Microbial Communities during Hot Composting as Detected by PCR–Single-Strand-Conformation Polymorphism-Based Genetic Profiles of Small-Subunit rRNA Genes. *Applied and Environmental Microbiology* 66(3): 930–936.

Pushpanathan, M., Jayashree, S., Gunasekaran, P., Rajendhran, J. 2014. Microbial Bioremediation: A Metagenomic Approach. In Das, S. (Ed) *Microbial Biodegradation and Bioremediation*, pp 407–419. Elsevier.

Radajewski, S., Ineson P., Parekh, N.R., Murrell J.C. 2000. Stable-Isotope Probing as a Tool in Microbial Ecology. *Nature* 403 (6770): 646–649.

Riesenfeld, C.S., Schloss, P.D., Handelsman, J. 2004. Metagenomics: Genomic Analysis of Microbial Communities. *Annual Review of Genetics* 38: 525–552.

Rincon-Florez, V.A., Carvalhais, L.C., Schenk, P.M. 2013. Culture-Independent Molecular Tools for Soil and Rhizosphere Microbiology. *Diversity* 5 (3): 581–612.

Robertson, D.E., Chaplin, J.A., DeSantis, G., Podar, M., Madden, M., Chi, E., Wong, K. 2004. Exploring Nitrilase Sequence Space for Enantioselective Catalysis. *Applied and Environmental Microbiology* 70 (4): 2429–2436.

Sabale, S.N., Suryawanshi, P.P., Krishnaraj, P.U. 2019. Soil Metagenomics: Concepts and Applications. In Hozzein W. N. (Ed) *Metagenomics—Basics, Methods and Applications*. IntechOpen. pp 1–27. London, United Kingdom, IntechOpen

Sar, P., Islam, E. 2012. Metagenomic Approaches in Microbial Bioremediation of Metals and Radionuclides. In Satyanarayana, T., Johri, B. N., Prakash, A. (Eds) *Microorganisms in Environmental Management*, pp 525–546. Dordrecht: Springer.

Sardrood, B.P., Goltapeh, E.M., Varma, A. 2013. An Introduction to Bioremediation. In Goltapeh, E. M., Danesh, Y. R., Varma, A. (Eds) *Fungi as Bioremediators,* pp 3–27. Springer Berlin, Heidelberg

Schuppler, M., Mertens, F., Schön, G., Göbel, U.B. 1995. Molecular Characterization of *Nocardioform actinomycetes* in Activated Sludge by 16S rRNA Analysis. *Microbiology* 141 (2): 513–521.

Sebat, J.L., Colwell, F.S., Crawford, R.L. 2003. Metagenomic Profiling: Microarray Analysis of an Environmental Genomic Library. *Applied and Environmental Microbiology* 69(8): 4927–4934.

Sharma, A., Mistry, V., Kumar, V., Tiwari, P. 2022. Production of Effective Phyto-Antimicrobials via Metabolic Engineering Strategies. *Current Topics in Medicinal Chemistry*. DOI: 10.2174/1568026622666220310104645

Sharma, S. 2012. Bioremediation: Features, Strategies and Applications. *Asian Journal of Pharmacy and Life Science* 2 (2): 202–213.

Sharma, V.K., Kumar, N., Prakash, T., Taylor, T.D. 2010. MetaBioME: A Database to Explore Commercially Useful Enzymes in Metagenomic Datasets. *Nucleic Acids Research* 38 (suppl 1): 468–472.

Shekhar, S.K., Godheja J., Modi, D.R. 2020. Molecular Technologies for Assessment of Bioremediation and Characterization of Microbial Communities at Pollutant-Contaminated Sites. In Bharagava, R. N., Saxena, G., (Eds) *Bioremediation of Industrial Waste for Environmental Safety*, Springer Nature: Singapore, pp 437–474.

Simon, C., Daniel, R. 2011. Metagenomic Analyses: Past and Future Trends. *Applied and Environmental Microbiology* 77 (4): 1153–1161.

Smith, E., Thavamani, P., Ramadass, K., Naidu, R., Srivastava, P., Megharaj, M. 2015. Remediation Trials for Hydrocarbon-Contaminated Soils in Arid Environments: Evaluation of Bioslurry and Biopiling Techniques. *International Biodeterioration & Biodegradation* 101: 56–65.

Suenaga, H., Ohnuki, T., Miyazaki, K. 2007. Functional Screening of a Metagenomic Library for Genes Involved in Microbial Degradation of Aromatic Compounds. *Environmental Microbiology* 9 (9): 2289–2297.

Thapa, B., Kumar, A. Kc., Ghimire, A. 2012. A Review on Bioremediation of Petroleum Hydrocarbon Contaminants in Soil. *Kathmandu University Journal of Science, Engineering and Technology* 8 (1): 164–170.

Torsvik V., Øvreås L., Thingstad T.F. 2002. Prokaryotic Diversity–Magnitude, Dynamics, and Controlling Factors. *Science* 296 (5570): 1064–1066.

Tyagi, M., Fonseca, M.M.R., Carvalho, C.C. 2011. Bioaugmentation and Biostimulation Strategies to Improve the Effectiveness of Bioremediation Processes. *Biodegradation* 22 (2): 231–241.

Uchiyama, T., Abe, T., Ikemura, T., Watanabe, K. 2005. Substrate-Induced Gene-Expression Screening of Environmental Metagenome Libraries for Isolation of Catabolic Genes. *Nature Biotechnology* 23 (1): 88–93.

Van Hellemond, E.W., Janssen, D.B., Fraaije, M.W. 2007. Discovery of a Novel Styrene Monooxygenase Originating from the Metagenome. *Applied and Environmental Microbiology* 73 (18): 5832–5839.

Varjani, S., Srivastava, V.K., Sindhu, R., Thakur, I.S., Gnansounou, E. 2018. Culture Based Approaches, Dependent and Independent, for Microbial Community Fractions in Petroleum Oil Reservoirs. *Indian Journal of Experimental Biology* 56(7): 451–459.

Vidali, M. 2001. Bioremediation. An Overview. *Pure and Applied Chemistry* 73 (7): 1163–1172.

Wagner, M., Erhart, R., Manz, W., Amann, R., Lemmer, H. 1994. Development of an rRNA-Targeted Oligonucleotide Probe Specific for the Genus *Acinetobacter* and Its Application for in Situ Monitoring in Activated Sludge. *Applied and Environmental Microbiology* 60 (3): 792–800.

Wu, Z., Lin, W., Li, B., Wu, L., Fang, C., Zhang, Z. 2015. Terminal Restriction Fragment Length Polymorphism Analysis of Soil Bacterial Communities under Different Vegetation Types in Subtropical Area. *PLoS One* 10 (6): 0129397.

Zhang, J.F. 2010. Records of Bizarre Jurassic Brachycerans in the Daohugou Biota, China (Diptera, Brachycera, Archisargidae and Rhagionemestriidae). *Palaeontology* 53 (2): 307–317.

7 Plant Growth-Promoting Microbial Consortia for the Alleviation of Abiotic Stress in Medicinal and Aromatic Plants

Suman Singh
University of Lucknow, Lucknow, India
Sucheta Singh
CSIR–National Botanical Research Institute, Lucknow, India
Ashutosh Awasthi
Teerthanker Mahaveer University, Moradabad, India

CONTENTS

DOI: 10.1201/9781003147091-7

7.1 INTRODUCTION

India is an agricultural country. About 80% of rural population of India depends on agriculture and its allied activities as a core source of livelihood. The agriculture sector contributes about 15% in GDP (gross domestic product) of India (Goyal et al. 2016). India produces a large quantity of food grains such as pulses, cereals and millets and exports a large quantity of medicinal and aromatic plants (MAPs) of economic importance (Yadav and Badola 2019). Cultivation of MAPs also boosts the Indian economy pertaining to their export potential, employment generation and poverty mitigation, mainly in the rural sector. The majority of the world population mainly relies on herbal drug and traditional medicines for their immunity boosting potential and primary health care necessities. In India, about 20,000 medicinal plants have been recognized and among these about 800 plants are used for curing diseases (Kamboj 2000). Hence, the high yielding biomass of MAPs is the current requisite for the growers, as well as the pharmaceutical industries. It has been estimated by the World Health Organization that the global market of MAPs will have reached US$5 trillion from US$62 billion over the next three decades. India shares merely 0.5% of the international export market of medicinal plants despite becoming major producers of MAPs (Purohit and Vyas 2004). Among the reasons behind the reduced production of these plants are environmental stresses and climate changes. The major plant productivity restraining factors are abiotic stresses, including heat, and drought, waterlogging and salinity, as well as heavy metals, and, according to FAO (2007), 96.5% of the world land area has been affected by these environmental stresses (Velthuizen 2007). The harmful impact of these abiotic stresses on the growth of plants is well recognized (Yadav et al. 2020). In addition, heavy metal stress also causes damage to the environment and humans (Tchounwou et al. 2012). Plants face various metabolic changes such as alteration in primary and secondary metabolite content, including changes in gene expression related to metabolite biosynthesis pathways under stress conditions. High levels of stress in medicinal and aromatic plants can affect the secondary metabolite production. Moreover, agro-ecological instabilities due to abiotic stresses significantly affect quantitative and qualitative loss in plant yield and global food security (Roberts and Mattoo 2018). As the world population is growing day by day, there are chances of augmenting food insecurity and increasing the demand for herbal drugs. Therefore, plant productivity must be doubled to meet the anticipated food and herbal drug demand, while maintaining the quality (Foley et al. 2011). In order to gain high-yield MAPs productivity, necessary nutrients should be provided in the soil for plant growth promotion and pathogen restriction. As plants are constantly using nutrients from the soil, the restock of lost nutrients needs to be continued. Therefore, chemical fertilizers are generally utilized to replenish plant nurturing (Santiago et al. 2017). Although chemical fertilizers are used for achieving enhanced food and fiber production, chemical fertilizers are very expensive and consistent use may reduce the soil quality and cause ecological damage (Khan et al. 2018).

The microbial inoculants involving the plant growth-promoting microorganisms (PGPM) could be an important eco-friendly and economical solution to enhance plant productivity by alleviating the abiotic stress (Pankaj et al. 2019; 2020). The PGPMs chiefly include plant growth-promoting rhizobacteria (PGPR), plant growth-promoting fungus (PGPF), arbuscular mycorrhizal fungi (AMF), etc. PGPMs play a beneficial role in agriculture by acting as a natural fertilizer. PGPM support plant growth through direct and indirect mechanisms combating abiotic stresses (Hashem et al. 2019). A review of the literature showed that PGPM helped to enhance plant growth, secondary metabolite production/phytochemical synthesis, bioavailability of nutrients, improved stress tolerance and restrained disease occurrence (Dojima and Craker 2016; Majeed and Muhammad 2019; Rana et al. 2012). It has been confirmed through various studies that mixed inoculants (termed microbial consortia) can perform better in different environmental conditions and soil types as compared to single inoculants due to their poor adaptability to varying environmental conditions (Khan et al. 2017; Vassilev et al. 2015). Microbial consortia have multiple activities like nitrogen fixation, siderophore production, phosphorous solubilization, disease resistance, metabolic activities, hormone and antibiotic synthesis which make them more reliable technologies for the augmenting of plant yield in comparison to other methods (Ma 2019; Upadhyay et al. 2012; Zahir et al. 2018). Selection of suitable carrier material and bio-inoculants, along with their shelf-life study for the development of microbial consortia formulation is also necessary for its successful commercialization. The current chapter illustrates the effects of abiotic stress on MAPs and role of microbial consortia in the alleviation of abiotic stress in MAPs with their adaptation mechanisms. This chapter encourages formulating highly efficient microbial consortia for sustainable production of MAPs.

7.2 ABIOTIC STRESS AND ITS IMPACT ON PLANT PRODUCTIVITY

7.2.1 Various Types of Abiotic Stress

The adverse effect of abiotic factors on the living population performance and physiology in the ecosystem is called abiotic stress. Abiotic stress comprises extreme temperatures (cold and heat), drought (water deficits), flood (waterlogging), high salinity, light intensity and heavy metals toxicity, which may cause damage to plants and animals worldwide. As plants are sessile, they have to tackle the adverse effect of abiotic stress in terms of signaling mechanisms and acuity in response to adaptation or resistance to harmful environmental stress (Gull et al. 2019). Plants show different modifications according to abiotic stress conditions. Generally, growth and development of plants are retarded under abiotic stress. However, production of secondary metabolites and essential oil content was found to be enhanced in MAPs to combat the impact of abiotic stress (Pradhan et al. 2017). The effect of abiotic stress on plants productivity and its metabolism has been described below.

7.2.2 Drought and Its Impact on Plants

Drought stress is a restraining factor for plant productivity that can cause an imbalance between food demand and supply for the global population (Abobatta 2019). Drought stress occurs due to high light intensity and limited water supply in soil that is not available to plant roots. The combination of high light and water deficit conditions induce stomatal closure in leaves. Consequently, carbon di oxide uptake notably reduces, which results in reduced consumption of $NADP^+$ for CO_2 fixation. Thus, high amount of $NADPH+H^+$ accumulates, which alter the metabolic processes and reduced synthesis of chemical components like phenols, alkaloids or isoprenoids (Selmar and Kleinwächter 2013). Drought stress is in fact regarded as a negative factor accountable for losses in agricultural yield; but, with reference to medicinal plants, the condition is dissimilar. Although this happens only under moderate drought conditions, under severe drought the biomass of MAPs reduces and secondary metabolites content increases (Kleinwächter et al. 2015). Drought stress disturbs the growth of roots and leaves and, due to minimum accessibility of water leaf water potential, relative water content also reduces (Blum 2017; Hasanuzzaman et al. 2014). Moreover, other physiological and biochemical responses also occur under drought stress. The growth of plants disturbs due to generation of reactive oxygen species. However, some osmo-protectants such as polyols, glycine betaine and proline are released by plants under drought conditions to maintain the metabolic functions. Additionally, plant growth hormones, including auxins, cytokinin, gibberellins, abscisic acid and various enzymes (act as antioxidants) also helps plants in reducing the adverse effect of drought (Farooq et al. 2009).

Literature on the growth and development of MAPs under drought stress have been less available. MAPs have displayed different responses under drought stress. For example, under drought stress, fresh herb weight was decreased in *Salvia miltiorrhiza* (Liu et al. 2011), *Ocimum basilicum (*Ekren et al. 2012), *Mentha spicata, Rosmarinus officinalis* (Delfine et al., 2005) and *Mentha piperita* (Khorasaninejad et al. 2011). Some investigators have reported a reduction in essential oil yield in Lamiaceae family plants like *Mentha arvensis* (Misra and Srivastava 2000), *Rosmarinus officinalis* (Zehtab-Salmasi et al. 2001), and *Salvia officinalis* (Govahi et al. 2000) under drought stress. However, other investigators have recorded an augment in essential oil yield in some Lamiaceae species, for instances, *Salvia officinalis* (Bettaieb et al. 2009) and *Lavandula angustifolia* (Karamzadeh 2003) under similar conditions. These contradictory results can be explained by their trichome density/oil glands analysis and the effects of level and duration of drought stress. There are three possible mechanisms described which explained that drought stress decreases the MAPs productivity: (1) Photosynthetic rate is decreased due to early leaf wilting or rolling or leaf senescence; (2) Radiation used efficiency is reduced by drought stress which ultimately reduce the biomass; (3) Harvesting index is reduced which may restrain MAP productivity (Farahani et al. 2009).

7.2.3 Salinity Stress and Its Impact on Plants

Salinity of soil is a major problem for agriculture, and two types of salinity have been detected: Primary and secondary. Primary salinity takes place naturally while secondary salinity occurs as a result of human activities. Excessive irrigation (increase in groundwater tables), utilization of contaminated water, or vegetation clearing are some common reasons for secondary salinity (Bharti et al. 2013). Soil salinity disturbs the ecological equilibrium by badly affecting soil quality and agricultural productivity (Srivastava and Kumar 2015). Salinity also badly affects plant development involving germination, growth and reproduction and also induces nutrient deficiency, ion toxicity, and osmotic stress in plants (Akbarimoghaddam et al. 2011). Moreover, salinity also imposes leaf area reduction, decreased chlorophyll content and stomatal conductance, thus affecting photosynthesis efficiency (Netondo et al. 2004). Growth and productivity of MAPs have been reported to be negatively affected by salinity stress (Banerjee and Roychoudhury 2017). For instance, productivity of medicinal plants such as *Trachyspermum ammi*, milk thistle, *Ammi majus*, fennel and cumin were reduced by salt stress (Ashraf and Orooj 2006).

Photosynthesis is an essential physiological process for growth of plants. Salt stress inhibits the chlorophyll metabolism and suppressed photosynthesis process; ultimately cause death of plant. Chlorophyll content (a, b and total) was found to be reduced in *Thymus vulgaris, Ziziphora multiflora, Satruja hortensis, Teucrium polium* (Said-Al Ahl and Omer 2011). Some essential oil-bearing plants like *Mentha pulegium, Mentha piperita* and *Mentha suaveolens* showed a reduction in essential oil yield under salt stress over the control (non-stress) (Aziz et al. 2008). The levels of major compounds were differently affected by salinity stress in various plant essential oil. Eugenol and linalool are the major component of *Ocimum basilicum* essential oil. When subjected to salt stress, eugenol content decreased, while linalool increased (Said-Al Ahl et al. 2010). Another study reported that *Origanum vulgare* essential oil components such as carvacrol, p-cymene and c-terpinene were reduced under salt stress (Said-Al Ahl and Hussein 2010). The reason behind the fluctuation in essential oil constituents may be the density of oil glands increased during stress and cause accumulation of essential oil in some plants. Another reason may be the downturn of primary metabolism of plants during stress, which affected the growth of plants and decreased the essential oil content (Said-Al Ahl et al. 2016). Phenolic compounds are released by plants when subjected to abiotic stress. Higher plants have been reported to produce phenolic compounds to scavenge the ROS during salt stress (Waskiewicz et al. 2012).

7.2.4 Heat Stress and Its Impact on Plants

Heat stress (high temperature) is another abiotic factor that can badly affect the growth and productivity of plants (Bita and Gerats 2013). Due to an increase in the greenhouse effect, the global temperature increases and alters the climatic conditions.

It is considered as an agrarian effect in various arid and semiarid regions. Heat stress also alters the biochemical and physiological processes in plants, which seriously affect the growth and development of plants (Ghasemi et al. 2016). The literature on the effect of heat stress on biological activities and secondary metabolite production is scarcely available. A study illustrated that chemical composition of *Mentha piperata* and *M. arvensis* varied during heat stress. Menthol is the major component of *Mentha* essential oil and found to be decreased under increasing temperature, whereas pulegone and menthyl acetate increased as the temperature enhances. Menthone and menthofuran also showed variation under heat stress (Heydari et al. 2018). Another study showed that *Mentha piperata* and *Catharanthus roseus* recorded significant reduction in the height and weight of plants under heat stress and conditions become more pronounced when subjected to the combined stress of drought and heat. Additionally, stress also stimulated the accumulation of osmoprotectants and secondary metabolites. Concentration of terpenoids, tannins and alkaloids increased while phenol, saponin and flavonoids content reduced in response to stress (Alhaithloul et al. 2019). Ibrahim et al. (2010) demonstrated that accumulation of terpenoids in *Betula pendula* and *Populus tremula* have increased at elevated temperatures. Escandón et al. (2018) illustrated that the accumulation of secondary metabolites such as flavonoids and terpenoids in *Pinus radiata* is regulated by zeatin riboside and isopentenyl adenosine.

7.2.5 Cold Stress and Its Impact on Plants

Among abiotic factors, cold is another harmful stress condition that constrains plant yield worldwide. Cold (low temperature) stress includes freezing (<0°C) and chilling (between 0°C and 10°C) stresses. Chilling stress generally affects productivity of plants by altering primary metabolism and destroying cell membrane stability, while freezing stress induces ice formation inside cells that cause severe membrane damage. Solute precipitation and denaturation of protein also occurs due to freezing injury (Ghassemi et al. 2021; Salinas 2002). Cold stress also negatively affected photosynthesis, respiration rate, nutrient uptake and development of reproductive organs. It also enhanced protein degradation, electrolyte leakage, lipid peroxidation, ROS (Cao et al. 2011; Islam et al. 2011). Consequently, these result in yellowing and wilting of leaves, necrosis, reduced germination, stunted growth and underdeveloped reproductive organs (Yadav 2010). All plants may not be able to generate stress tolerance while some plants have specific adaptation responses. For example, many MAPs such as *Satureja thymbra, Cistus incanus, Thymus sibthorpii, Teucrium polium* and *Phlomis fruticose* showed chemical defenses and generate mechanical barriers against cold stress. Various biochemical, physiological and molecular defense mechanisms are included in cold acclimation. During this process, other modification such as accumulation of compatible solutes, phytohormones, enzymatic and non-enzymatic changes occur (Boscaiu et al. 2008).

7.2.6 Heavy Metal Stress and Its Impact on Plants

Heavy metals have high density relative atomic weight with an atomic number more than 20. Excessive quantity of heavy metals can be deleterious to organisms. Some heavy metals such as Pb, Cd, Hg and As are highly detrimental and some such as Cu, Co, Mn, Fe, V, Ni and Zn are harmful for living beings when present in high amounts. In nature heavy metals are present in soil but anthropogenic activities like use of pesticides, chemical fertilizers in agriculture, sewage sludge, municipal waste disposal, burning of fossil fuels, etc. enhance the concentration of these metals to a level that is detrimental to plants and animals (Chibuike and Obiora 2014). Various studies have recorded plant growth reduction on heavy metal contaminated soil (Ashfaque et al. 2016; Chibuike and Obiora 2014). Heavy metal toxicity directly inhibits the cytoplasmic enzymes synthesis and destroys the cell structures due to ROS. Indirect heavy metal toxicity disturbs the community structure of soil microorganisms and also replaces beneficial nutrient content at cation exchange sites of plants. Both toxic effects decrease plant growth and development and under extreme conditions the death of the plant takes place (Jamla et al. 2021; Taiz and Zeiger 2002). MAPs have the capability to accumulate substantial numbers of toxic metals and it has been reported that heavy metal stress may change the secondary metabolite content of plants. Nonetheless, the effect of stress is variable in yield, essential oil content and secondary metabolite contents (Verpoorte et al. 2002; Zheljazkov and Warman 2003). A study has showed that *Mentha piperata* and *Ocimum basilicum* productivity are not much influenced by heavy metal treatments (Cu, Pb, Cd) (Zheljazkov et al. 2006). Higher concentration of copper reduced the yield of *Anethum graveolens* (Zheljazkov et al. 2006). Another study shows that essential oil content was found to be increased in *M. piperata*, while decreased in *M. arvensis* and *M. citrata* (Prasad et al. 2010). In *Salvia officinalis* plant, the essential oil components α- and β-thujones and its oil quality was found to be deteriorated in heavy metal polluted soil (Stancheva et al. 2009). Similarly, secondary metabolites also exhibit variation when subjected to heavy metal stress. Eugenol, the main component of *Ocimum tenuiflorum* essential oil was increased by chromium contamination (20 μM) exposure of Cr showed higher production of eugenol (25%) (Rai et al. 2004). In a similar way, phyllanthin and hypophyllanthin found in *Phyllanthus amarus* were increased by cadmium stress (Rai et al. 2005). However, heavy metal accumulated MAPs were found thorough examination to avoid increased concentration of heavy metals reaching the buyer (Shahid et al. 2016).

7.2.7 Waterlogging

Under flooding/or waterlogged conditions plants face various challenges like gas exchange limitations, micronutrient toxicities, nutrient deficiencies, etc. (Setter et al. 2006). Hypoxic or O_2 limiting situations in plant roots occur under flooding which lead to higher accumulation of 1-aminocyclopropane 1-carboxylate synthase (ACS) enzyme and other stress proteins. As a result of stress, more ACC synthesizes its roots; however, the more ACC accumulated into the roots due to a failure in conversion of ethylene, extra ACC had to be moved to the shoots where there is an aerobic atmosphere. In the presence of oxygen ACC gets converted to ethylene, and due to an accumulation of ethylene, plants show stunted growth, chlorosis, wilting (epinasty), and necrosis (Vwioko et al. 2017). It is essential to improve the oxygen diffusion in the flooded plants to get away from the hypoxic environment. Plants suffer various morphological changes like wilting, reduced size of stomatal pore, shoot elongation, hyponastic growth, suberized exodermis, aerenchyma formation, adventitious root, lenticels hypertrophy development (Voesenek & Bailey-Serres 2015). For example, in *Mentha arvensis* rapid shoot elongation has been seen (Phukan et al. 2016) to escape from waterlogging stress.

7.3 DETAILED MECHANISMS OF PLANT-MICROBE INTERACTIONS FOR BETTER MANAGEMENT OF VARIOUS STRESSES

A variety of biochemical and physiological modifications with respect to abiotic and biotic stress in plants have been reported. Various mechanisms have been recognized behind microbial activated stress tolerance in plants (Figure 7.1).

7.3.1 Siderophores

Iron is an essential element for the growth and development of a plant. It is found in soil as an unsolvable form of Fe^{3+} that is not available to the plants. To overcome this situation, several microbes produce specific ferric ion chelating agents commonly known as siderophores (Saha et al. 2016). Siderophore-forming bacteria were able to improve iron nutrition, suppress diseases, and improve plant health (Shen et al. 2013). Many siderophores have the ability to elicit induced systemic resistance (ISR)

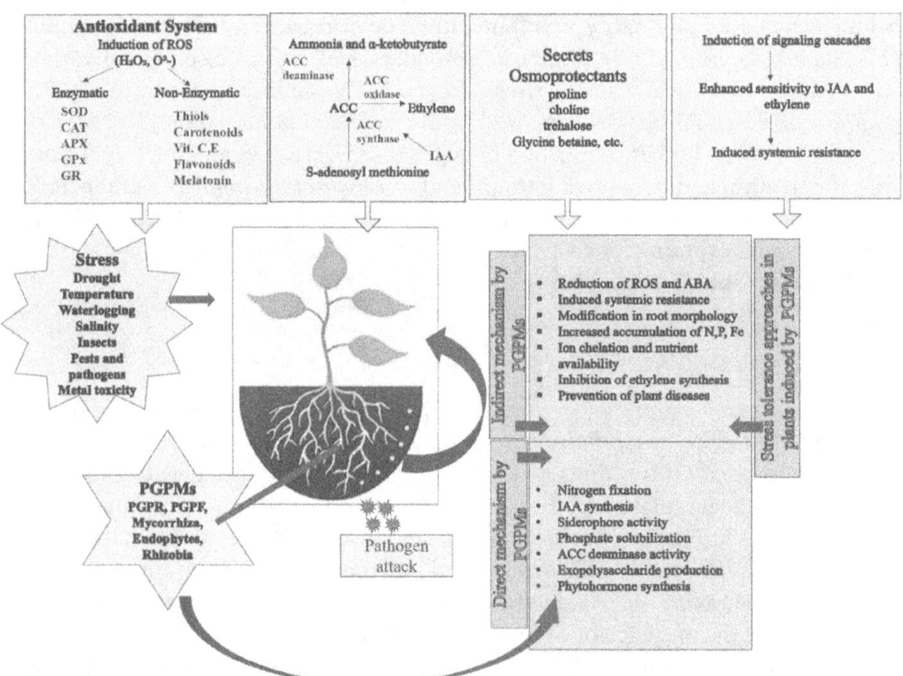

FIGURE 7.1 Plant-microbe interaction mechanism under environmental stress.

in host plants, which depends on several factors such as PGPR, soil type, selection of crop, availability of iron, affinity of siderophore for iron, specific pathogen (Bashan and de Bashan 2005).

7.3.2 Phosphate Solubilization

Phosphorous is an important plant nutrient found in about 50% of soil as insoluble form. The insoluble forms of P are calcium phosphate and inorganic phosphate with iron and aluminum while the soluble types are monobasic and the dibasic phosphates (HPO_4^{2-} and H_2PO^{4-}) easily assimilated by plants (Jha and Saraf 2015). Microorganisms use different mechanism to solubilize the insoluble forms of phosphate. *Pseudomonas, Bacillus* and *Rhizobia* have been recognized as effective phosphate solubilizers (Goswami et al. 2014a). The mechanism involves secretion of organic acid by microbes due to sugar metabolism. These acids act as good chelators of divalent Ca^{2+} cations associated with the release of phosphates from insoluble phosphatic compounds by lowering the pH in the soil (Goswami et al. 2014b; 2015). Many studies have revealed that phosphate solubilizing PGPR possess more than one mechanism degrading complex form of phosphate in to simple form (Hariprasad and Niranjana 2009).

7.3.3 Nitrogen Fixation

Nitrogen is an essential element for growth and development of plants and it is also not available to plants unless reduced to ammonia by rhizobia and diazotrophic bacteria. PGPR are reported to convert atmospheric N_2 into ammonia and ammonium using nitrogenase enzymes in soil (Riggs et al. 2001). A variety of PGPRs such as *Azotobacter* spp. *Bacillus* spp. and *Beijerinckia* spp. have been recognized to fix gaseous nitrogen symbiotically (Barea et al. 2005; Esitken et al. 2006). Furthermore, some free living diazotrophs such as *Azoarcus, Azospirillum, Burkholderia, Gluconacetobacter* and *Pseudomonas* fix gaseous N_2 non-symbiotically (Niranjana and Hariprasad 2014).

7.3.4 Induced Systemic Response (ISR)

The defense mechanism is activated in particular parts of the plants at which stimulation has occurred and causes ISR. Induction is caused due to the attack of pathogens or insects, and ISR support the plant to regulate occurrence of diseases. The stimulation of defense mechanism in the form of ISR consists of jasmonate and ethylene signaling pathways within the plant body against pathogens (Verhagen et al. 2004).

7.3.5 ACC Deaminase Activity

In ethylene biosynthetic pathway, 1-aminocyclopropane-1-carboxylate (ACC) which is a precursor of ethylene is synthesized by S-adenosylmethionine (SAM) with the help of 1-aminocyclopropane-1-carboxylate synthase (ACS). It has been reported that ACC deaminase possessing bacteria were able to hydrolyze ACC into ammonia and α-ketobutyrate (Bal et al. 2013). Therefore, ACC deaminase-synthesizing PGPR increases the plant tolerance to abiotic stress by regulating ACC content and minimizing ethylene synthesis in plant tissues (Glick et al. 2007; Saleem et al. 2018). A number of studies have reported that ACC deaminase-producing microbes were able to improve plant growth and alleviate abiotic stresses in various plants (Brotman et al. 2013; Chang et al. 2014; Glick 2012; 2015; Naveed et al. 2014a, 2014b; Pankaj et al. 2020; Pinedo et al. 2015). A study confirmed the role of ACC deaminase and indole acetic acid (IAA) production by beneficial microbes competent to increase defense mechanisms in plants against various environmental stresses such as high salt, drought and heavy metal stress (Barra et al. 2016; Hermosa et al. 2012).

7.3.6 Secretion of Phytohormones

Under various environmental stresses, PGPR with multifunctional activities directly control plant growth and development by phytohormones synthesis such as auxins (IAA), abscisic acid (ABA), gibberellins, ethylene, and cytokinins (Egamberdieva 2013b; Fahad et al. 2015). Various literature has reported that IAA-producing microbes play an important role in root development and enhance tolerance against abiotic stresses (Creus et al. 2005; Dimkpa et al. 2009; Molina-Favero et al. 2008). *Lavandula dentate* plants inoculated with IAA-producing *B. thuringiensis* improved plant growth, nutrient uptake and phyciochemical activities in plants under drought conditions (Armada et al. 2014). *P. putida* with gibberellins synthesizing properties improved soybean plants growth under drought conditions (Sang-Mo et al. 2014).

7.3.7 Release of Antioxidant Enzymes

Under abiotic and biotic stresses plants generates reactive oxygen species (ROS) comprising of hydrogen peroxide (H_2O_2), superoxide radical ($O_2^{\bullet -}$), hydroxyl radical ($OH^{\bullet}$), singlet oxygen ($^1O_2$) and alkoxy radicals (RO) (Gupta et al. 2016; Kalia et al. 2017). ROS causes oxidative injury to plants by altering membranes, DNA, proteins and lipids structure, as well as functions (Tiwari 2016). Various studies have reported that PGPM inoculation in plants helps to regulate the antioxidant enzymes synthesis mechanism, thus protecting plants against abiotic stress (Enebe and Babalola 2018). A study reported that AM fungi (*Glomus versiforme*) inoculation in citrus enhanced activities of superoxide dismutase (SOD), peroxidase (POD) and catalase (CAT) resulting in improved osmotic adjustment under drought stress (Wu and Xia. 2006). Likewise, co-inoculation of *Pseudomonas mendocina* and *G. intraradices* or *G. mosseae* inoculation in lettuce ameliorated oxidative damage in plants by enhancing catalase activity under severe drought conditions (Kohler 2008).

7.3.8 Exopolysaccharides

Exopolysaccharides (EPS) are known for bioremediation, microbial aggregation, protection from desiccation and are commercially used as a thickening, stabilizing, gelling and coagulating agents by many industries. Many PGPRs produce biofilm around the roots that sustain sufficient moisture inside the polymeric extracellular matrix formed by the bacterial populations. Plants inoculated with EPS-producing bacteria like *Pseudomonas* showed tolerance against drought stress due to EPS production (Sandhya et al. 2009). Under drought stress, inoculation of *Azospirillum* strain which was rich EPS producer showed ameliorative effects against low moisture content of soil on wheat and maize by bringing about many biochemical changes in the plant system (Alvarez et al. 2008; El-Komy et al. 2003).

7.4 EXPLOITATION OF PLANT GROWTH-PROMOTING MICROBES FOR THE ALLEVIATION OF ABIOTIC STRESS

Plants give raw materials for various industries like pharmaceuticals, food, cosmetics, flavoring and fragrances. Thus, conservation and management of plant genetic resources is required to meet the demand of food, fodder, fiber, etc. Plant associated beneficial microorganisms, including AM fungi, PGP fungi, PGP rhizobacteria, endophytes, algae and cyanobacteria have major roles in improving plant productivity with increasing resistance against abiotic and biotic stress (Etesami and Beattie 2017; Meena et al. 2017). A review of several reports shows that under different stress conditions, PGPR have potential to increase productivity by numerous direct and indirect mechanisms (Etesami 2018; Etesami and Beattie 2017; Kaushal and Wani 2016; Shrivastava and Kumar 2015; Ullah and Bano 2015). Researchers have showed their interest in the application of PGPRs for agricultural productivity many times previously, but the use of multi-strain microbial formulations under abiotic stress conditions is acquiring increasing importance in present conditions. The current research has moved in the direction of multi-strain formulations development (bacterial-bacterial/fungal-bacterial/fungal-fungal) for their better performance in comparison to single strains (Vorholt et al. 2017; Woo and Pepe 2018).

7.4.1 Alleviation of Salt Stress by PGPM

Many studies reported that symbiosis of AM Fungi in plants increased nutrient availability, osmoprotectants activity and photosynthetic efficiency under salt stress alleviation. For example, inoculation of *Glomus intraradices* in *Ocimum basilicum* (Zuccarini and Okurowska 2008), *Glomus mosseae* and *Glomus intraradices* in *Mentha arvensis* (Bharti et al. 2013) under salt stress and *G. mosseae* and *Bacillus subtilis* in *Artemisia annua* (Awasthi et al. 2011) under normal conditions showed better performances in terms of improved biomass, secondary metabolites content and nutrient uptake compared to the uninoculated control. Karthikeyan et al. (2012) had reported that application of endophytic bacteria *Achromobacter xylosoxidans* in *Catharathus roseus* showed reduced level of ethylene under highly salt stress conditions. Another similar study reported that treatment with *Trichoderma harzianum*, *Bacillus coagulans* and AM fungi *Glomus aggregatum* in *Glycyrrhiza glabra* plants showed a significant increase in biomass, nutrient content and secondary metabolites (Selvaraj and Sumithra 2011). PGPM symbiosis in plants can improve antioxidative defense mechanisms by enhancing the activities of enzymes like catalase, ascorbate peroxidase, superoxide dismutase ameliorating the oxidative damage created by salinity in *Trigonella foenum graceum* (Evelin and Kapoor 2014) and *Mentha arvensis* (Bharti et al. 2013). They have been recognized to enhance plant growth directly as they can improve and mobilize the supply of nutrients, such as nitrogen (Dobbelaere et al. 2003) and phosphorous and make them available for plant uptake or by production of phytohormones and growth regulators, as well as by the suppression of pathogens through mechanisms of antibiosis, iron sequestration by siderophores, secreting cell wall degrading enzymes like chitinase, cellulose, etc. and improving soil structure by production of polysaccharide (Kachhap et al. 2015). It has been reported in some studies that under a low salt level, secondary metabolites and essential oil contents increased in plants and further decreased when a high level of salt stress was applied. However, under high salt stress it was constant in treatment with PGPM. For example, the essential oil content increased with increasing salinity (10g/L NaCl) in *Rosmarinus officinalis* but decreased with further increases in salt stress. When beneficial bacterium *Pseudomonas fluorescens* was applied in these salt affected plants, the essential oil content was not reduced by salinity stress (Bidgoli et al. 2019). The synergistic effects of mixed cultures have been exhibited in a number of MAPs, for instance, *Alpinia galanga* and *Coleus amboinicus* (Mani 2004), *Calendula officinalis* (Hosseinzadah et al. 2011), *Ocimum basilicum* (Hemavathi et al. 2006), *Phyllanthus amarus* (Earanna and Bagyaraj 2004), *Mentha arvensis* (Singh et al. 2019) *Saraca asoca* (Lakshmipathy et al. 2001), and *Silybum marianum* (Egamberdieva et al. 2013a).

7.4.2 Alleviation of Drought Stress by PGPM

Application of ACC deaminase-producing PGPR *Bacillus licheniformis* in pepper plants showed drought resistance activities as compared to control plants (Lim and Kim 2013). Another study reported that inoculation of *Bacillus pumilus* in *Glycyrrhiza uralensis* enhanced growth, chlorophyll content, and improved the water state, reducing the effect of drought stress (Zhang et al. 2021). In another case, the effect of biofertilizers under drought stress has been studied in Valerian and it was found that a combination of *Azospirillum,* mycorrhiza and *Azotobacte*r and another combination of *Azotobacter* and mycorrhiza enhanced the biomass, oil content, relative water content, chlorophyll content and nutrient uptake in comparison to other treatments and control under drought stress (Mina et al. 2019). Another similar study showed that a consortium involving *Bacillus lentus, Pseudomonas* sp and *Azospirillum brasilens* application in *Ocimum basilicum* improved plant growth, antioxidant activity and photosynthetic efficiency pointing toward resistance against water-induced stress (Heidari and Golpayegani 2012). In a study, the effect of PGPRs (*Azospirillm brasilense, Azotobacter chrococum, Pseudomonas fluorescent, Azotobacter + Azospirillum + Pseudomonas)* and vermicompost in *Melissa officinalis* plants has been examined under drought stress. The biofertilizer application increased the total chlorophyll and relative water content (Kazeminasab et al. 2016).

7.4.3 Alleviation of Temperature Stress (Heat and Cold) by PGPM

The bio-inoculants can mitigate the harmful effect of temperature stress in plants. Heat stress tolerance had been increased by application of *Pseudomonas, Bacillus* and mycorrhizal fungi in tomatos which provide strength to plants producing ROS-degrading enzymes (catalase, peroxidases, superoxide dismutase), lipid peroxidation and a reduction in H_2O_2 levels (Duc et al. 2018). Another study showed that ACC deaminase-producing microbes have great potentiality in reducing heat stress by degrading the ethylene precursor. For example, *Paraburkholderia phytofirmans* possessing ACC deaminase activity promotes the growth of potatos under heat stress (Bensalim et al. 1998).

Growth of plants can be hindered by low temperature but the presence of beneficial microorganisms provides tolerance against cold temperatures to plants by performing various mechanisms (Amara et al. 2015). The *Serratia marcescens* produces some substances like indole acetic acid, which is induced by cold temperatures (4°C and 15°C) in the *Pantoea dispersa* plant (Selvakumar et al. 2008).

Plant growth-promoting psychrotolerant strains PGERs and NARs that were applied in Wheat seeds showed higher yields and nutrient uptake under cold stress (Mishra et al. 2008). Another study reported that the application of ACC deaminase-producing *Paraburkholderia phytofirmans* PsJN in grapevines increased cold resistance by decreasing cell membrane damage (Barka et al. 2006; Theocharis et al. 2012). Another study showed that inoculation of *Trichoderma harzianum* and *Achromobacter xylosoxidans* in *Ocimum sanctum* improved the fresh herb yield, photosynthetic efficiency, nutrient uptake, proline, starch and phenolics content, as compared to non-inoculated plants. Accumulation of ACC content was reduced in the consortium treated plants over control (Singh et al. 2020). It has been reported by Glick (2014) that ACC deaminase-producing bacteria can improve plant growth and reduce environmental stress. Some other PGPRs, such as *Brevibacterium frigoritolerans, Pseudomonas fragi, P. fluorescens, P. proteolytica* and *P. chlororaphis* possessing ACC deaminase activity, showed decreased ROS production, lipid peroxidation and frost injury in bean plants (Tiryaki et al. 2019).

7.4.4 Alleviation of Waterlogging Stress by PGPM

Microbial inoculants reported to protect plants from waterlogging/flooding injury. A study described that ACC deaminase-producing PGPR *Achromobacter xylosoxidans* when inoculated in *Ocimum sanctum* plants recorded improved growth and minimum ethylene content as compared to non-inoculated waterlogged plants (Barnawal et al. 2012). Another study exhibited that *P. fluorescens* having ACC deaminase activity increased rice root elongation *in vitro* and colonization efficiency of the endophyte significantly over the control under flooding conditions (Etesami et al. 2014). It has been proved by Grichko and Glick (2001) that inoculation of ACC deaminase-producing bacteria in plants showed higher growth, photosynthetic rate and considerably reduced ethylene production in plant under flooding stress.

Symbiotic association of AM fungi and rhizobia (*B. japonicum*) has been found to enhance nutrient acquisition in soy crops improving yield and growth of this crop under water stress (Hattori et al. 2013). Another study showed that inoculation of AM fungi (*Glomus intraradices*) in *Pterocarpus officinalis* seedlings has significantly improve P acquisition and plant growth development in leaves under flooding conditions (Fougnies et al. 2007). Colonization of *Bradyrhizobium elkanii* and

Bradyrhizobium japonicum in Soybean crops improved plant growth by enhancing N_2 fixation in plants under soil flooding conditions (Beutler et al. 2014; Scholles and Vargas 2004).

7.4.5 Alleviation of Heavy Metal Stress by PGPM

Plant-microbe interaction plays a vital role in regulating heavy metal stresses in soil and plants. They have ability to amass, alter or detoxify heavy metals. Therefore, they provide additional room for rhizoremediation and phytoremediation uses. The conversion of heavy metal into less toxic forms has been done through redox reaction mediated by beneficial microbe interaction (Amstaetter et al. 2010). Specific enzymes play a key role in the occurrence of redox reactions. For instance, for conversion of CueO/CuiD/CopR, multicopper oxidases are required in Cu efflux. While in reduction of Cr^{+6} to Cr^{+3} chromate reductase of ChrA required. In case of Hg, Hg^{2+} converted in to Hg^0 using the protein MerA (Mishra et al. 2017). Arsenic oxidizing bacteria *Bacillus* sp. and *Geobacillus* sp. has been reported to biotransform toxic As^{3+} to less toxic As^{5+} (Majumder et al. 2013). Another mode of action performs by rhizospheric bacteria/fungi involve reduction of heavy metal toxicity by producing metal-binding peptides and chelators, such as metallothioneins (MT) and PC (glutathione-derived peptides) (Miransari 2011). Murthy et al. (2011) have showed that MT biosynthesis was enhanced in *Bacillus cereus* under Pb exposure. Another similar study exhibited that *Providencia vermicola* (SJ2A) have heavy metal tolerance activity by releasing MT-assisted periplasmic Pb sequestration (Sharma et al. 2017). Mycorrhizal fungi such as *Rhizophagus irregularis* also help in detoxification of Zn, Cu and Cd (González-Guerrero et al. 2008). They act by entering in to the cell and start sequestration or compartmentalization of heavy metals in to different subcellular organelles (Hložková et al. 2016).

Another mechanism performed by PGPR in enhancing plant growth under heavy metal stress is the production of ACC deaminase enzyme (Han et al. 2015; Zhang et al. 2011). ACC deaminase helps in regulation of ethylene synthesis and facilitates plant tolerance to stress conditions. Phytohormones produced by microbes under heavy metal stress also trigger tolerance or resistance approaches to overcome heavy metal toxicity in plants. For example, *B. subtilis* improved growth of *Brassica juncea* by producing IAA, which induces rapid root elongation in plants in Ni-contaminated soil (Zaidi et al. 2006).

In addition to ACC deaminase and IAA activity, siderophore producers, phosphate solubilizers, and nitrogen fixing microbes under heavy metal stress also help in plant growth inducing nutrient uptake ability and altering the bioavailability of heavy metal (Gupta et al. 2014; Wu et al. 2010).

7.5 NEED FOR MICROBIAL CONSORTIA DEVELOPMENT: THE CURRENT TRENDS

The use of microbial consortia has turned out to be a recent research trend in the field of biotechnology, bioprocess engineering and applied microbiology. It is apparent that mixed cultures can perform efficiently the complex course of actions, augmenting the productivity more in comparison to a single culture. The most important thing in the development of mixed consortia is the compatibility among microbial members, which determines the long-term stability and industrial application. Non compatibility among microbial members of microbial consortia may destroy the community structure as well as their performance. Thus, optimization of compatible association among microbes, carrier material and their storage play a crucial role in the best performance of the microbial consortia formulation for crop productivity (Ghosh et al. 2016). For instance, two different strains of *Escherichia coli* are known to metabolize glucose and xylose individually, but their synergistic association metabolizes the sugars more efficiently in comparison to their single culture (Eiteman et al. 2008). This technology will take off the extra burden from a single culture to perform. Extensive research on plant microbe interaction, microbe-microbe interaction is necessary to design, optimize and develop bioformulation. Opting for appropriate PGPM, carrier material and large-scale preparation and preservation of inoculants for the long term are the necessary steps involved in the development of bioformulation for commercial use. For production of good quality bioformulation, carrier material should be cheap, eco-friendly, easy to handle, have water-holding capacity > 50% and should be easily available (Malik et al. 2005). Many types of carrier materials like kaolinite, zeolite, alginate, talc, turf, peat, pyrophyllite, montmorillonite, press mud, vermiculite, vermicompost and sawdust are used for the development of bioformulation (El-Fattah et al. 2013; Zayed 2016). Generally, dry bioformulations are preferred in comparison to wet bioformulations due to their long shelf life, simple to store and transport (Omer 2010). Various types of supplementary materials such as hormones, mineral nutrients, fungicides and carbon sources are also used to increase the adhesive capacity of microbes to host and enhance efficiency of bioformulation. In modern times, encapsulation technologies have been generally used to produce microbial inoculants with variations in composition and morphology (John et al. 2011; Schoebitz et al. 2012). Another option in advanced technology is the use of pure lyophilized cultures, which can be employed directly or in combination with a solid carrier material (Malusá et al. 2012).

7.6 FUNCTIONING OF MICROBIAL CONSORTIA IN ALLEVIATING ABIOTIC STRESS AND ENHANCING PLANT GROWTH

Global climate change and environmental stresses substantially restrict plant growth and development. Abiotic stresses like nutrient deficiency, salinity, heavy metal toxicity, high light intensity, low light intensity, ozone, drought, temperature (high and low) and UV-B radiation are environmental factors which cause adverse effect on plants (Hasanuzzaman et al. 2012; Smekalova et al. 2014). Continuous changes in the environment cause numerous morphological, biochemical and physiological changes in plants resulting in a decreased crop yield and productivity of pharmaceutically important secondary metabolites (Seki et al. 2003). Here, the mechanisms underlying microbes mediated enhancements of plant resistance to abiotic stress are explained (Table 7.1).

Cold stress or increasing temperature causes major crop damaging stresses (Koini et al. 2009; Pareek et al. 2010). A study of transcriptomic and metabolism

TABLE 7.1
Responses of Various Bio-Inoculants in Medicinal Plants under Abiotic Stress

Microbes	Medicinal Plant	Type of Abiotic Stress	Responses	References
Achromobacter xylosoxidans	*Catharnthus roseus*	Salt stress	ACC deaminase, reduced ethylene level	Karthikeyan et al. 2012
Azospirillum brasilense and *Pseudomonas fluorescens*	*Catharanthus roseus*	Salt stress	Increased biomass and ajmalicine content	Karthikeyan et al. 2009
Halomonas desiderata and *Exiguobacterium oxidotolerans*	*Mentha arvensis*	*Catharanthus roseus*	Increased herb yield	Bharti et al. 2014
Bacillus coagulans	*Glycyrrhiza glabra*	*Catharanthus roseus*	Increased biomass and secondary metabolites content	Selvaraj and Sumithra 2011
Bacillus pumilus and *Exiguobacterium oxidotolerans*	*Bacopa monnieri*	*Catharanthus roseus*	Plant weight, bacoside-A content	Bharti et al. 2013
Brachybacterium paraconglomeratum	*Chlorophytum borivilianum*	Salinity stress	ACC Deaminase, reduced ethylene level, IAA and ABA level affected	Barnawal et al. 2016
Pseudomonas extremorientalis	*Silybum marianum*	Salinity stress	Increased biomass	Egamberdieva et al. 2013a
Pseudomonas sp. and *Bacillus lentus*	*Ociumum basilicum* L	Salinity stress	Increased enzyme activity, including GR (glutathione reductase) and APX (ascorbate peroxidase)	Golpayegani and Tilebeni 2011
Azotobacter chroococcum	*Trigonella foenum-graecum* L. (fenugreek)	Drought	Increased shoot biomass and uptake of nutrients	Tank and Saraf 2003
Glomus mosseae and *Trichoderma harzianum*	*Andrographis paniculata*	Grown at different P level	Andrographolide improved, P level increased	Arpana and Bagyaraj. 2007
PGPR *Achromobacter xylosoxidants*	*Ocimum sanctum*	Waterlogged stress	ACC-deaminase activity, ethylene level decreased, maximum growth and yield of the basil	Barnawal et al. 2012
Exiguobacterium oxidotolerans	*Mentha arvensis*	Salt stress	ACC-deaminase activity, plant growth, oil content and physiological status	Bharti et al. 2014

of cold stressed plants showed direct inhibition of metabolic enzyme, reprogramming of gene expression and raises phenolics production and increased assimilation of lignin and suberin in cell walls (Chinnusamy et al. 2007; Moura et al. 2010). Psychrotolerant endophytic *Pseudomonas vancouverensis* (OB155) and *P. frederiksbergensis* (OS261) having induced cold acclimation functioning genes *LeCBF1* and *LeCBF3* resulting in an ability to better cope with cold stress (10–12°C) with reducing membrane damage and improved antioxidant activity of plants (Subramanian et al., 2015). In similar reports, treatment with endophytes *Burkholderia phytofirmans* strain (PsJN) on plants had increased plant growth and provided strengthening of cell walls under cold stress conditions (Su et al. 2015). This indicates that endophytes sense physiological changes in plants and modulate various gene expressions for adaptation in the new environments. High temperature of soil is a major problem to the survival and colonization of microbes with plants in which protein denaturation and aggregation types of cellular damage occurred. In sudden increasing temperature, organisms respond to increasing synthesis of heat shock proteins (HSPs) consisting of chaperons (like *DnaJ, DnaK, GroES, GroEL, ClpA, ClpB, ClpX*) (Munchbach et al. 1999). A well-known example is *P. aeruginosa* strain AMK-P6, thermotolerant in nature, induced the synthesis of HSPs in high temperature (Ali et al. 2009).

Drought stress is directly linked to water deficit conditions and indirectly to increasing temperature and solar heat (Xu et al. 2010). Under drought conditions, peroxidation activity may be triggered resultant to cause negative effects on antioxidant metabolism. Superoxide dismutase (SOD) plays a significant role in antioxidant metabolism and its activity affected the expression of FeSOD and Cu/ZnSOD (Foyer and Noctor 2005; Naya et al. 2007). Endophytic *B. phytofirmans* strain PsJN showed a diverse range of functionalities involved in ROS detoxification transcriptional regulation and cellular homeostasis under drought conditions (Sheibani-Tezerji et al. 2015). Rhizobacterium *Paenibacillus polymyxa* confers drought tolerance in plants by inducing expression of the drought responsive gene, *ERD15* (Early Response to Dehydration 15) (Timmusk and Wagner 1999).

In the case of flooding conditions, water imposes a critical problem in terms of oxygen insufficiency leading to wilting. As a result, water submerged roots subsequently influence nutrient uptake by reducing the potassium/sodium ion uptake and preventing the translocation of potassium ions to the plant shoot (Armstrong and Drew 2002; Sairam et al. 2008). Under waterlogged conditions, plants stressed with hypoxia resulted in the production of ACC in excess amounts through over expressed ACC synthase activity compared to well aerated plant roots. ACC deaminase activity evaluated in two strains of *Pseudomonas putida*, namely wild type and an isogenic (mutant) strain that lack of enzyme. The result showed that plant produced less ethylene when inoculated with wild strain under waterlogged conditions (Ravanbakhsh et al. 2017).

Salinity stress causes cell dehydration due to osmotic imbalance and cytoplasmic water elimination has the result of decreasing the volume of vacuole and cytoplasme, which is indirectly reduced *in planta* secondary metabolite content (Ramakrishna and Ravishanker. 2011). For example, anthocyanins content increased in salt stressed conditions (Parida and Das 2005) whereas salt sensitive species showed a minimum level of anthocyanin under salinity stress (Daneshmand et al. 2010). It is mostly observed that aromatic amino acid such as arginine, cysteine and methionine level decreased in plants grown under salinity. However, generation of nitric oxide (NO), modulation of hormones, proline accumulation, accumulation of polyols and glycine betaine increased in plant cells and alleviates salinity stress (Gupta and Huang 2014; Matysik et al. 2002).

A number of remedies are involved in avoiding abiotic stresses on plants. One of the most eco-friendly methods to cure plant responses to environmental stresses is plant growth-promoting microorganisms. A plant-microbe interaction strategy significantly alleviates abiotic stress on plants. In a *Trichoderma*-plant interaction study, it is confirmed that the putative sequence of ACC-deaminase found in *Trichoderma* genome, which is confirmed by gene silencing through RNAi (Kubicek et al. 2011; Viterbo et al. 2010). Kumar et al (2015), reported Induced Systemic Tolerance (IST) to salinity in plants by *Pseudomonas* sp. (AK-1) and *Bacillus* sp. (SJ-5) treatments resulting in increased proline accumulation and lipoxygenase activity.

Next-generation RNA sequencing analysis of two strains of *Sinorhizobium meliloti* (wild-type strain-1021 and an IAA-overproducing derivative RD64) demonstrated that sigma factor *Rpo*H1 and other stress responses genes were upregulated to overproduction of IAA (Defez et al. 2016). Transcriptome analysis of different plants showed that a number of noncoding miRNA worked in regulatory roles like transcriptional regulation, ion transport, auxin homeostasis under drought, salinity and cold stresses (Trindade et al. 2010). For example, the regulatory role of miR393 involved for salinity stress (Gao et al. 2011), miR169 for salinity and drought stress by modulating the expression of nuclear transcription factor YA (NF-YA) (Zhao et al. 2009) and miR169c controlling expression of stomatal activity involving genes conferred drought tolerance (Zhang et al. 2011).

In nutritional status, plant nutritional status and nutrient availability from the environment exhibited significant effects on plant-microbe interaction. It is well known that plant and phosphate-acquiring AMF is symbiotically regulated by phosphate status in the soil and in plant. Castrillo et al (2017) discussed that plant phosphate level and immunity affected root microbiome composition under phosphate limiting conditions. To conquer this problem, colonization of *Colletotrichum tofieldiae* in plants activated the phosphate starvation response (PSR) gene to enhance uptake phosphate along with affecting *PHR*1 function involved in repression of immune related gene expression and enhanced root endophytic microbiome assembly for better plant performance. In the case of nitrogen deficient conditions, miR2111 synthesized in shoots and transported through phloem to roots for post transcriptional silencing of *TML* (positive regulator for nodulation) involved in increasing host susceptibility to *Rhizobium* for nodulation. In another case, it has been shown that under iron deficient conditions, pathogens directly affect plants and further reduced the availability of iron to plants. This is analyzed in the *Arabidopsis* plant's increased activity of *MYB72* and *BGLU42* (β-glucosidase) mediated production and exudation of scopoletin involved in iron solubility and its availability, as well as increased rhzospheric microbiomes to protect plants from pathogen attack via ISR (Zamioudis et al. 2014).

7.7 CONCLUSION AND FUTURE PROSPECTS

Agriculture has been carrying the burden of plant based pharmaceuticals on planet Earth. Low availability of agricultural land is a serious problem. Recently, wasteland utilization for planting and cultivation of pharmaceutically important medicinal plants have shown to be additional major problems worldwide. If these problems are resolved, the pressure on agricultural land availability will be reduced. Another problem, due to the availability of a large number of species with diverse climatic/seasons, is that medicinal and aromatic plants are very difficult to grow in existing cropping systems or are grown as sole crops in different types of agricultural lands. However, overdependence of the human population on synthetic and semi-synthetic chemical fertilizers and pesticides has led to the circulation and spreading of life-threatening chemicals in plants and their useful products, which is not only harmful for human consumption but can also interrupt the ecological balance. Moreover, such disturbances can alter plant–microbe interactions by modifying microbial colonization property and biogeochemical cycles. In this chapter, we have focused on how various abiotic stresses are mitigated by microbes on plants. Plant growth-promoting microbes are involved in improving plant growth, remediating and managing contaminated soil and degraded wastelands. Various strategies on sound grounds of physiological, biochemical, molecular and ultra structural parameters proved that microorganisms improved plant growth, health and pharmaceutically important secondary metabolites, as well as improved soil microbial communities. Using various omics approaches like genomics, metagenomics, proteomics, metatranscriptomics, metaproteomics and metabolomics and their involvement in study will strengthen our knowledge of the plant microbe's interaction. Furthermore, it will also give gene cascades, biosynthetic pathways and the accumulation of various beneficial metabolites with effect of microbes. These studies improve new existing protocol for plant-microbes interaction under stress conditions. However, the development of stress tolerant plant genotypes through genetic manipulation and molecular breeding is a long term process and unstable in nature. However, microbial inoculation to mitigate stresses in plants in an environmentally friendly manner is a short term process. Furthermore, some microorganisms have broad host ranges, giving consistent results under field conditions. It is also shown that variations in the performance of microbes depended on many factors, including plant genotypes, age, climate, weather conditions, soil characteristics, etc. According to deep studies regarding complex dynamics between plant microbes, the interaction is poorly understood. We need to identify subtleties that govern plant microbes' associations at the molecular level. In another way, application of modern tools like nanomaterials and nanofertilizers of microorganisms are required in improving soil characteristics and lead to direct impacts on plant health. In recent years, nano based technologies for agriculture growth have been applied in developed countries. Apart from modern tools, promising microbial strains can also inoculate through organic and inorganic carriers, prepared by solid or liquid fermentation technologies. A number of delivering protocol can be used through seedling dips, seed treatments, biopriming, soil application, sucker treatment, foliar spray, hive insert, fruit spray and PGPR formulations. The development of PGPR formulations with compatible promising strains have synergy between strains that lead to production of various compounds at site of colonization for better adjustments and adaptations in plants grown in abiotic stressed conditions. Advantages of PGPR formulation are the broad spectrum of action, reliability, enhanced efficacy and they also have survived under variable environmental field conditions without genetic engineering. The application of PGPR formulations of promising strain mixtures is a better strategy for the management of abiotic stress, as well as plant growth promotion. Using these strategies, we can minimize inputs of chemical and maximize outputs (crop yields). The development of superior microorganism strains by using genetic manipulations can also be exploited as low input and are sustainable for the management of plant abiotic stresses. These strategies should also be implemented for enhancing the crop productivity to meet the needs of the growing population. In sum, the application of these techniques for enhancement of microbes can serve as key to sustainable agriculture by maintaining a balanced nutrient cycling and improving soil fertility, plant tolerance and crop productivity.

ACKNOWLEDGMENTS

Suman Singh is grateful to the Department of Science and Technology, New Delhi, India, for financial support (Grant No. DST/DISHA/SoRf/PM-016/2013). Sucheta Singh acknowledges the SERB–National Post Doctoral Fellowship (SERB-NPDF) for financial support (PDF/2019/001174).

REFERENCES

Abobatta, W.F. 2019. Drought adaptive mechanisms of plants – a review. *Advances in Agriculture and Environmental Science* 2(1): 62–65. DOI: 10.30881/aaeoa.00022.

Akbarimoghaddam, H., Galavi, M., Ghanbari, A., Panjehkeh, N. 2011. Salinity effects on seed germination and seedling growth of bread wheat cultivars. *Trakia Journal of Sciences* 9: 43–50.

Alhaithloul, H.A., Soliman, M.H., Amta, K.L., ElEsawi, M.A., Elkelish, A. 2019. Changes in ecophysiology, osmolytes, and secondary metabolites of the medicinal plants of *Mentha piperita* and *Catharanthus roseus* subjected to drought and heat stress. *Biomolecules* 10(1): 43.

Ali, Sk.Z., Sandhya, V., Grover, M., Kishore, N., Rao, L.V., Venkateswarlu, B. 2009. *Pseudomonas* sp. strain AKM-P6 enhances tolerance of sorghum seedlings to elevated temperatures. *Biology and Fertility of Soil* 46: 45–55

Alvarez, S., Marsh, E.L., Schroeder, S.G., Schachtman, D.P. 2008. Metabolomic and proteomic changes in the xylem sap of maize under drought. *Plant, Cell and Environment* 31: 325–340.

Amara, U., Rabia, K., Hayat, R. 2015. Soil bacteria and phytohormones for sustainable crop production. In *Bacterial Metabolites in Sustainable Agroecosystem. Sustainable Development and Biodiversity*; Maheshwari, D. (Ed.); Springer: Cham, Switzerland, 12: 87–103.

Amstaetter, K., Borch, T., Larese-Casanova, P., Kappler, A. 2010. Redox transformation of arsenic by Fe (II)-activated goethite (a-FeOOH). *Environmental Science and Technology* 44: 102–108. doi: 10.1021/es901274s.

Armada, E., Roldan, A., Azcon, R. 2014. Differential activity of autochthonousbacteria in controlling drought stress in native *Lavandula* and *Salvia* plants species under drought conditions in natural arid soil. *Microbial Ecology* 67: 410–420.

Armstrong, W., Drew, M.C. 2002. Root growth and metabolism under oxygen deficiency. *Plant Roots: The Hidden Half.* Waisel, Y., Eshel, A., and Kafkaf U., Eds.; Marcel Dekker: New York, 729–761.

Arpana, J., Bagyaraj, D. 2007. Response of kalmegh to an arbuscular mycorrhizal fungus and a plant growth promoting rhizomicroorganism at two levels of phosphorus fertilizer. *Am.-Eurasian Journal of Agriculture and Environmental Sciences* 2: 33–38.

Ashfaque, F., Inam, A., Sahay, S., Iqbal, S. 2016. Influence of heavy metal toxicity on plant growth, metabolism and its alleviation by phytoremediation—A promising technology. *Journal of Agriculture and Ecology Research International* 6(2): 1–19.

Ashraf, M., Orooj, A. 2006. Salt stress effects on growth, ion accumulation and seed oil concentration in an arid zone traditional medicinal plant ajwain (*Trachyspermum ammi* [qL.] Sprague). *Journal of Arid Environments* 64: 209–220.

Awasthi, A., Bharti, N., Nair, P., Singh, R., Shukla, A.K., Gupta, M.M., Darokar, M.P., Kalra, A. 2011. Synergistic effect of *Glomus mosseae* and nitrogen fixing *Bacillus subtilis* strain Daz26 on artemisinin content in *Artemisia annua* L. *Applied Soil Ecology* 49: 125–130.

Aziz, E.E., Al-Amier, H., Craker, L.E. 2008. Influence of salt stress on growth and essential oil production in peppermint, pennyroyal, and apple mint. *Journal of Herbs Spices and Medicinal Plants* 14: 77–87.

Bal, H.B. 2013. Isolation of ACC deaminase producing PGPR from rice rhizosphere and evaluating their plant growth promoting activity under salt stress. *Plant and Soil* 366: 93–105.

Banerjee, A., Roychoudhury, A. 2017. Effect of salinity stress on growth and physiology of medicinal plants. *Medicinal Plants and Environmental Challenges.* Ghorbanpour, M. and Varma, A., Eds.; Springer, Cham. https//doi.org/10.1007/978-3-319-68717-9_10.

Barea, J.M., Pozo, M.J., Azcon, R., Azcon-Aguilar, C. 2005. Microbial co-operation in the rhizosphere. *Journal of Experimental Botany* 56:1761–1778.

Barka, E.A., Nowak, J., Clement, C. 2006. Enhancement of chilling resistance of inoculated grapevine plantlets with a plant growth-promoting rhizobacterium, *Burkholderia phytofirmans* strain PsJN. *Applied and Environmental Microbiology* 72:7246–7252.

Barnawal, D., Bharti, N., Maji, D., Chanotiya, C.S., Kalra, A. 2012. 1-aminocyclopropane-1-carboxylic acid (ACC) deaminase-containing rhizobacteria protect *Ocimum sanctum* plants during waterlogging stress via reduced ethylene generation. *Plant Physiology and Biochemistry* 58: 227–235.

Barnawal, D., Bharti, N., Tripathi, A. et al. 2016. ACC-deaminase-producing endophyte *Brachybacterium paraconglomeratum* strain SMR20 ameliorates *Chlorophytum* salinity stress via altering phytohormone generation. *Journal of Plant Growth Regulator* 35: 553–564.

Barra, P.J., Inostroza, N.G., Acuña, J.J., Mora, M.L., Crowley, D.E., Jorquera, M.A. 2016. Formulation of bacterial consortia from avocado (*Persea americana* Mill.) and their effect on growth, biomass and superoxide dismutase activity of wheat seedlings under salt stress. *Applied Soil Ecology* 102: 80–91. doi: 10.1016/j.apsoil.2016.02.014.

Bashan, Y., de-Bashan, L.E. 2005. Bacteria. In *Encyclopaedia of Soils in the Environment*; Hillel, D., Ed.; Elsevier: Oxford, UK, 103–115.

Bensalim, S., Nowak, J., Asiedu, S.K. 1998. A plant growth promoting rhizobacterium and temperature effects on performance of 18 clones of potato. *American Journal of Potato Research* 75: 145–152.

Bettaieb, I., Zakhama, N., Wannes, W.A., Kchouk, M.E., Marzouk, B. 2009. Water deficit effects on *Salvia officinalis* fatty acids and essential oils composition. *Scientia Horticulturae* 120: 271–275.

Beutler, A.N., Giacomeli, R., Albertom, C.M., Silva, V.N., da Silva, Neto, G.F., Machado, G.A., Santos, A.T.L. 2014. Soil hydric excess and soybean yield and development in Brazil. *Australian Journal of Crop Science* 8: 1461–1466.

Bharti, N., Barnawal, D., Awasthi, A., Yadav, A., Kalra, A. 2014. Plant growth promoting rhizobacteria alleviate salinity induced negative effects on growth, oil content and physiological status in *Mentha arvensis*. *Acta Physiologiae Plantarum* 36: 45–60.

Bharti, N., Yadav, D., Barnawal, D., Maji, D., Kalra, A. 2013. *Exiguobacterium oxidotolerans*, a halotolerant plant growth promoting rhizobacteria, improves yield and content of secondary metabolites in *Bacopa monnieri* (L.) Pennell under primary and secondary salt stress. *World Journal of Microbiology and Biotechnology* 29: 379–387.

Bidgoli, R.D., Azarnezhad, N., Akhbari, M., Ghorbani, M. 2019. Salinity stress and PGPR effects on essential oil changes in *Rosmarianum officinalis* L. *Agriculture and Food Security* 8: 2.

Bita, C.E., Gerats, T. 2013. Plant tolerance to high temperature in a changing environment: Scientific fundamentals and production of heat stress-tolerant crops. *Frontiers in Plant Science* 4: 273. doi:10.3389/fpls.2013.00273.

Blum, A. 2017. Osmotic adjustment is a prime drought stress adaptive engine in support of plant production. *Plant, Cell and Environment* 40: 4–10. doi: 10.1111/pce.12800.

Boscaiu, M., Lull, C., Lidon, A., Bautista, I., Donat, P., Mayoral, O. 2008. Plant responses to abiotic stress in their natural habitats. *Bulletin of the University of Agricultural Sciences and Veterinary Medicine Cluj-Napoca Horticulture* 65: 53–58.

Brotman, Y., Landau, U., Cuadros-Inostroza, Á., Takayuki, T., Fernie, A.R., Chet, I. 2013. *Trichoderma*-plant root colonization: escaping early plant defense responses and activation of the antioxidant machinery for saline stress tolerance. *PLoS Pathogens.* 9:e1003221. doi: 10.1371/journal.ppat.1003221.

Cao, H.X., Sun, C.X., Shao, H.B., Lei, X.T. 2011. Effects of low temperature and drought on the physiological and growth changes in oil palm seedlings. *African Journal of Biotechnology* 10: 2630–2637.

Castrillo, G., Teixeira, P.J.P.L., Paredes, S.H., Law, T.F., de Lorenzo, L., Feltcher, M.E., Finkel, O.M., Breakfield, N.W., Mieczkowski, P., Jones, C.D. et al. 2017. Root microbiota drive direct integration of phosphate stress and immunity. *Nature* 543: 513–518.

Chang, P., Gerhardt, K.E., Huang, X.D., Yu, X.M., Glick, B.R., Gerwing, P.D. 2014. Plant growth-promoting bacteria facilitate the growth of barley and oats in salt impacted soil: Implications for phytoremediation of saline soils. *International Journal of Phytoremediation* 16: 1133–1147.

Chibuike, G.U., Obiora, S.C. 2014. Heavy metal polluted soils: Effects on plants and bioremediation methods. *Applied and Environmental Soil Science.* Article ID: 752708. https://doi.org/10.1155/2014/752708.

Chinnusamy, V., Zhu, J., Zhu, J.K. 2007. Cold stress regulation of gene expression in plants. *Trends in Plant Sciences* 12: 444–451.

Creus, C.M., Graziano, M., Casanovas, E.M., Pereyra, M.A., Simontacchi, M., Puntarulo, S., Barassi, C.A., Lamattina, L. 2005. Nitric oxide is involved in the *Azospirillum brasilense*-induced lateral root formation in tomato. *Planta.* 221: 297–303

Daneshmand, F., Arvin, M.J., Kalantari, K.M. 2010. Physiological responses to NaCl stress in three wild species of potato *in vitro. Acta Physiologiae Plantarum* 32: 91–101.

Defez, R., Esposito, R., Angelini, C., Bianco, C. 2016. Overproduction of incole-3-acetic acid in free living rhizobia induces transcriptional changes resembling those occurring in nodule bacteroids. *Molecular Plant Microbe Interactions* 29(6): 484–495.

Delfine, S., Loreto, F., Pinelli, P., Tognetti, R., Alvino, A. 2005. Isoprenoids content and photosynthetic limitations in rosemary and spearmint plants under water stress. *Agriculture, Ecosystems and Environment* 106: 243–252.

Dimkpa, C., Weinand, T., Asch, F. 2009. Plant rhizobacteria interactions alleviate abiotic stress conditions. *Plant, Cell and Environment* 32: 1682–1694.

Dobbelaere, S., Croonenborghs, A., Thys, A., Ptacek, D., Vanderleyden, J., Dutto, P., Labandera-Gonzalez, C., Caballero-Mellado, J., Anguirre, J.F., Kapulnik, Y., Brener, S., Burdman, S., Kadouri, D., Sarig, S., Okon, Y. 2003. Response of agronomically important crops to inoculation with *Azospirillum*. *Australian Journal of Plant Physiology* 28: 871–879.

Dojima, T., Craker, L.E. 2016. Potential Benefits of Soil Microorganisms on Medicinal and Aromatic Plants. In *Medicinal and Aromatic Crops: Production, Phytochemistry, and Utilization;* Jeliazkov V.D.Z. and Cantrell C.L., Eds; American Chemical Society Publications: Washington, DC, 75–90.

Duc, N.H., Csintalan, Z., Posta, K. 2018. Arbuscular mycorrhizal fungi mitigate negative effects of combined drought and heat stress on tomato plants. *Plant Physiology and Biochemistry* 132: 297–307.

Earanna, N., Bagyaraj, D.J. 2004. Influence of AM fungi and growth promoting rhizomicroorganisms on growth and herbage yield of *Phyllanthusamarus* Schum. and Thom. *J Geobios* 31: 117–20.

Egamberdieva, D. 2013b. The role of phytohormone producing bacteria in alleviating salt stress in crop plants. In *Biotechnological techniques of Stress Tolerance in Plants.* Miransari, M., Ed., Stadium Press LLC: USA, 21–39.

Egamberdieva, D., Jabborova, D., Mamadalieva, N. 2013a. Salt-tolerant *Pseudomonas extremorientalis* able to stimulate growth of *Silybum marianum* under salt stress condition. *Journal of Medicinal and Aromatic Plant Science and Biotechnology* 7: 65–75.

Eiteman, M.A., Lee, S.A., Altman, E. 2008. A co-fermentation strategy to consume sugar mixtures effectively. *Journal of Biological Engineering* 2: 3.

Ekren, S., Sönmez, Ç., Özçakal, E., Kurttas, Y.S.K., Bayram, E., Gürgülü, H. 2012. The effect of different irrigation water levels on yield and quality characteristics of purple basil (*Ocimum basilicum* L.). *Agricultural Water Management* 109: 155–161.

El-Fattah, D.A.A., Wedad, E. E., Mona, S.Z., Mosaad, K.H. 2013. Effect of carrier materials, sterilization method, and storage temperature on survival and biological activities of *Azotobacter chroococcum* inoculants. *Annals of Agricultural Science* 58(2): 111–118.

El-Komy, H.M., Hamdia, M.A., ElBaki, G.A. 2003. Nitrate Reductase in wheat plants grown under water stress and inoculated with *Azospirillum* spp. *Journal of Plant Biology* 46: 281–7.

Enebe, M.C., Babalola, O.O. 2018. The influence of plant growth-promoting rhizobacteria in plant tolerance to abiotic stress: A survival strategy. *Applied Microbiology and Biotechnology* 102: 7821–7835.

Escandón, M., Meijón, M., Valledor, L., Pascual, J., Pinto, G., Cañal, M.J. 2018. Metabolome integrated analysis of high-temperature response in *Pinus radiata*. *Frontiers in Plant Science* 9: 485.

Esitken, A., Pirlak, L., Turan, M., Sahin, F. 2006. Effects of floral and foliar application of plant growth promoting rhizobacteria (PGPR) on yield, growth and nutrition of sweet cherry. *Scientia Horticulturae* 110: 324–327.

Etesami, H. 2018. Can interaction between silicon and plant growth promoting rhizobacteria benefit in alleviating abiotic and biotic stresses in crop plants? *Agriculture, Ecosystem and Environment* 253: 98–112. 10.1016/j.agee.2017.11.007.

Etesami, H., Beattie, G.A. 2017. Plant-microbe interactions in adaptation of agricultural crops to abiotic stress conditions. In *Probiotics and Plant Health*; Kumar V., et al. Eds.; Springer: Berlin, 163–200. 10.1007/978-981-10-3473-2_7.

Etesami, H., Mirseyed, H.H., Alikhani, H.A. 2014. Bacterial biosynthesis of 1-aminocyclopropane-1-caboxylate (ACC) deaminase, a useful trait to elongation and endophytic colonization of the roots of rice under constant flooded conditions. *Physiology and Molecular Biology of Plants* 20(4): 425–434. doi:10.1007/s12298-014-0251-5.

Evelin, H., Kapoor, R., Giri, B. 2009. Arbuscular mycorrhizal fungi in alleviation of salt stress, a review. *Annals of Botany* 104: 1263–1280.

Fahad, S., Hussain, S., Bano, A., Saud, S., Hassan, S., Shan, D. 2015. Potential role of phytohormones and plant growth-promoting rhizobacteria in abiotic stresses: Consequences for changing environment. *Environmental Science and Pollution Research* 22: 4907–4921.

Farahani, H.A., Valadabadi, S.A., Daneshian, J., Shiranirad, A.H., Khalvati, M.A. 2009. Medicinal and aromatic plants farming under drought conditions. *Journal of Horticulture and Forestry* 1: 86–92.

Farooq, M., Wahid, A., Kobayashi, N., Fujita, D., Basra, S.M.A. 2009. Plant drought stress: Effects, mechanisms and management. *Agronomy for Sustainable Development* 29:153–188.

Foley, J.A., Ramankutty, N., Brauman, K.A., Cassidy, M.S., Gerber, J.S., Johnston, M., Mueller, N.D., O'Connell, C., Ray, D.K., West, P.C. et al. 2011. Solutions for a cultivated planet. *Nature* 478: 337–342.

Fougnies, L., Renciot, S., Muller, F., Plenchette, C., Prin, Y., de Faria, S.M., Bouvet, J.M., Sylla, S.N., Dreyfus, B., Ba, A.M. 2007. Arbuscular mycorrhizal colonization and nodulation improve tolerance in *Pterocarpus officinalis* Jacq. seedlings. *Mycorrhiza* 17: 159–166.

Foyer, C.H., Noctor, G. 2005. Oxidant and antioxidant signaling in plants: A re-evaluation of the concept of oxidative stress in a physiological context. *Plant, Cell and Environment* 28: 1056–1071.

Gao, P., Bai, X., Yang, L., Lv, D., Pan, X., Li, Y., Cai, H., Ji, W., Chen, Q., Zhu, Y. 2011. osa-MIR393: A salinity- and alkaline stress-related microRNA gene. *Molecular Biology Reports* 38(1): 237–42.

Ghasemi, M., Modarresi, M., Babaeian, J.N., Bagheri, N., Jamali A. 2016. The evaluation of exogenous application of salicylic acid on physiological characteristics, proline and essential oil content of chamomile (*Matricaria chamomilla* L.) under normal and heat stress conditions. *Agriculture.* 6: 31. doi:10.3390/agriculture6030031.

Ghassemi, S., Delangiz, N., Asgari, L.B. et al. 2021. Review and future prospects on the mechanisms related to cold stress resistance and tolerance in medicinal plants. *Acta Ecologica Sinica.* https://doi.org/10.1016/j.chnaes.2020.09.006.

Ghosh, S., Chowdhury, R. & Bhattacharya, P. 2016. Mixed consortia in bioprocesses: Role of microbial interactions. *Applied Microbiology and Biotechnology* 100: 4283–4295. https://doi.org/10.1007/s00253-016-7448-1.

Glick, B.R. 2012. Plant growth-promoting bacteria: Mechanisms and applications. *Scientifica (Cairo)* 2012: pp. 1–16, 963401. https://doi.org/10.6064/2012/963401.

Glick, B.R. 2014. Bacteria with ACC deaminase can promote plant growth and help to feed the world. *Microbiological Research* 169(1): 30–39.

Glick, B.R. 2015. *Beneficial Plant-Bacterial Interactions.* Springer: Berlin.

Glick, B.R., Cheng, Z., Czarny, J., Duan, J. 2007. Promotion of plant growth by ACC deaminase-producing soil bacteria. *European Journal of Plant Pathology*, 119:329–339.

Golpayegani, A., Tilebeni, H.G. 2011. Effect of biological fertilizers on biochemical and physiological parameters of basil (*Ocimum basilicm* L.) medicine plant. *American-Eurasian Journal of Agricultural and Environmental Sciences* 11: 411–416.

González-Guerrero, M., Melville, L.H., Ferrol, N., Lott, J.N., Azcon-Aguilar, C. and Peterson, R.L. 2008. Ultrastructural localization of heavy metals in the extraradical mycelium and spores of the arbuscular mycorrhizal fungus *Glomus intraradices. Canadian Journal of Microbiology* 54: 103–110. doi: 10.1139/w07-119.

Goswami, D., Parmar, S., Vaghela, H., Dhandhukia, P., Thakker, J. 2015. Describing *Paenibacillus mucilaginosus* strain N3 as an efficient plant growth promoting rhizobacteria (PGPR). *Cogent Food & Agriculture* 1(1): 1000714.

Goswami, D., Patel, K., Parmar, S., Vaghela, H., Muley, N., Dhandhukia, P., Thakker, J.N. 2014a. Elucidating multifaceted urease producing marine *Pseudomonas aeruginosa* BG as a cogent PGPR and bio-control agent. *Plant Growth Regulation*.75:253–263. doi:10.1007/s10725-014-9949-1.

Goswami, D., Pithwa, S., Dhandhukia, P., Thakker, J.N. 2014b. Delineating *Kocuria turfanensis* 2M4 as a credible PGPR: A novel IAA-producing bacteria isolated from saline desert. *Journal of Plant Interaction*, 9: 566–576.

Govahi, M., Ghalavand, A., Nadjafi, F., Sorooshzadeh, A. 2015. Comparing different soil fertility systems in Sage (*Salvia officinalis*) under water deficiency. *Ind. Crops Prod.* 74: 20–27.

Goyal, S.K., Prabha Rai, J.P., Singh, S.R. 2016. Indian agriculture and farmers—Problems and reforms. *Indian Agriculture and Farmers* 246:79–87.

Grichko, V.P., Glick, B. R. 2001. Amelioration of flooding stress by ACC-deaminase containing plant growth promoting bacteria. *Plant Physiology and Biochemistry* 39: 11–17.

Gull, A., Lone, A.A., Wani, N.U.I. 2019. Biotic and abiotic stresses in plants. DOI: http://dx.doi.org/10.5772/intechopen.85832

Gupta, B., Huang, B. 2014. Mechanism of salinity tolerance in plants: Physiological, biochemical, and molecular characterization. *International Journal of Genomics*, Article ID 701596. https://doi.org/10.1155/2014/701596

Gupta, D.K., Chatterjee, S., Datta, S., Vee, V. and Walther, C. 2014. Role of phosphate fertilizers in heavy metal uptake and detoxification of toxic metals. *Chemosphere* 108: 134–144. doi: 10.1016/j.chemosphere.2014.01.030.

Gupta, D.K., Pena, L.B., Romero-Puertas, M.C. 2016. NADPH oxidases differentially regulate ROS metabolism and nutrient uptake under cadmium toxicity. *Plant, Cell and Environment* 40(4): 509–526.

Han, Y., Wang, R., Yang, Z., Zhan, Y., Ma, Y., Ping, S. et al. 2015. 1-aminocyclopropane-1-carboxylate deaminase from *Pseudomonas stutzeri* a1501 facilitates the growth of rice in the presence of salt or heavy metals. *Journal of Microbiology and Biotechnology* 25: 1119–1128.

Hariprasad, P., Niranjana, S.R. 2009. Isolation and characterization of phosphate solubilizing rhizobacteria to improve plant health of tomato. *Plant Soil* 316: 13–24.

Hasanuzzaman, M., Hossain, M.A., da Silva, J.A.T., Fujita, M. 2012. Plant response and tolerance to abiotic oxidative stress: Antioxidant defense is a key factor. In: Venkateswarlu B., Shanker A., Shanker C., Maheswari M. (eds), *Crop Stress Management: Perspectives and Strategies*. Springer: New York, 261–315.

Hasanuzzaman, M., Nahar, K., Fujita, M. 2014. Regulatory role of polyamines in growth, development and abiotic stress tolerance in plants. In: *Plant Adaptation to Environmental Change: Significance of Amino Acids and Their Derivatives*, 157–193.

Hashem, A., Tabassum, B., Fathi, Abd Allah E. 2019. *Bacillus subtilis*: A plant growth promoting rhizobacterium that also impacts biotic stress. *Saudi Journal of Biological Science* 26: 1291–1297.

Hattori, R., Matsumura, A., Yamawaki, K., Tarui, A., Daimon, H. 2013. Effects of flooding on arbuscular mycorrhizal colonization and root-nodule formation in different roots of soybeans. *Journal of Agricultural Science* 4: 673–677.

Heidari, M., Golpayegani, A. 2012. Effects of water stress and inoculation with plant growth promoting rhizobacteria (PGPR) on antioxidant status and photosynthetic pigments in basil (*Ocimum basilicum* L.). *Journal of Saudi Society of Agricultural Sciences* 11(1): 57–61.

Hemavathi, V.N., Sivakumar, B.S., Suresh, C.K., Earanna, N. 2006. Effect of *Glomus fasciculatum* and plant growth promoting rhizobacteria on growth and yield of *Ocimum basilicum. Karnataka Journal of Agricultural Science* 19: 17–20.

Hermosa, R., Viterbo, A., Chet, I., Monte, E. 2012. Plant-beneficial effects of *Trichoderma* and of its genes. *Microbiology* 158: 17–25.

Heydari, M., Zanfardino, A., Taleei, A., Bushehri, A.A.S. et al. 2018. Effect of heat stress on yield, monoterpene content and antibacterial activity of essential oils of *Mentha* x *piperita* var. Mitcham and *Mentha arvensis* var. *piperascens*. *Molecules* 23 (8): 1903. doi:10.3390/molecules23081903.

Hložková, K., Matenová, M., Žáˇcková, P., Strnad, H., Hršelová, H., Hroudová, M., et al. 2016. Characterization of three distinct metallothionein genes of the Ag-hyperaccumulating ectomycorrhizal fungus *Amanita strobiliformis*. *Fungal Biology* 120: 358–369. doi: 10.1016/j.funbio.2015.11.007.

Hosseinzadah, F., Satei, A., Ramezanpour, M. 2011. Effects of mycorrhiza and plant growth promoting rhizobacteria on growth, nutrient uptake and physiological characteristics in *Calendula officinalis* L. *Middle East Journal of Scientific Research* 8(5): 947–53.

Ibrahim, M.A., Maenpaa, M., Hassinen, V., Kontunen-Soppela, S., Malec, L., Rousi, M., Pietikainen, L., Tervahauta, A., Karenlampi, S., Holopainen, J.K, et al. 2010. Elevation of night-time temperature increases terpenoid emissions from *Betula pendula* and *Populus tremula*. *Journal of Experimental Botany* 61: 1583–1595.

Islam, S., Ezakor, E., Garner, J.O. 2011. Effect of chilling stress on the chlorophyll fluorescence, peroxidase activity and other physiological activities in *Ipomoea batatas* L. genotypes, *American Journal of Plant Physiology* 6(2): 72–82.

Jamla, M., Khare, T., Joshi, S., Patil, S., Penna, S., Kumar, V. 2021. Omics approaches for understanding heavy metal responses and tolerance in plants. *Current Plant Biology* 27: 100213.

Jha C.K., Saraf M. 2015. Plant growth promoting rhizobacteria (PGPR): A Review. *Journal of Agricultural Research Development* 5: 108–119. http://www.e3journals.org

John, R.P., Tyagi, R.D., Brar, S.K., Surampalli, R.Y., Pre´vost, D. 2011. Bio–encapsulation of microbial cells for targeted agricultural delivery. *Critical Reviews in Biotechnology* 31: 211–226.

Kachhap, S., Chaudhary, A., Singh, S.D. 2015. Response of plant growth promoting rhizobacteria (PGPR) in relation to elevated temperature conditions in groundnut (*Arachis hypogaea* L.). *The Ecoscan* 9: 771–778.

Kalia, R., Sareen, S., Nagpal, A. 2017. ROS-induced transcription factors during oxidative stress in plants: A tabulated review. In *Reactive Oxygen Species and Antioxidant Systems in Plants: Role and Regulation under Abiotic Stress*; Khan, M. and Khan, N., Eds.; Springer: Singapore, 129–158.

Kamboj, V.P. 2000. Herbal medicine. *Current Science* 78: 35–43.

Karamzadeh, S. 2003. Drought and production of second metabolites in medicinal and aromatic plants. *Drought Journal* 7: 90–95.

Karthikeyan, B., Jaleel, A.C., Azooz, M.M. 2009. Individual and combined effects of *Azospirillum brasilense* and *Pseudomonas fluorescens* on biomass yield and ajmalicine production in *Catharanthus roseus*. *Academic Journal of Plant Sciences* 2: 69–73.

Karthikeyan, B., Joe, M.M., Islam, R.M., Sa, T. 2012. ACC deaminase containing diazotrophic endophytic bacteria ameliorate salt stress in *Catharanthus roseus* through reduced ethylene levels and induction of antioxidative defense systems. *Symbiosis*. 56(2):77–86. Doi 10.1007/s13199-012- 0162-6.

Kaushal, M., Wani, S.P. 2016. Rhizobacterial-plant interactions: Strategies ensuring plant growth promotion under drought and salinity stress. *Agriculture, Ecosystem and Environment* 231: 68–78. 10.1016/j.agee.2016.06.031.

Kazeminasab, A., Yarnia, M., Lebaschy, M.H., Mirshekari, B., Rejali, F. 2016. The effect of vermicompost and PGPR on physiological traits of lemon balm (*Melissa officinalis* L.) plant under drought stress. *Journal of Medicinal Plants By-Products* 2: 135–144.

Khan, M.N., Mobin, M., Abbas, Z.K., Alamri, S.A. 2018. Fertilizers and their contaminants in soils, surface and groundwater. In *The Encyclopedia of the Anthropocene*; DellaSala, D.A. and Goldstein, M.I. Eds., 5: 225–240 Elsevier, Oxford.

Khan, M.Y., Zahir, A.Z., Asghar, H.N., Waraich, E.A. 2017. Preliminary investigations on selection of synergistic halotolerant plant growth promoting rhizobacteria for inducing salinity tolerance in wheat. *Pakistan Journal of Botany* 49(4): 1541–1551.

Khorasaninejad, S., Mousavi, A., Soltanloo, H., Hemmati, K., Khalighi, A. 2011. The effect of drought stress on growth parameters, essential oil yield and constituent of peppermint (*Mentha piperita* L.). *Journal of Medicinal Plants Research* 5: 5360–5365.

Kleinwächter, M., Paulsen, J., Bloem, E., Schnug, E., Selmar, D. 2015. Moderate drought and signal transducer induced biosynthesis of relevant secondary metabolite in thyme (*Thymus vulgaris*), greater celandine (*Chelidonium majus*) and parsley (*Petroselinum crispum*). *Industrial Crops and Products* 64: 158–166.

Kohler, J. 2008. Plant growth promoting rhizobacteria and arbuscular mycorrhizal fungi modify alleviation biochemical mechanisms in water stressed plants. *Functional Plant Biology* 35: 141–151.

Koini, M.A., Alvey, L., Allen, T., Tilley, C.A., Harberd, N.P., Whitelam, G.C. et al. 2009. High temperature-mediated adaptations in plant architecture require the bHLH transcription factor PIF4. *Current Biology* 19: 408–413.

Kubicek, C.P., Herrera-Estrella, A., Seidl-Seiboth, V., Martinez, D.A., Druzhinina, I.S., Thon, M. et al. 2011. Comparative genome sequence analysis underscores mycoparasitism as the ancestral life style of *Trichoderma*. *Genome Biology* 12: 40.

Kumar, S., Vaishnav, A., Jain, S., Varma, A., Choudhary, D.K. 2015. Bacterial-mediated induction of systemic tolerance to salinity with expression of stress alleviating enzymes in soybean (*Glycine max* L. Merrill). *Journal of Plant Growth Regulation* 34: 558–573.

Lakshmipathy, R., Chandrika, K., Gowda, B., Balakrishna, A.N., Bagyaraj, D.J. 2001. Response of *Saraca asoca* (Roxb.) de Wilde to inoculation with *Glomus mosseae*, *Bacillus coagulans* and *Trichoderma harzianum*. *Journal of Soil Biology and Ecology* 21: 76–80.

Lim, J.H., Kim, S.D. 2013. Induction of drought stress resistance by multi-functional PGPR Bacillus licheniformis K11 in pepper. *Plant Pathology Journal* 29(2): 201–208.

Liu, H., Wang, X., Wang, D., Zou, Z., Liang, Z. 2011. Effect of drought stress on growth and accumulation of active constituents in *Salvia miltiorrhiza* Bunge. *Industrial Crops Products*. 33: 84–88.

Ma, Y. 2019. Seed coating with beneficial microorganisms for precision agriculture. *Biotechnology Advances* 37(7): 107423.

Majeed, A., Muhammad, Z. 2019. Salinity: A major agricultural problem—causes, impacts on crop productivity and management strategies. In *Plant Abiotic Stress Tolerance*. Hasanuzzaman M., Hakeem K., Nahar K. and Alharby H. Eds.; Springer: Cham. https://doi.org/10.1007/978-3-030-06118-0_3.

Majumder, A., Bhattacharyya, K., Bhattacharyya, S., Kole, S.C. 2013. Arsenic-tolerant, arsenite-oxidising bacterial strains in the contaminated soils of West Bengal, India. *Science of the Total Environment* 46: 1006–1014.

Malik, B.S., Paul, S., Sharma, R.K., Sethi, A.P., Verma, O. P. 2005. Effect of *Azotobacter chroococcum* on wheat (*Triticum aestivum*) yield and its attributing components. *Indian Journal of Agricultural Science* 75: 600–602.

Malusa,´ E., Sas–Paszt, L., Ciesielska, J. 2012. Technologies for beneficial microorganisms inocula used as biofertilizers. *Scientific World Journal*, 2012: 491206. doi: 10.1100/2012/491206.

Mani, N. 2004. Phytochemical and antimicrobial studies on *Alpinia galangal* and *Coleus amboinicus* as influenced by native AM fungi. PhD thesis, Bharathidasan University, India.

Matysik, J., Alia, A., Bhalu, B., Mohanty, P. 2002. Molecular mechanisms of quenching of reactive oxygen species by proline under stress in plants. *Current Science* 82: 525–532.

Meena, K.K., Sorty, A.M., Bitla, U.M., Choudhary, K., Gupta, P., Pareek, A. et al. 2017. Abiotic stress responses and microbe-mediated mitigation in plants: The omics strategies. *Frontiers in Plant Science* 8: 172.

Mina, J.G., Mehrdad, Y., Abdollah, H.G., Farhad, F., Amir, M.D. 2019. Evaluating effects of drought stress and bio-fertilizer on quantitative and qualitative traits of valerian (*Valeriana officinalis* L.), *Journal of Plant Nutrition* 42(13): 1417–1429.

Miransari, M. 2011. Hyperaccumulators, arbuscular mycorrhizal fungi and stress of heavy metals. *Biotechnology Advances* 29: 645–653. doi: 10.1016/j.biotechadv.2011.04.006.

Mishra, J., Singh, R., Arora, N.K. 2017. Alleviation of heavy metal stress in plants and remediation of soil by rhizosphere microorganisms. *Frontiers in Microbiology* 8: 1706.

Mishra, P.K., Mishra, S., Selvakumar, G., Bisht, S.C., Bisht, J.K., Kundu, S., Gupta, H.S. 2008. Characterisation of a psychrotolerant plant growth promoting *Pseudomonas* sp. strain PGERs17 (MTCC 9000) isolated from North Western Indian Himalayas. *Annals of Microbiology* 58: 561–568.

Misra A., Srivastava N.K. 2000. Influence of water stress on Japanese mint. *Journal of Herbs Spices and Medicinal Plants* 7: 51–58.

Molina-Favero, C., Creus, C.M., Simontacchi, M., Puntarulo, S., Lamattina, L. 2008. Aerobic nitric oxide production by *Azospirillum brasilense* Sp245 and its influence on root architecture in tomato. *Molecular Plant Microbe Interactions* 2: 1001–1009.

Moura, J.C.M.S., Bonine, C.A.V., Viana, J.O.F., Dornelas, M.C., Mazzafera, P. 2010. Abiotic and biotic stresses and changes in the lignin content and composition in plants. *Journal of Integrative Plant Biology* 52(4): 360–376.

Munchbach, M., Nocker, A., Narberhaus, F. 1999. Multiple small heat shock proteins in rhizobia. *Journal of Bacteriology* 181: 83–90.

Murthy, S., Bali, G., Sarangi, S. 2011. Effect of lead on metallothionein concentration in lead-resistant bacteria *Bacillus cereus* isolated from industrial effluent. *African Journal of Biotechnology* 10: 15966–15972.

Naveed, M., Hussain, M.B., Zahir, Z.A., Mitter, B., Sessitsch, A. 2014a. Drought stress amelioration in wheat through inoculation with *Burkholderia phytofirmans* strain PsJN. *Journal of Plant Growth Regulation* 73: 121–131.

Naveed, M., Mitter, B., Reichenauer, T.G., Wieczorek, K., Sessitsch, A. 2014b. Increased drought stress resilience of maize through endophytic colonization by *Burkholderia phytofirmans* PsJN and *Enterobacter* sp FD17. *Environmental and Experimental Botany* 97: 30–39.

Naya, L., Ladrera, R., Ramos, J., González, E.M., Arrese-Igor, C., Minchin, F.R. et al. 2007. The response of carbon metabolism and antioxidant defenses of alfalfa nodules to drought stress and to the subsequent recovery of plants. *Plant Physiology* 144: 1104–1114.

Netondo, G.W., Onyango, J.C., Beck, E. 2004. Sorghum and salinity, gas exchange and chlorophyll fluorescence of sorghum under salt stress. *Crop Science* 44: 806–811.

Niranjana, S.R., Hariprasad, P. 2014. Understanding the mechanism involved in PGPR-mediated growth promotion and suppression of biotic and abiotic stress in plants. In *Future Challenges in Crop Protection 59 Against Fungal Pathogens, Fungal Biology*; Goyal, A. and Manoharachary, C. Eds., Springer, pp 59–108. DOI 10.1007/978-1-4939-1188-2_3.

Omer, A.M. 2010. Bioformulations of *Bacillus* spores for using as biofertilizers. *Life Science Journal* 7(4): 124–131.

Pankaj, U., Singh, D.N., Mishra, P., Gaur, P., Vivekbabu, C.S., Shanker, K., Verma, R.K. 2020. Autochthonous halotolerant plant growth promoting rhizobacteria promote bacoside A yield of *Bacopa monnieri* (L) Nash and phytoextraction of salt-affected soil. *Pedosphere* 30(5): 671–683.

Pankaj, U., Singh, D.N., Singh, G., Verma, R.K. 2019. Microbial inoculants assisted growth of *Chrysopogon zizanioides* promotes phytoremediation of salt affected soil. *Indian Journal of Microbiology* 59(2): 137–146.

Pareek, A., Sopory, S.K., Bohnert, H.K., Govindjee. 2010. (Eds). *Abiotic Stress Adaptation in Plants: Physiological, Molecular and Genomic Foundation.* Springer: Dordrecht, 526.

Parida, A.K., Das, A.B. 2005. Salt tolerance and salinity effects on plants: A Review. *Ecotoxicology and Environmental Safety* 60: 324–349.

Phukan, U.J., Mishra, S., Shukla, R.K. 2016. Waterlogging and submergence stress: Affects and acclimation. *Critical Reviews in Biotechnology* 36(5): 956–66.

Pinedo, I., Ledger, T., Greve, M., Poupin, M.J. 2015. *Burkholderia phytofirmans* PsJN induces long-term metabolic and transcriptional changes involved in *Arabidopsis thaliana* salt tolerance. *Frontiers in Plant Science* 6: 466. doi:10.3389/flps.2015.00466.

Pradhan, J., Sahoo, S.K., Lalotra, S., Sarma, R.S. 2017. Positive impact of abiotic stress on medicinal and aromatic plants. *International Journal of Plant Science* 12(2): 309–313.

Prasad, A., Singh, A.K., Chanotiya C.S., Patra, D.D. 2010. Effect of chromium and lead on yield, chemical composition of essential oil, and accumulation of heavy metals of mint species. *Communications in Soil Science and Plant Analysis* 41: 2170–2186.

Purohit, S.S., Vyas, S.P. 2004. *Medicinal Plant Cultivation—A Scientific Approach, Including Processing and Financial Guidelines.* Agrobios, Jodhpur, India, 1-3.

Rai, V., Khatoon, S., Bisht, S., Mehrotra, S. 2005. Effect of cadmium on growth, ultramorphology of leaf and secondary metabolites of *Phyllanthus amarus* Schum. and Thonn. *Chemosphere* 61: 1644–1650.

Rai, V., Vajpayee, P., Singh, S.N., Mehrotra, S. 2004. Effect of chromium accumulation on photosynthetic pigments, oxidative stress defense system, nitrate reduction, proline level and eugenol content of *Ocimum tenuiflorum* L. *Plant Science Journal* 167: 1159–1169.

Ramakrishna, A., Ravishankar, G.A. 2011. Influence of abiotic stress signals on secondary metabolites in plants. *Plant Signaling and Behavior* 6(11): 1720–1731.

Rana, A., Saharan, B., Nain, L., Prasanna, R., Shivay, Y.S. 2012. Enhancing micronutrient uptake and yield of wheat through bacterial PGPR consortia. *Soil Science and Plant Nutrition* 58: 573–582.

Ravanbakhsh, M., Sasidharan, R., Voesenek, L.A., Kowalchuk, G.A., Jousset, A. 2017. ACC deaminase-producing rhizosphere bacteria modulate plant responses to flooding. *Journal of Ecology* 105: 979–986.

Riggs, P.J., Chelius, M.K., Iniguez, A.L., Kaeppler, S.M., Triplett, E.W. 2001. Enhanced maize productivity by inoculation with diazotrophic bacteria. *Australian Journal of Plant Physiology* 28: 829–836.

Roberts, D.P., Mattoo, A.K. 2018. Sustainable agriculture—enhancing environmental benefits, food nutritional quality and building crop resilience to abiotic and biotic stresses. *Agriculture.* 8(1): 8. https://doi.org/10.3390/agriculture8010008.

Saha, M., Sarkar, S., Sarkar, B., Sharma, B.K., Bhattacharjee, S., Tribedi, P. 2016. Microbial siderophores and their potential applications: A Review. *Environmental Science and Pollution Research* 23(5): 3984–3999.

Said-Al Ahl, H.A.H., Abou-Ellail, M., Omer, E.A. 2016. Harvest date and genotype influences growth characters and essential oil production and composition of *Petroselinum crispum* plants. *Journal of Chemical and Pharmaceutical Research* 8: 992–1003.

Said-Al Ahl, H.A.H., Hussein, M.S. 2010. Effect of water stress and potassium humate on the productivity of oregano plant using saline and fresh water irrigation. *Ozean Journal of Applied Science* 3: 125–141.

Said-Al Ahl, H.A.H., Meawad, A.A., Abou-Zeid, E.N., Ali, M.S. 2010. Response of different basil varieties to soil salinity. *International Agrophysics* 24: 183–188.

Said-Al Ahl, H.A.H., Omer, E.A. 2011. Medicinal and aromatic plants production under salt stress. A review. *Herba Polonica* 57: 72–87.

Sairam, R.K., Kumutha, D., Ezhilmathi, K., Deshmukh, P.S., Srivastava, G.C. 2008. Physiology and biochemistry of waterlogging tolerance in plants. *Biologia Plantarum.* 52(3): 401–412.

Saleem, A.R., Brunetti, C., Khalid, A., Della Rocca, G., Raio, A., Emiliani, G. 2018. Drought response of *Mucuna pruriens* (L.) DC. Inoculated with ACC deaminase and IAA producing rhizobacteria. *PLoS ONE* 13(2): e019121.

Salinas, J. 2002. Molecular mechanisms of signal transduction in cold acclimation. In *Plant Signal Transduction*; Scheel, D. and C. Wasternack, C. Eds.; Oxford University Press: Oxford, UK,116–139.

Sandhya, V.Z., Grover, M., Reddy, G., Venkateswarlu, B. 2009. Alleviation of drought stress effects in sunflower seedlings by the exopolysaccharides producing *Pseudomonas putida* strain GAP-P45. *Biology and Fertility of Soils* 46: 17–26.

Sang-Mo, K., Radhakrishnan, R., Khan, A.L., Min-Ji, K., Jae-Man, P., Bo-Ra, K., Dong-Hyun, S., In-Jung, L. 2014. Gibberellin secreting rhizobacterium, *Pseudomonas putida* H-2-3 modulates the hormonal and stress physiology of soybean to improve the plant growth under saline and drought conditions. *Plant Physiology and Biochemistry* 84: 115–124.

Santiago, C.D., Yagi, S., Ijima, M., Nashimoto, T., Sawada, M., Ikeda, S., Asano, K., Orikasa, Y., Ohwada, T. 2017. Bacterial compatibility in combined inoculations enhances the growth of potato seedlings. *Microbes and Environments* 32(1): 14–23.

Schoebitz, M., Simonin, H., Poncelet, D. 2012. Starch filler and osmoprotectants improve the survival of rhizobacteria in dried alginate beads. *Journal of Microencapsulation* 29: 532–538.

Scholles, D., Vargas, L.K. 2004. Viability of soybean inoculation with Bradyrhizobium strains in flooded soil. *Revista Brasileira de Ciência do Solo* 28: 973–979.

Seki, M., Kamei, A., Yamaguchi-Shinozaki, K., Shinozaki, K. 2003. Molecular responses to drought, salinity and frost: Common and different paths for plant protection. *Current Opinion in Biotechnology* 14: 194–199.

Selmar, D., Kleinwächter, M. 2013. Stress enhances the synthesis of secondary plant products: The impact of stress-related over-reduction on the accumulation. *Plant and Cell Physiology* 54(6): 817–826.

Selvakumar, G., Kundu, S., Joshi, P., Nazim, S., Gupta, A.D., Mishra, P.K., Gupta, H.S. 2008. Characterization of a cold-tolerant plant growth-promoting bacterium *Pantoea dispersa* 1A isolated from a sub-alpine soil in the North Western Indian Himalayas. *World Journal of Microbiology and Biotechnology* 24: 955–960.

Selvaraj, T., Sumithra, P. 2011. Effect of *Glomus aggregatum* and plant growth promoting rhizomicroorganisms on growth, nutrition and content of secondary metabolites in *Glycyrrhiza glabra* L. *Indian Journal of Pure and applied Biosciences* 26: 283–290.

Setter, T.L.H., Khabaz-Saberi, W.I., Singh, K.N., Kulshreshtha, N., Sharma, S.K. 2006. Review of waterlogging tolerance in wheat in India: Involvement of element/microelement toxicities, relevant to yield plateau and opportunities for crop management. In *International Symposium on Balanced Fertilization*; Ludhiana, D.K. Benbi, M.S. Brar, S.K. Bansal. Eds.; International Potash Institute and Dept. of Soils, Punjab Agricultural University: Ludhiana, 513–520.

Shahid, M., Dumat, C., Khalid, S., Schreck, E., Xiong, T. et al. 2016. Foliar heavy metal uptake, toxicity and detoxification in plants: A comparison of foliar and root metal uptake. *Journal of Hazardous Materials.* 325: 36–58.

Sharma, J., Shamim, K., Dubey, S.K., Meena, R.M. 2017. Metallothionein assisted periplasmic lead sequestration as lead sulfite by *Providencia vermicola* strain SJ2A. *Science of Total Environment* 579: 359–365.

Sheibani-Tezerji, R., Rattei, T., Sessitsch, A., Trognitz, F., Mitter, B. 2015. Transcriptome profiling of the endophyte *Burkholderia phytofirmans* PsJN indicates sensing of the plant environment and drought stress. *mBio.* 6(5): e00621–15.

Shen, X., Hu, H., Peng, H., Wang, W., Zhang, X. 2013. Comparative genomic analysis of four representative plant growth-promoting rhizobacteria in *Pseudomonas. BMC Genomics* 14: 271. doi:10.1186/1471-2164-14-271.

Shrivastava, P., Kumar, R. 2015. Soil salinity, A serious environmental issue and plant growth promoting bacteria as one of the tools for its alleviation. *Saudi Journal of Biological Science* 22: 123–131.

Singh, S., Tripathi, A., Chanotiya, C.S., Barnawal, D., Singh, P., Patel, V.K., Vajpayee, P., Kalra, A. 2020. Cold stress alleviation using individual and combined inoculation of ACC deaminase producing microbes in *Ocimum sanctum. Environmental Sustainability* 3: 289–301.

Singh, S., Tripathi, A., Maji, D., Awasthi, A., Vajpayee, P., Kalra, A. 2019. Evaluating the potential of combined inoculation of *Trichoderma harzianum* and *Brevibacterium halotolerans* for increased growth and oil yield in *Mentha arvensis* under greenhouse and field conditions. *Industrial Crops and Products* 131: 173–181.

Smékalová, V., Dosčosilová, A., Komis, G., Samaj, J. 2014. Crosstalk between secondary messengers, hormones and MAPK modules during abiotic stress signalling in plants. *Biotechnology Advances* 32(1): 2–11.

Stancheva, I., Geneva, M., Hristozkova, M., Boychinova, M., Markovska, Y. 2009. Essential oil variation of *Salvia officinalis* (L.) grown on heavy metals polluted soil. *Biotechnology and Biotechnology Equipment* 23: 373–376

Stancheva, I., Geneva, M., Hristozkova, M., Boychinova, M., Markovska, Y. 2009. Essential oil variation of *Salvia Officinalis* (L.). *Grown on Heavy Metals Polluted Soil, Biotechnology & Biotechnological Equipment* 23(supl): 373–376.

Su, F., Jacquard, C., Villaume, S., Michel, J., Rabenoelina, F., Clément, C. et al. 2015. *Burkholderia phytofirmans* PsJN reduces impact of freezing temperatures on photosynthesis in *Arabidopsis thaliana. Frontiers in Plant Science* 6: 810.

Subramanian, P., Mageswari, A., Kim, K., Lee, Y., Sa, T. 2015. Psychrotolerant endophytic *Pseudomonas* sp. strains OB155 and OS261 induced chilling resistance in tomato plants (*Solanum lycopersicum* Mill.) by activation of their antioxidant capacity. *Molecular Plant-Microbe Interactions* 28: 1073–1081.

Taiz, L., Zeiger, E. 2002. *Plant Physiology*, Sinauer Associates: Sunderland, Mass, USA.

Tank, N., Saraf, M. 2010. Salinity-resistant plant growth promoting rhizobacteria ameliorates sodium chloride stress on tomato plants. *Journal of Plant Interaction.* 5: 51–58.

Tchounwou, P.B., Yedjou, C.G., Patlolla, A.K., Sutton, D.J. 2012. Heavy metals toxicity and the environment. *EXS* 101: 133–164.

Theocharis, A., Bordiec, S., Fernandez, O. et al. 2012. *Burkholderia phytofirmans* PsJN primes *Vitis vinifera* L. and confers a better tolerance to low nonfreezing temperatures. *Molecular Plant-Microbe Interactions* 25(2): 241–249.

Timmusk, S., Wagner, E.G.H. 1999. The plant growth-promoting rhizobacterium *Paenibacillus polymyxa* induces changes in *Arabidopsis thalianan* gene expression: A possible connection between biotic and abiotic stress responses. *Molecular Plant Microbe Interactions* 12(11): 951–9.

Tiryaki, D., Aydın, İ., Atıcı, Ö. 2019. Psychrotolerant bacteria isolated from the leaf apoplast of cold-adapted wild plants improve the cold resistance of bean (*Phaseolus vulgaris* L.) under low temperature. *Cryobiology* 86: 111–119.

Tiwari, S. 2016. *Pseudomonas putida* attunes morphophysiological: Biochemical and molecular responses in *Cicer arietinum* L. during drought stress and recovery. *Plant Physiology and Biochemistry* 99: 108–117.

Trindade, I., Capitao, C., Dalmay, T., Fevereiro, M.P., Santos, D.M. 2010. miR398 and miR408 are up-regulated in response to water deficit in *Medicago truncatula*. *Planta* 231: 705–716.

Ullah, S., Bano, A. 2015. Isolation of plant growth-promoting rhizobacteria from rhizospheric soil of halophytes and their impact on maize (*Zea mays* L.) under induced soil salinity. *Canadian Journal of Microbiology* 61: 307–313.

Upadhyay, S.K., Singh, J.S., Saxena, A.K., Singh, D.P. 2012. Impact of PGPR inoculation on growth and antioxidant status of wheat under saline conditions. *Plant Biology* 14: 605–611.

Vassilev, N., Vassileva, M., Lopez, A., Martos, V., Reyes, A., Maksimovic, I., Eichler-Löbermann, B., Malusa, E. 2015. Unexploited potential of some biotechnological techniques for biofertilizer production and formulation. *Applied Microbiology* 99(12): 4983–96.

Velthuizen, V.H. 2007. *Mapping Biophysical Factors That Influence Agricultural Production and Rural Vulnerability*; Food & Agriculture Organization: Rome, Italy.

Verhagen, B.W., Glazebrook, J., Zhu, T., Chang, H.S., van Loon, L.C. 2004. The transcriptome of rhizobacteria-induced systemic resistance in arabidopsis. *Molecular Plant Microbe Interactions* 17: 895–908.

Verpoorte, R., Contin, A., Memelink, J. 2002. Biotechnology for the production of plant secondary metabolites. *Photochemistry Reviews* 1: 13–25.

Viterbo, A., Landau, U., Kim, S., Chernin, L., Chet, I. 2010. Characterization of ACC deaminase from the biocontrol and plant growth-promoting agent *Trichoderma asperellum* T203. *FEMS Microbiology Letters* 305: 42–48.

Voesenek, L.A., Bailey-Serres, J. 2015. Flood adaptive traits and processes: An overview. *New Phytologist*. 206: 57–73.

Vorholt, J.A., Vogel, C., Carlström, C.I., Muller, D.B. 2017. Establishing causality: Opportunities of synthetic communities for plant microbiome research. *Cell Host Microbe*. 22: 142–155.

Vwioko, E., Adinkwu, O., El-Esawi, M.A. 2017. Comparative physiological, biochemical, and genetic responses to prolonged waterlogging stress in Okra and Maize given exogenous ethylene priming. *Frontiers in Physiology* 8: 632.

Waskiewicz, A., Muzolf-Panek, M., Golinski, P. 2012. Phenolic content changes in plants under salt stress. In: Ahmad P. et al (Eds.), *Ecophysiology and Responses of Plants under Salt Stress*. Springer: Germany, 283–314

Woo, S.L., Pepe, O. 2018. Microbial consortia: Promising probiotics as plant biostimulants for sustainable agriculture. *Frontiers in Plant Science* 9: 1801.

Wu, G., Kang, H., Zhang, X., Shao, H., Chu, L., Ruan, C. 2010. A critical review on the bio-removal of hazardous heavy metals from contaminated soils: Issues, progress, eco-environmental concerns and opportunities. *Journal of Hazardous Materials* 174: 1–8.

Wu, Q.S., Xia, R.X. 2006. Arbuscular mycorrhizal fungi influence growth, osmotic adjustment and photosynthesis of citrus under well-watered and water stress conditions. *Journal of Plant Physiology* 163(4): 417–425.

Xu, Z., Zhou, G., Shimizu, H. 2010. Plant responses to drought and re-watering. *Plant Signaling and Behavior* 5: 649–654.

Yadav, P.K., Badola, S. 2019. Newsletter on wildlife trade in India Post Traffic special issue on Medicinal Plants challenges in conservation and sustainable trade of caterpillar fungus in India. Issue 31. https://www.academia.edu/40050643/Trade_in_Medicinal_and_Aromatic_plants_of_India_An_overview.

Yadav, S., Modi, P., Dave, A., Vijapura, A., Patel, D., Patel, M. 2020. Effect of abiotic stress on crops. *Sustainable Crop Production* [Internet]. Hasanuzzaman, M., Filho, M.C.M.T., Fujita, M. and Nogueira, T.A.R., Eds; IntechOpen: London. https://www.intechopen.com/chapters/68945 doi: 10.5772/intechopen.88434

Yadav, S.K. 2010. Cold stress tolerance mechanisms in plants. A review. *Agronomy for Sustainable Development* 30(3): 515–527.

Zahir, Z.A., Ahmad, M., Hilger, T.H., Dar, A., Malik, S.R., Abbas, G., Rasche, F. 2018. Field evaluation of multistrain biofertilizer for improving the productivity of different mungbean genotypes. *Soil and Environment* 37(1): 45–52.

Zaidi, S., Usmani, S., Singh, B.R. and Musarrat, J. 2006. Significance of *Bacillus subtilis* strain SJ 101 as a bioinoculant for concurrent plant growth promotion and nickel accumulation in *Brassica juncea*. *Chemosphere* 64: 991–997.

Zamioudis, C., Hanson, J., Pieterse, C.M.J. 2014. β-Glucosidase BGLU42 is a MYB72-dependent key regulator of rhizobacteria-induced systemic resistance and modulates iron deficiency responses in *Arabidopsis* roots. *New Phytologist* 204: 368–379.

Zayed, M.S. 2016. Advances in formulation development technologies. In *Microbial Inoculants in Sustainable in Sustainable Productivity, Vol. 2: Functional Applications;* Singh D.P., Singh H.B., and Prabha R. Eds.; Springer. Front Microbiol. 2016 Dec 23;7:2105. doi: 10.3389/fmicb.2016.02105.

Zehtab-Salmasi, S., Javanshir, A., Omidbaigi, R., Aly-Ari, H., Ghassemi-Golezani, K. 2001. Effects of water supply and sowing date on performance and essential oil production of anise (*Pimpinella anisum* L.). *Acta Agronomica Hungarica* 49: 75–81.

Zhang, X., Xia, H., Li, Z., Zhuang, P., Gao, B. 2011. Identification of a new potential Cd-hyperaccumulator *Solanum photeinocarpum* by soil seed bank-metal concentration gradient method. *Journal of Hazardous Materials* 189: 414–419.

Zhang, X., Xie, Z., Lang, D., Chu, Y., Cui, G., Jia, X. 2021. *Bacillus pumilus* improved drought tolerance in *Glycyrrhiza uralensis* G5 seedlings through enhancing primary and secondary metabolisms. *Physiologia Plantarum*. doi: 10.1111/ppl.13236.

Zhang, X., Zou, Z., Gong, P., Zhang, J., Ziaf, K., Li, H., Xiao, F., Ye, Z. 2011. Over-expression of microRNA169 confers enhanced drought tolerance to tomato. *Biotechnology Letters* 33: 403–409.

Zhao, B., Ge, L., Liang, R. et al. 2009. Members of miR-169 family are induced by high salinity and transiently inhibit the NF-YA transcription factor. *BMC Molecular Biology* 10: 29.

Zheljazkov, V.D., Craker, L.E., Xing, B. 2006. Effects of Cd, Pb, and Cu on growth and essential oil contents in dill, peppermint, and basil. *Environmental and Experimental Botany* 58: 9–16.

Zheljazkov, V.D., Warman, P.R. 2003. Application of high Cu compost to Swiss chard and basil. *Science of Total Environment* 302: 13–26.

Zuccarini, P., Okurowska, P. 2008. Effects of mycorrhizal colonization and fertilization on growth and photosynthesis of sweet basil under salt stress. *Journal of Plant Nutrition* 31: 497–513.

8 Symbiotic Effect of Mycorrhizal Fungi and Native Plants on Phytoremediation of Contaminated Soils

Laila Midhat
Cadi Ayyad University, Marrakech, Morocco
and
High School of Engineering and Innovation
Marrakech, Morocco

Naaila Ouazzani, Laila Mandi, Mohamed Radi, and Lahcen Ouahmane
Cadi Ayyad University, Marrakech, Morocco

Elmehdi Ouatiki, Soumia Amir, and Abdessamad Tounsi
Sultan Moulay Slimane University, Témara, Morocco

CONTENTS

DOI: 10.1201/9781003147091-8

8.1 INTRODUCTION

The Industrial Revolution, countless wars, extensive mining and long-standing agricultural activities have had major consequences on soils and have left many polluted soils and sites worldwide. Traditionally, after the closure of the mine, soils were generally abandoned in huge surfaces without any treatment. Diverse chemical contaminants can have a harmful effect on soil quality and, therefore, on ecosystem and human health. Among them, heavy metals are a major problem due to their persistence, long distance transport, their ability to accumulate in tissues and their high toxic character, even at low concentrations (Ali et al. 2013; Mahar et al. 2016). Moreover, these toxic metals leach into the surrounding environment through water and or wind. Worldwide, ecological rehabilitation of these wastes is a great challenge and presents a major concern of governments and researchers.

Biological methods have been considered to be easier to implement in comparison to other conventional cleanup methods. Biological techniques exploit the properties of a living organism with the aim of cleaning up a given environment. The living organism used can be a microorganism (bacteria, fungi), a plant (algae, plants, trees) or an animal (earthworms, for example). These organisms act on pollutants by different mechanisms such as absorption, accumulation, digestion, transformation, degradation or evapotranspiration. These mechanisms make the pollutant less toxic, dilute it, immobilize it or extract it. There are different biological techniques of soil remediation, which can be applied alone or in combination with other physicochemical processes.

Phytoremediation is among the most promising biological techniques for the rehabilitation of metal-contaminated soils. It uses native green plants to remove metals from the environment and accumulates them in their organs (Mahar et al. 2016).

Phytoremediation is an innovative, economical, eco-friendly and technically feasible technique. Phytoremediation efficiency is based on the choice of the plant species. Native and spontaneous plants are promising candidates in terms of growth, tolerance and accumulation under stressful environmental conditions (Yoon et al. 2006).

This assisted natural remediation can be improved by adding a certain type of microorganism through a symbiosis phenomenon called mycorrhization. Indeed, some fungi maintain a symbiotic relationship with their host plants by forming a new organism that renders available water and mineral nutrition and helps defend roots against pathogens as well as enhance resistance to various types of stress (Garbaye 1991; Bâ et al. 2011; Mukhopadhyay and Maiti 2011). This chapter provides an overview of the industrial activity's issues, bioremediation techniques and the efficiency of plant-mycorrhizal fungi symbiotic systems on the rehabilitation of contaminated soils. Biological techniques are classified into two main categories: Bioremediation techniques that mainly use bacteria and the phytoremediation techniques that exploit the properties of plants.

8.2 BIOREMEDIATION TECHNIQUES

Bioremediation is based on the natural activity and metabolism of microorganisms, mainly bacteria and fungi. These microorganisms can be endogenous or exogenous to the contaminated site. They have depolluting properties which allow degrading, transforming, immobilizing or solubilizing pollutants (Smith 1990; Garbisu and Alkorta 2003). Mostly, bioremediation techniques can be applied *in situ*. These techniques can be subdivided into several categories according to the biological principle or method of decontamination implemented (Garbisu and Alkorta 2003). For example, biodegradation mechanisms use certain microorganisms to rend pollutants less harmful for the environment (Vidali 2001). Bioleaching involves microorganisms that allow the extraction of pollutants fixed or trapped in the soil or in certain minerals (e.g. leaching of sulfide ores) (Rohwerder et al. 2003). Finally, bio-sorption through microorganisms is based on the modification of the mobility and fixation of metals (Barkay and Schaefer 2001).

Generally, bioremediation techniques are low cost but require a long-time treatment to visualize an effective result, about 5 to 10 years. In addition, it is important to adapt each technique to the environmental conditions of the target area.

8.3 PHYTOREMEDIATION TECHNIQUES

Phytoremediation is a term which groups various techniques that use the potential of higher plants and their symbiotic microorganisms to extract, degrade or immobilize pollutants contained in sediments, soils, sludges, water surfaces and groundwater (Ali et al. 2013; Mahar et al. 2016; Singh et al. 2020). Pollutants treated are metals, solvents, polycyclic aromatic hydrocarbons, explosives and phytosanitary products (Bert. 2013). It is new, inexpensive technology, effective, applicable *in situ*, non-destructive and environmentally friendly. It impacts positively the functions and structure of the soil, improving its fertility with inputs organic matter (Mulligan 2001; Ghosh and Singh 2005; Pilon-Smits 2005; Odjegba and Fasidi 2007; Ahmadpour et al. 2012; Ali et al. 2013; Mahar et al. 2016). Phytoremediation combines two remediation strategies: (i) phytostabilization (or phytorestoration) based on reducing metals mobility in contaminated soils or sediments; (ii) phytodecontamination based on reducing metal concentrations in the environment. It groups together several techniques depending on the environment to be treated and the nature of the pollutants. Phytoremediation includes several techniques: phytoextraction, phytostabilization, phytovolatilization, phytodegradation and rhizofiltration (Ghosh and Singh 2005; Pilon-Smits 2005; Ali et al. 2013; Mahar et al. 2016).

- ***Phytodegradation*** is based on the use of plants to transform certain organic pollutants in by-products less toxic for the plant and, therefore, does not concern heavy metals

- ***Rhizodegradation*** is based on the same principle of degradation of organic pollutants but involves soil microorganisms whose degradation activity is stimulated by root exudates
- ***Phytovolatization*** uses specific plants that absorb organic or inorganic pollutants, transforming them into volatile forms that are less toxic and releasing them in the atmosphere through their leaves
- ***Rhizofiltration*** is based on the use of terrestrial or aquatic plants able to develop their root system in polluted water and to concentrate large quantities of pollutants due to their high root biomass
- ***Phytoextraction*** (phytoaccumulation) is an *in situ* technology, using plants with high ability to uptake metals from soil and accumulate them in their organs, which will then be harvested (Lasat 2002; Mahar et al. 2016). After harvest, plant tissues with high metal concentrations will be treated by thermal, chemical or microbiological processes. Metals can also be recovered and reused for economic purposes (phytomining) (Brooks 1998; Chaney et al. 2007; Ali et al. 2013). This process is mainly used for trace elements and is rarely done for organic pollutants. There are two strategies of phytoextraction : Phytoextraction-assisted (induced) and continuous phytoextraction (Salt et al. 1998; Ali et al. 2013; Mahar et al. 2016)
- ***Phytostabilization*** or phyto-immobilization is using plants to reduce metal mobility and bioavailability in rhizospheric soil or immobilize them chemically by precipitation, stabilization and/or absorption (Ali et al. 2013; Mahar et al. 2016). The plants used in phytostabilization are species metallotolerant adopting an exclusion strategy (Berti and Cunningham 2000). Indeed, the metals are uptaken and stabilized in the root system to prevent their dispersion in the environment (Ghosh and Singh 2005; Pilon-Smits 2005; Krämer 2010)

The technique of phytoremediation is used when rapid immobilization is needed. This strategy is not a decontamination technique in the strict sense but a method of managing polluted sites intended to immobilize pollutants in the soil. Indeed, it ensures better landscape integration for sites with significant contamination and for which other methods are not applicable (Vangronsveld et al. 2009).

8.4 NATIVE PLANT SPECIES: METALLOPHYTES

Plant choice is one of the impact parameters on the successful phytoremediation process (Mahar et al. 2016). The selected plant species should be native and indigenous to the target region. Native plant species usually include plants that are found in specific areas without any human intervention and have naturally grown accustomed to the local climate. Generally, they have a good ability to support and accumulate high levels of contaminants in their tissues without suffering from toxicity (Baker 1981; Yoon et al. 2006; Mahar et al. 2016). Generally, they efficiently adapt to environmental conditions, requiring reduced agricultural and maintenance resources.

Native plants have specific and unique mechanisms that give them a high ability for uptake, translocation and storage of toxic metals and they can behave as accumulators and/or excluders plants (Sinha et al. 2004).

Native plants are a group of species that can thrive on metal-rich soils with a great ability to avoid metal absorption by root cells thought various extracellular mechanisms, such as root sorption, precipitation, and exclusion of metals (Dalvi and Bhalerao 2013). Some plants failed in the first defense system by limiting the metal uptake by the roots. At this moment, the second type of defense is an activated tolerance strategy. After entering root cells, metals, especially ions forms, are accumulated inside the cytosol and can establish complexes with organic and inorganic chelators (Ali et al. 2013). Metal chelation by complexation make metals immobile and reduce their toxic effects. After chelation, the complexes formed are transported and stored in inactive compartments, where they present less harmful effects to plants (Yan et al. 2020).

Screening of native plant species is considered a crucial objective of studies carried out for the rehabilitation of abandoned contaminated sites. Many native plant species were collected, identified and used in phytoremediation (phytoextraction/phytostabilization) strategies (Table 8.1). Most of these plant species were characterized by a great capacity to tolerate and accumulate various metals in their organs.

TABLE 8.1
Concise List of Plant Species Applied in Phytoremediation Strategies

Family	Plant species	References
Apocynaceae	*Alyxia rubricaulis*	Chaney et al. (2010)
Azollaceae	*Azolla pinnata*	Rai (2008)
Brassicaceae	*Thlaspi caerulescens*	Lasat (2002)
	Thlaspi rotundifolium	
	Lyssum caricum	Li et al. (2003)
	Alyssum murale	
	Alyssum bertolonii	
	Thlaspi caerulescens	Chaney et al. (2010)
	Arabidopsis halleri	Reeves and Brooks (1983); Reeves and Baker (2000)
Convolvulaceae	*Ipomea alpina*	Lasat (2002)
Euphorbiaceae	*Euphorbia cheiradenia*	Chehregani and Malayeri (2007)
Lamiaceae	*Aeolanthus biformifolius*	Chaney et al. (2010)
	Haumaniastrum robertii	Chaney et al. (2010)
Pteridaceae	*Pteris ryukyuensis*	Kalve et al. (2011)
	Pteris vittata	Srivastava et al. (2006)
Theaceae	*Schima superba*	Chaney et al. (2010)

The effectiveness of phytoremediation can also be improved by adding specific microorganisms (bacteria, fungi), which will immobilize contaminants and fertilize the soil. This is referred to as assisted phytoremediation.

8.5 MYCORRHIZATION

Mycorrhizae have existed for over 450 million years, when plants first colonized the earth (Bâ et al. 2011). Paleontological research carried out provides an idea on how old this symbiotic relationship is. Taylor et al. (1995) presented fossils that resemble endomycorrhizae arbuscules. In addition, Beimforde et al. (2011) presented optical photomicrographs of ectomycorrhizal structures of the *Eomelanomyces cenococcoides* species from the Eocene amber of India.

The word mycorrhiza is of Greek origin (mukês means fungi and rhiza means root) and is a symbiotic complex created from a mutually beneficial relationship between the roots of almost 90% of ground flora ((Smith and Read 1997) and certain mycorrhizal fungi (Kaufman 2020). In nature there are seven types of mycorrhizae, of which two are the most common. These differ in the morphological aspects of the hyphae and their relationship with the infected plant host as well as in anatomical characters related to the combination (Nouaim and Chaussod 1996). The fungal hyphae that don't extend into the root structure of the cell body are called ectomycorrhizae, whereas the fungal hyphae which penetrate into the cell membrane and infiltrate the cell wall are called endomycorrhizae (Huey et al. 2020). The other five types are ectendomycorrhizal, arbutoid, monotropoid, orchidoid and ericoid.

8.5.1 Ectomycorrhizal and Endomycorrhizal Associations

Ectomycorrhizal: Mutual contact between fungi and host plants is accomplished by the mycelia of the fungus, which develop in such a way as not to infiltrate the interior of living root cells by forming a three-part structure mentioned below:

i. The *fungal mantle*, which can be up to 40 µm thick on the outer part of the root, acts as a barrier against pathogens
ii. The *Hartig network* (Smith and Read 1997), where symbiotic exchanges take place is a set of hyphae that grows inward between the epidermal and the root cortical cells
iii. The *external hyphae*, the mycelium, which diffuse outside the root, house the sporophores

The most ectomycorrhizal fungi are ascomycetes and usually basidiomycetes forming carpophores on soils and some of them are edible. Hosts are frequently of the resinous type (Pinaceae) (Smith and Read 1997).

Endomycorrhizal: Fungal hyphae extend into the root cells cortex pushing back their plasmalemma without crossing it and that is why they cannot be seen without staining (Oldroyd 2013; Javeria et al. 2017; Nouria 2018). Generally, based on the morphology of the ends of the hyphae, we can distinguish two groups of endomycorrhiza: The mycorrhizae with clusters and the mycorrhizae with arbuscules.

8.5.2 Mycorrhizae with Clusters

Their name is determined according to the peloton aspect of the tip of the hyphae of the mycorrhizal fungi (Barman et al. 2016). These hyphae repel the plasmalemma from the cortical cells of the host plant root, forming clumps(Peterson et al. 2004; Walker 2013). This group also contains other types of mycorrhizae:

- *Ericaceous mycorrhizae* or Ericoids are a unique type of mycorrhizae restricted to many families of the great order of ericaceous angiosperms. A common characteristic of those plants growing these mycorrhizae is the development of highly characterized roots, the so-called the "hair roots" (Peterson et al. 2004; Walker 2013)
- *Orchid mycorrhizae* are found only in the orchid family, which is one of the biggest flowering plant families (Peterson, Massicotte, and Melville 2004; Walker 2013)

8.5.3 Arbuscular Mycorrhizae

Arbuscular mycorrhizae (AM), originally named vesicular-arbuscular mycorrhizae, are symbiotic mutualistic relationships among most vascular plant roots and a few species of fungi of the novel phylum Glomeromycota (Schüßler et al. 2001) More than 80% of land plant communities form mutualistic relationships with AM species, playing a major role in the performance of the plants (Munkvold et al. 2004). There are two main types of AM: The Paris type and the Arum type (Smith and Read 1997; Peterson et al. 2004; Matsuda et al. 2021). Unlike ectomycorrhizae, AMs don't make evident morphological changes in their root structure so they can be observed only by using various microscopic methods (Duponnois et al. 2005). This is the most common type, especially in herbaceous plants and many woody species (Ouahmane 2007)

8.5.4 Roles of Mycorrhizal Symbioses

Plants are immobile living beings (Greppin et al. 1986) fixed by their roots in the soil, which limits enormously the resources of water and nutrients, the quantities of which vary according to bioclimatic conditions and the living environment of the plant. To increase the exchange surface, the majority of plants have formed significant symbiotic partnerships with microbes that live on the rhizosphere part of soil. This symbiosis is achieved through the elongation of mycelial filaments that can reach up to one kilometer in length (Fortin et al. 2008). These mycelial filaments act essentially to transfer water and mineral elements necessary for the development of the host plants (Smith and Read 1997; Sim and Eom 2006).

8.5.5 Improvement of Phosphate Nutrition

Phosphorus (P) is an essential macronutrient for the growth and the evolution of plants, representing around 0.2% of their dry weight. The phosphate group is used in the construction of various molecules such as nucleic acids (DNA, RNA), enzymes, phosphoproteins and phospholipids, which gives it a fundamental structural role. Without a dependable source of this nutrient) which they take up as orthophosphate) $H_2PO_4^-$ and HPO_4^{2-}) (Nahar et al. 2021), plants cannot thrive (Schachtman et al. 1998). Numerous works and several bibliographical syntheses have been published on the beneficial effect of mycorrhiza in the improvement of mineral nutrition (phosphate, nitrogen and trace elements) and water nutrition. The endomycorrhizal symbiotic relationship changes the uptake pathways of Pi present in the host plant through the establishment of a "mycorrhizal pathway" that can provide 20–100% of Pi entry into endomycorrhized plants (Plassard et al. 2015).

8.5.6 Improving Nitrogen Nutrition

Nitrogen (N) is an essential element for the growth of plants (Zheng et al. 2013). However, it also is a major component of nucleic acid (DNA) proteins (chlorophyll), amino acids and enzymes, in addition to pigments needed for photosynthesis (Chapin and Kedrowski 1983). It is a limiting element for the productivity of the plant (Clarholm 1993). In general, specially ectomycorrhized mycorrhizal forest trees absorb ammonium (NH_4^+) more rapidly than nitrate (NO_3) (Botton and Chalot 1999). Furthermore, the study of ammonium uptake in the ectomycorrhizal fungus *Paxillus involutus* has shown that it occurs through a proton-dependent symport (Javelle and Chalot 1999).

8.5.7 Improving Trace Element Nutrition

Trace elements are involved in various enzymatic reactions and play an essential function in the metabolism of proteins, lipids and carbohydrates (Kabata-pendias and Pendias 2001). These elements are not very mobile in the soil and their uptake is mainly due to a better soil penetration by the extra-root mycorrhizae hyphae (Nouaim and Chaussod 1996). You et al. (2021) showed that infection of *Phragmites australis* with arbuscular mycorrhizal fungi enhanced Zn and Cd extraction.

8.5.8 Improvement of Hydric Nutrition

The poverty of arid soils and seedlings in nutrients and also the long dry period that can last for several months, cause a restriction in the nutritive and hydric resources of the plants. These stressful conditions cause the plants to seek these resources away from the rhizosphere; this is done with the help of the hyphae of the mycorrhizal symbiotic association. Püschel et al. (2020) proved that inoculation of the *Medicago truncatula* plant with the AM fungus *Rhizophagus irregularis* increased its water uptake.

8.5.9 Tolerance to Environmental Stress

The evolution of plants is influenced by different factors in their natural environment that can disrupt the different resources favorable to their functioning; however, mycorrhized plants have the capacity to resist these changes. The results of many studies have demonstrated that plant tolerance to water stress is improved by colonization with mycorrhizal fungi (Al-Karaki and Clark 1998). The role of mycorrhizal symbiosis is crucial in promoting the growth of plants and protecting them from the harmful effects of different stresses, such as salinity (Hameed et al. 2014). To cope with salinity, different strategies of associating plants with beneficial soil microorganisms may play an especially significant function in adapting these plants to adverse ecological conditions.

8.5.10 Mycorrhizal Fungi as Bioprotection Agents

The various interactions between the constituents of nature can be beneficial or detrimental, and plants themselves can also be disrupted by pathogens present in their environment, particularly the soil. To protect themselves from these threats, plants have developed protective systems, including the establishment of the mycorrhizal symbiotic association which can predispose the plant to respond more quickly to pathogen attacks (Harrier and Watson 2004). The use of mycorrhizae in agriculture should therefore be considered as a proactive strategy to biological control. According to much research to date, the application of arbuscular mycorrhizae (AM) is a promising, easy-to-use and environmentally friendly prevention method, and the presence of AM allows for a substantial reduction of external chemical inputs, fertilizers and pesticides (Dalpé 2005; Singh et al. 2019). Among the researchers who have carried out work in this sense, Reyes-Tena et al. (2017) showed the existence of an associative effect with co-inoculation by mycorrhizal fungi and actinomycetes in plant growth promotion and that bioprotection from *Phytophthora capsici* caused wilt in pepper plants. In addition, Akema and Futai (2005) demonstrated the importance of ECM symbiosis in controlling the globally threatened pine wilt disease.

8.6 CONCLUSION AND FUTURE PROSPECTS

Rehabilitation of contaminated soils is becoming an urgent preoccupation in order to overcome the environmental risk and ensure sustainability of resources. Biological methods are the most environmentally friendly and least expensive techniques, compared to other traditional methods. Indeed, phytoremediation, which represents one of the most widely used biological techniques, is essentially based on using green plant species to accumulate, stabilize and/or degrade the various pollutants dispersed in nature. In addition to their natural ability to accumulate, researchers have evolved this method by emphasizing the soil constituents present in the rhizosphere of plants, such as the symbiotic relationship (mycorrhizae) with microorganisms. This technique revealed that it could be used as a promoting strategy for the rehabilitation of metal-contaminated soils.

REFERENCES

Ahmadpour, P., Ahmadpour, F., Mahmud, T.M.M., Abdu, A., Soleimani, M., Hosseini Tayefeh, F. 2012. Phytoremediation of heavy metals: A green technology. African Journal of Biotechnology 11(76): 14036–14043.

Akema, T., Futai, K. 2005. Ectomycorrhizal development in a *Pinus thunbergiistand* in relation to location on a slope and effect on tree mortality from pine wilt disease. Journal of Forest Research 10(2): 93–99. doi: 10.1007/s10310-004-0101-3.

Ali, H., Khan, E., Sajad, M.A. 2013. Phytoremediation of heavy metals-concepts and applications. Chemosphere 91: 869–881. doi.org/10.1016/j.chemosphere.2013.01.075.

Al-Karaki, G.N., Clark, R.B. 1998. Growth, mineral acquisition, and water use by mycorrhizal wheat grown under water stress. Journal of Plant Nutrition 21(2): 263–276.

Bâ, A., Duponnois, R., Diabaté, M. and Dreyfus, B. 2011. Les champignons ectomycorhiziens des arbres forestiers en Afrique de l'Ouest: Méthodes d'étude, diversité, écologie, utilisation en foresterie et comestibilité. IRD Editions. IRD. p.252.

Baker, A.J.M. 1981. Accumulators and excluders-strategies in the response of plants to heavy metals. Journal of Plant Nutrition 3: 643–654.

Barkay, T., Schaefer, J. 2001. Metal and radionuclide bioremediation: issues, considerations and potentials. Current Opinion in Microbiology 4: 318–323.

Barman, J., Samanta, A., Saha, B., Datta, S. 2016. Mycorrhiza the oldest association between plant and fungi. Resonance 21(12): 1093–1104. https://doi.org/10.1007/s12045-016-0421-6.

Beimforde, C., Schäfer, N., Dörfelt, H., Nascimbene, P.C., Singh, H., Heinrichs, J., Reitner, J., Rana, R.S., Schmidt, A.R. 2011. Ectomycorrhizas from a lower eocene angiosperm forest. New Phytologist 192(4): 988–996.

Bert V. 2013. Les phytotechnologies appliquées aux sites et sols pollués. EDP Sciences, Paris, 100 p <ineris-00969552>.

Berti, W.R., Cunningham, S.C. 2000. Phytostabilization of Metals. In: Raskin, I., Ensley, B.D. (Eds.): Phytoremediation of Toxic Metals: Using Plants to Clean up the Environment. John Wiley & Sons, New York: 71–88.

Botton, B., Chalot, M. 1999. Nitrogen assimilation: Enzymology in ectomycorrhizas. In: Varma A. & Hock B. (Eds.), Mycorrhiza: Structure, Function, Molecular Biology, and Biotechnology, 2nd edition, Springer, Berlin: 333–372.

Brooks, R.R. 1998. Geobotany and Hyperaccumulators. In: Brooks, R.R. (Ed.): Plants that Hyperaccumulate Heavy Metals: Their Role in Phytoremediation, Microbiology, Archaeology, Mineral Exploration and Phytomining. CAB International, Oxford, UK: 55–94.

Chaney, R.L., Angle, J.S., Broadhurst, C.L., Peters, C.A., Tappero, R.V., Sparks, D.L. 2007. Improved understanding of hyperaccumulation yields commercial phytoextraction and phytomining technologies. Journal of Environmental Quality. 36: 1429–1443.

Chaney, R.L., Broadhurst, C.L., Centofanti, T. 2010. Phytoremediation of Soil Trace Elements. In: Hooda, P.S. (Ed.): Trace Elements in Soils. John Wiley & Sons, Chichester: 311–352.

Chapin, F.S., Kedrowski, R.A. 1983. Seasonal changes in nitrogen and phosphorus fractions and autumn retranslocation in evergreen and deciduous Taiga trees. Ecology 64(2): 376–391.

Chehregani, A., Malayeri, B.E. 2007. Removal of heavy metals by native accumulator plants. International Journal of Agriculture and Biology. 9: 462–465.

Clarholm, M. 1993. Microbial biomass P, labile P, and acid phosphatase activity in the humus layer of a spruce forest, after repeated additions of fertilizers. Biology and Fertility of Soils 16: 287–292.

Dalpé, Y. 2005. Les mycorhizes: Un outil de protection des plantes mais non une panacée. Phytoprotection 86(1): 53–59.

Dalvi, A. A., Bhalerao, S.A. 2013. Response of plants towards heavy metal toxicity: An overview of avoidance, tolerance and uptake mechanism. Annals of Plant Sciences. 2: 362–368.

Duponnois, R., Colombet, A., Hien, V., Thioulouse, J. 2005. The mycorrhizal fungus *Glomus intraradices* and rock phosphate amendment influence plant growth and microbial activity in the rhizosphere of *Acacia holosericea*. Soil Biology & Biochemistry 37: 1460–68. doi: 10.1016/j.soilbio.2004.09.016.

Fortin, JA., Plenchette, C., Piché, Y. 2008. Les Mycorhizes: La Nouvelle Révolution Verte. In: Les mycorhizes, la nouvelle révolution verte. Édition Multimondes, Québec, Canada.131.

Garbaye, J. 1991. Biological interactions in the myeorrhizosphere. Experientia 47: 370–375.

Garbisu, C., Alkorta, I. 2003. Basic concepts on heavy metal soil bioremediation. The European Journal of Mineral Processing and Environmental Protection. 3: 58–66.

Ghosh, M., Singh, S.P. 2005. A review on phytoremediation of heavy metals and utilization of its byproducts. Applied Ecology and Environmental Research. 3: 1–18.

Greppin, H., Auderset, G.U.Y., Bonzon, M., Degli agosti, R., Lenk, R., Penel, C. 1986. Le mécanisme de l ' induction florale. Saussurea 17: 71–84.

Hameed, A., Dilfuza, E., Abd-allah, E.F., Hashem, A., Kumar, A., Ahmad, P. 2014. Salinity stress and arbuscular mycorrhizal symbiosis in plants. Use of Microbes for the Alleviation of Soil Stresses, Springer, New York: 139–159.

Harrier, L.A., Watson, C.A. 2004. The potential role of arbuscular mycorrhizal (AM) fungi in the bioprotection of plants against soilborne pathogens in organic and/or other sustainable farming systems. Pest Management Science 60: 149–157.

Huey, C.J., Gopinath, S.C.B., Uda, M.N.A., Zulhaimi, H.I., Jaafar, M.N., Kasim, F.H., Yaakub, A.R.W. 2020. Mycorrhiza: A natural resource assists plant growth under varied soil conditions. Biotech 10(5): 204. doi: 10.1007/s13205-020-02188-3.

Javelle, A., Chalot, M. 1999. Ammonium and methylamine transport by the ectomycorrhizal fungus *Paxillus involutus* and ectomycorrhizas. FEMS Microbiology Ecology 30(4): 355–366.

Javeria, S., Kumar, V., Sharma, P., Prasad, L., Kumar, M., Varma, A. 2017. Mycorrhizal symbiosis: Ways underlying plant–fungus interactions. Mycorrhiza—Eco-Physiology, Secondary Metabolites, Nanomaterials, Fourth Edition, Springer, Cham: 183–207.

Kabata-pendias, A., Pendias, H. 2001. Trace Elements in Soils and Plants Trace Elements in Soils and Plants. 3rd Edition, CRC Press, Boca Raton, 403pp.

Kalve, S., Sarangi, B.K., Pandey, R.A., Chakrabarti, T. 2011. Arsenic and chromium hyperaccumulation by an ecotype of *Pteris vittata*—prospective for phytoextraction from contaminated water and soil. Current Science 100: 888–894.

Kaufman, M. 2020. An organic-based food system: A voyage back and forward in time. In Diet for a Sustainable Ecosystem. Springer, Berlin: 375–395.

Krämer, U. 2010. Metal hyperaccumulation in plants. Annual Review of Plant Biology 61: 517–534.

Lasat, M.M. 2002. Phytoextraction of toxic metals: A review of biological mechanisms. Journal of Environmental Quality 31: 109–120.

Li, Y.M., Chaney, R., Brewer, E., Roseberg, R., Angle, J.S., Baker, A., Reeves, R., Nelkin, J. 2003. Development of a technology for commercial phytoextraction of nickel: economic and technical considerations. Plant Soil 249: 107–115.

Mahar, A., Wang, P., Ali, A., Awasthi, M.K., Lahori, A.H., Wang, Q., Li, R., Zhang, Z. 2016. Challenges and opportunities in the phytoremediation of heavy metals contaminated soils: A review. Ecotoxicology and Environmental Safety 126: 111–121.

Matsuda, Y., Kohei, K., Yudai, K., Toko, T. 2021. Colonization status and community structure of arbuscular mycorrhizal fungi in the coniferous tree, *Cryptomeria japonica*, with special reference to root orders. Plant and Soil 468(1): 423–38.

Mukhopadhyay, S., Maiti, S.K. 2011. Trace metal accumulation and natural mycorrhizal colonisation in an afforested coalmine overburden dump: A case study from India. International Journal of Mining, Reclamation and Environment 25(2): 187–207.

Mulligan, C.N., Yong, R.N., Gibbs, B.F. 2001. Remediation technologies for metal-contaminated soils and groundwater: An evaluation. Engineering Geology 60(1–4): 193–207.

Munkvold, L., Kjøller, R., Vestberg, M., Rosendahl, S., Jakobsen, I. 2004. High functional diversity within species of arbuscular mycorrhizal fungi. New Phytologist 164(2): 357–364.

Nahar, K., Bovill, B., Mcdonald, G. 2021. Mycorrhizal colonization in bread wheat varieties differing in their response to phosphorus. Journal of Plant Nutrition 44(1): 29–45.

Nouaim, R., Chaussod, R. 1996. Rôle des mycorhizes dans l ' alimentation hydrique et minérale des plantes, notamment des ligneux de zones arides. Options Mediterraneennes 20: 9–26.

Nouria, D. 2018. Evaluation Du Potentiel Mycorhizogène Sous Acacia Saligna Introduite Pour La Revégétalisation de La Sablière de Terga.

Odjegba, V.J., Fasidi, I.O. 2007. Phytoremediation of heavy metals by *Eichhornia crassipes*. Environmentalist. 27: 349–355.

Oldroyd, Giles E.D. 2013. Speak, friend, and enter: Signalling systems that promote beneficial symbiotic associations in plants. Nature Reviews Microbiology 11(4): 252–63.

Ouahmane, L. 2007. Rôles de la mycorhization et des plantes associées (lavande et thym) dans la croissance du cyprès de l'Atlas (*Cupressus atlantica* G.) : conséquences sur la biodiversité rhizosphérique et la réhabilitation des milieux dégradés [unpublished].

Peterson, R.L., Massicotte, H.B., Melville, L.H. 2004. Mycorrhizas: Anatomy and Cell Biology. CABI Publishing, Wallingford, UK.

Pilon-Smits, E. 2005. Phytoremediation. Annual Review of Plant Biology 56: 15–39.

Plassard, C., Robin, A., Le Cadre, E. et al. 2015. Améliorer la biodisponibilité du phosphore : comment valoriser les compétences des plantes et les mécanismes biologiques du sol ?. Innovations Agronomiques 43(43): 115–138.

Püschel, D., Bitterlich, M., Rydlová, J. and Jansa, J. 2020. Facilitation of plant water uptake by an arbuscular mycorrhizal fungus: A Gordian knot of roots and hyphae. Mycorrhiza 30: 219–330.

Rai, P.K. 2008. Phytoremediation of Hg and Cd from industrial effluents using an aquatic free-floating macrophyte Azollapinnata. International Journal of Phytoremediation 10: 430–439.

Reeves, R.D., Baker, A.J.M. 2000. Metal-accumulating plants. In: Raskin, I., Ensley, B.D. (Eds.): Phytoremediation of Toxic Metals: Using Plants to Clean Up the Environment. John Wiley & Sons, New York, NY: 193–229.

Reeves, R.D., Brooks, R.R. 1983. Hyperaccumulation of lead and zinc by two metallophytes.

Reyes-Tena, A., Gabriel R., Luis L., Evangelina E.Q.A. 2017. Effect of Mycorrhizae and Actinomycetes on growth and bioprotection of *Capsicum annuum* L. against *Phytophthora capsici*. Pakistan Journal of Agricultural Sciences 54(3): 513–22.

Rohwerder, T., Gehrke, T., Kinzler, K., Sand, W. 2003. Bioleaching review part A: Progress in bioleaching: fundamentals and mechanisms of bacterial metal sulfide oxidation. Applied Microbiology and Biotechnology 63: 239–248.

Salt, D.E., Smith, R.D., Raskin, L. 1998. Phytoremediation. Annual Review Plant Physiology and Plant Molecular Biology 49(1): 643–668.

Schachtman, D.P., Reid, R.J., Ayling, S.M. 1998. Update on phosphorus uptake phosphorus uptake by plants : From soil to cell. Plant Physiology Journal 116: 447–453.

Schüßler, A., Schwarzott, D., Walker, C. 2001. A New Fungal Phylum, the Glomeromycota. Phylogeny and Evolution 105: 1413–1421.

Sim, M., Eom, A. 2006. Effects of ectomycorrhizal fungi on growth of seedlings of *Pinus densiflora*. Mycobiology 34(4): 191–195.

Singh, G., Pankaj, U., Ajayakumar, P.V., Verma, R.K. 2020. Phytoremediation of sewage sludge by *Cymbopogon martinii* (Roxb.) Wats. var. motia Burk. grown under soil amended with varying levels of sewage sludge. International J of Phytoremediation 22(5): 540–550.

Singh, G., Pankaj, U., Chand, C., Verma, R.K. 2019. Arbuscular mycorrhizal fungi-assisted phytoextraction of toxic metals by *Zea mays* L. from tannery sludge. Soil and Sediment Contamination: An International Journal 28(8): 729–746.

Sinha, R.K, Herat, S, Tandon, P.K. 2004. Phytoremediation: Role of plants in contaminated site management. Book of Environmental Bioremediation Technologies. Springer, Berlin, Germany: 315–330.

Smith, M.R. 1990. The biodegradation of aromatic hydrocarbons by bacteria. Biodegradation 1(2–3): 191–206. doi: 10.1007/BF00058836. PMID: 1368147.

Smith, S.E., Read, D.J. (1997). Mycorrhizal Symbiosis. 2nd edition, Academic Press, London.

Srivastava, M., Ma, L.Q., Santos, J.A.G. 2006. Three new arsenic hyperaccumulating ferns. Science of The Total Environment 364: 1–3.

Taylor, T.N., Remy, W., Hass, H., Kerp, H. 1995. Fossil arbuscular mycorrhizae from the Early Devonian. Mycologia 87(4): 560–573.

Vangronsveld, J., Herzig, R., Weyens, N., Boulet, J., Adriaensen, K., Ruttens, A., Thewys, T., Vassilev, A., Meers, E., Nehnevajova, E., Van der Lelie, D., Mench, M. 2009. Phytoremediation of contaminated soils and groundwater: Lessons from the field. Environ. Sci. Pollut. Res. 16: 765–794.

Vidali, M. 2001. Bioremediation. An overview. Pure and Applied Chemistry 73: 1163–1172.

Walker, C. 2013. Arbuscular mycorrhiza in the living collections at the Royal Botanic Garden Edinburgh. Sibbaldia: The International Journal of Botanic Garden Horticulture 11: 143–57. doi: 10.24823/sibbaldia.2013.57.

Yan, A., Wang, Y., Tan, S. N., Mohd Yusof, M. L., Ghosh, S., Chen, Z. 2020. Phytoremediation: A promising approach for revegetation of heavy metal-polluted land. Frontiers in Plant Science 11. doi:10.3389/fpls.2020.00359.

Yoon, J., Cao, X., Zhou, Q., Ma, L.Q. 2006. Accumulation of Pb, Cu, and Zn in native plants growing on a contaminated Florida site. Sci. Total Environ. 368: 456–464.

You, Y., Wang, L., Ju, C., Wang, G., Ma, F., Wang, Y., Yang, D. 2021. Effects of arbuscular mycorrhizal fungi on the growth and toxic element uptake of *Phragmites australis* (Cav.) Trin. ex Steud under zinc/cadmium stress. Ecotoxicology and Environmental Safety 213: 112023.

Zheng, H., Zhenyu, W., Xia D., Stephen, H., Baoshan, X. 2013. Impacts of adding biochar on nitrogen retention and bioavailability in agricultural soil. Geoderma 206: 32–39.

9 Biopesticide Usage

An Alternative to the Menace of Chemical Hazards for Sustainable Plant Protection

M. Soniya Devi, Usha, and Vijay Kumar Mishra
Rani Lakshmi Bai Central Agricultural University
Jhansi, India

Rakesh Yonzone
Uttar Banga Krishi Viswavidyalaya, Majhian, India

CONTENTS

DOI: 10.1201/9781003147091-9

9.1 INTRODUCTION

With the upsurge in number of the harmful pest, farmers tend to suppress those injurious species in a rapid way. The only route for quick killing the pests is the utilization of chemical pesticides. Even though pesticides play a vital role in meeting the demand of people through increasing crop productivity and income generation of farmers over the decades, there have been many disadvantages as per the safety point of view. The pesticides not only demolish the harmful pests but also the non-target organisms (Hashimi et al. 2020). In addition to these, it hampers the ecological balance leading to pest resurgence, resistance development and environmental pollution. It is not advisable to use such chemicals until the pests reach economic threshold level. A few decades earlier, various rules and regulations were imposed on the import of the chemical pesticide products and the acceptable amount of pesticide residues on the commodity. Owing to the above limitations there has been increased social pressure to replace with safer alternatives. The alternative way to combat such issues is by using naturally occurring substances or its derivatives. There are certain living organism's viz., plant, nematodes and microbes which act as protectants for the agricultural crops. These useful organisms and their derivatives serves as safer purpose for combating the injurious pests in eco-friendly way thereby helping the ecosystem and human health (Gupta and Dikshit 2010).

Biopesticides are the pesticides based on microbes and natural products. This includes biochemical pesticides (BCP), microbial pesticides (MCP) and plant incorporated protectants (PIPs) (Gupta and Dikshit 2010; Jones 2010). These pesticides are derivatives of organisms including plants, fungus, bacteria, virus and other microbes, nematodes, etc. The earliest agricultural biopesticides which were used were plant extracts in the early 17th century. There was also a report on experiments with mineral oils as plant protectants in the 19th century. Thereafter, several studies were carried out on biopesticides. The spores of the bacteria, *Bacillus thuringiensis* were one of the oldest and still widely used biopesticides. Nonetheless, there was slow development and there remained a preference for chemical pesticides. However, in recent decades, it has become a major concern to people throughout the world to produce safe and quality food. Since then, there has been a visible change in the attitude toward biopesticides over chemicals from a safety point of view. This chapter will highlight the role played by biopesticides in sustaining the environment, their availability as commercial formulations and the challenges faced for their adoption.

9.2 PRESENT SCENARIO OF BIOPESTICIDE USAGE

9.2.1 Biopesticide Production

Agriculture farming worldwide is facing huge problems due to a variety of reasons. Among various factors is pest appearance and its management. This pest pressure in crop production has caused the agriculture sector to depend upon synthetic pesticides that have resulted in various negative effects, not only in the crops at the time of harvest but also on the development of pest resistance; thereby increasing the cost of cultivation for farmers. With the implementation of policies such as the Food Quality Protection Act,1996, due to the overuse of chemical pesticides in protection of food from various pests and diseases and their risk to human health, the use of pesticides in crop production has been reduced. With this Act, they now have new toxicity and residue standards which have forced the agriculture sector to reduce their dependence on synthetic pesticides and instead rely upon other eco-friendly chemicals, such as biopesticides and growth-promoting substances. Many documents cite plant growth regulators (PGRs) not as a biopesticides but many reviews and sources placed them as biopesticides. Some of the PGRs such as auxins, gibberellins, ethylene, cytokinins, abscisic acids, brassinostreoids, polyamines, jasmonates, etc., are important in regulating various physiological process in plant. They act as chemical messengers and enable plants to thrive under various stress conditions; thereby helping combat various adverse effects (Jan et al. 2020). Under pesticide stress conditions, they played a great role in the recovery of plants by detoxifying the pesticides. Salicylic acid aids in mitigating the pesticide toxicity improving pigment content, increasing photosynthesis leading to antioxidant enzyme activity and decreasing the effect of pesticides residue in various crops *viz.* rice, wheat, maize and sunflowers. It also elevates the glutathione S-transferase (GST) activity in the tomato plant under pesticide stress condition (Yüzbaşıoglu and Dalyan 2019). In rice, tea, cucumber, tomato and broccoli, brassinosteroids help in detoxifying various pesticides, including organochlorine, organophosphorus and carbamate (Zhou et al. 2015). Their importance has been continuously increasing in agriculture markets and among producers.

The biopesticide market has been growing continuously since its importance was realized and produced during the mid-1990. It reached about $3.2 billion by 2017. Some of the important factors that lead international companies to produces more biopesticides are the growth of organic farming and residue-free crops. There was an increase in biopesticide registration from 175 in 1998 to 452 in 2009 (Regnault-Roger 2012). The biopesticide market in recent times has increased due to more approval of biopesticides as compared to synthetic pesticides by the approving authorities. The registration of 13 different pesticides and 2 nematicides in China and 8 biopesticides and 4 botanicals in the United States was reported (Agrow 2013).

The important biopesticides include bacterial, fungal, protozoa, nematicidal and viral that specifically target pathogens and pests. Among various leading companies around the world, some of the important biopesticide-producing companies include Syngenta International AG, Valent Biosciences Corporation, Kumiai Chemical Industry Co. Ltd., Certis USA LLC, AgraQuest Inc., AG Biotech Australia Pty Ltd., BioWorks Inc, Greeneem, Bayer Crop Protection, GmBh, BionTech Inc., Koppert B.V., Marrone Bio Innovations, Troy Biosciences Inc., Isagro SpA, Sumitomo Chemical Co. Ltd., Prophyta Biologischer Pflanzenschutz GmbH and San Jacinto Environmental Supplies. There are various bacterial, fungal and other bioagent products available in the market and they hold about 65% of the total global market. Among them, bioagents *Bacillus thuringiensis* and its various species have the most dominant position in the market. Some other bioagents include species of the *Tirchoderma* such as *Trichoderma gamsi* and *Trichoderma harzianum*. The leading producers of biopesticides of these fugal and bacterial species are Russia, the United States, China and India (Leng et al. 2011).

China alone has 134 companies that produce biopesticides and have sales with an income of 6.2 billion Yuan, which accounted for about 12% of total pesticide sales in 2010. They have in total registered 25 microbial pesticides, which include 9 bacterial, 7 fungal and 7 viral biopesticide (Agrow 2013). In India, the total number of biopesticides registered was 12 with a production of 410 different biopesticide manufacturing units, among which 130 are governed by the private sector (Leng et al. 2011). The important botanical biopesticides that are found to be effective against various pests and diseases belong to families such as Compositae, Lamiaceae, Poaceae, Chenopodiaceae, Asteraceae and Meliaceae. Their products include rotenone, azadirachtin, pyrethrins and essential oils, which are commonly used and are widely accepted for organic cultivation. They are safe and eco-friendly and have been under use since the realization of their importance centuries ago. Although they still exist, their use had been restricted due to limitations in their efficacy, rapid degradation and the requirement of repeated application. Currently, their importance has grown rapidly. Due to identification of different modes of action and effectiveness, their production and commercialization has increased to a new height. Since microbial pesticides have dominated the market as compared to botanical pesticides new, active chemicals compounds are expected in the near future which helps in stimulating the pesticides industries (El-Wakeil 2013). More than 17 different botanical biopesticides have been registered with 200 active products in China. The sales of biopesticides in Europe have risen to about 20% and reached about US$100 million during 2011 (Agrow 2013).

The dominant botanical biopesticides are the azadirachtin, pyrethrum, essential oils and rotenone that have captured the market in larger quantities. The leading countries producing pyrethrin are Kenya and Australia. Kenya alone produces 70% of world production of pyrethrin These are also produced in other countries such as Tanzania and Ecuador but in lesser quantities (Attia et al. 2013). Among global botanical production, the sale of pyrethrum alone holds about 80% of the total sale (Isman 2006). Another important botanical pesticide includes the oil compound azadirachtin extracted from Neem trees (*Azadirachta indica*), which are the native trees from Burma and India. India alone produces neem oil of around 2.5 lakh tonnes and accounts for exports of about US$5.73 million products based on neem production in 2012. Besides India, Japan, Italy, Spain and the United States are among the leading importers.

There are various technologies to extract neem oil apart from traditional techniques such as steam distillation, which are time and labor intensive. Some of the technologies include SFE, MAE, which are effective and less time consuming. Some other oils which also have gained interest in controlling insect pests are the clove oil, thyme oil, mint oil and rosemary oil which are effective in controlling stored grain insect pests and diseases of some plants (Isman and Machial 2006). Since aromatic plants are grown all around the world, it's very difficult to estimate the production because of no accurate total harvest, making it ever harder to estimate the market share of biopesticides. But it has been estimated that the essential oil production in the world are valued around US$700 million. In China, 14 essential oil have been registered with production of 20 different products by 13 manufacturing units (Agrow 2013).

9.2.2 Biopesticide Consumption

Even though biopesticides contribute only a small percentage of the world pesticide market, their growth has been considerably faster than the synthetic pesticides. The total global market of pesticides was around US$37.5 billion, where biopesticides alone contributed about US$1.3 billion of the market. The biopesticide share of the global market is expected around US$2.1 billion and 3.7 billion in 2012 and 2017, respectively, where the synthetic pesticide market was predicted to be around US$44 billion and 61.5 billion in 2012 and 2017, respectively (BCC Research 2012). Shifting of chemical-based agriculture toward organic farming, there has been increase in the area around the world by 22 million hectares thereby increasing the demand of biopesticides globally (Regnault-Roger 2012). The global organic agricultural product has increased by US$29.8 billion in 2005 and has been growing steadily thereafter. The highest global biopesticide usage were recorded in North America and Europe with US$497.3 and 541.4 million during 2010 and 2011, followed by Latin America and Asia-Pacific market accounting for 20% and 10%, respectively (Thakore 2006).

Out of a 63% global biopesticide market, *Bacillus thuringiensis* and its sub-species occupies the most with 2% and 0.5% use of *Bt*-based products and botanical pesticides in China (Zhang et al. 2011). North America is the leading biopesticide consumer as compared to other countries, consuming a level of around US$139.8 million in 2010. In the United States, around 202 biopesticides were registered of which 102 were microbial based, whereas in the EU, 38 biopesticides were registered of which 34 were bioagent based (Bailey et al. 2010). However, there was a sharp decline of *Bt*-based product and an increase in *Bacillus subtilis* and fungal-based product sale. Furthermore, bioagent-based biopesticide total sales in Europe were recorded around 13.6% in 2011, with a turnover of US$73.4 million (Agrow 2013).

The use of botanical biopesticides has increased steadily over the years and has been recorded as the second highest sale in the European market with income of 18.5% shares. It has been used since time immemorial for minimizing pests and diseases (Ananthi 2019). Among the various biopesticides, essential oil based on azadirachtin was the fastest selling bioproduct on the market. There was a sale of Pyrethrum of about 1000 metric tonnes in the United States; however, its 90% use was in fields other than agriculture. They are very effective in controlling mites, insect pests, etc. The use of rotenone has sharply declined due to withdrawal for the European Union plant protection market and after their restricted use in the United States. Even though their market fluctuates, neem oil and azadrachtin hold the second position in biopesticide use in California. The use of nicotine is completely banned in Europe; however, it is used to treat rice and potatos in Bolivia and China (Regnault-Roger 2012).

One of the important attributes of biopesticide usages is its nature of decomposition without leaving any residues behind. Also, they are not harmful toward non-target organisms and are eco-friendly. They will help in reducing residues in foods thereby providing safer products to the consumer. They can be used in the long run as an alternative to synthetic pesticides as they are the only product that increases strength and competitiveness with chemicals in the global agriculture market.

9.3 IMPORTANCE OF BIOPESTICIDE USAGE IN AN AGRICULTURE CONTEXT

Agriculture, being the backbone of the Indian economy, contributes more than 18% of the total GDP and provides food security to the more than 1.27 billion inhabitants of India (Bhardwaj and Sharma 2013). One of the important factors in providing this food security is by using pesticides to prevent crop loss due to pests, pathogens and weeds. The application of synthetic pesticides has increased tremendously since the Green Revolution, causing harm to the environment and human health. Given that agriculture represents a major source of income to more than 65% of the Indian people, the population is continuously exposed to the adverse effect of pesticides. Due to their solubility in water and soil adsorb capacity, their residual impact has affected the environment adversely.

The use of pesticides is unregulated and uncontrolled, which is quite an alarming situation in present times. Small land holdings among the farmers, low irrigation, irregularity in monsoons and, most importantly, little knowledge about pesticides and its impact have contributed to creating a major problem in pesticide usage.

In India, the consumption patterns of the widely used chemicals are the insecticides. India accounts for 76% of the global pesticide usage. (Mathur 1999). The use of herbicides and fungicides are comparatively lesser than insecticides. Huge amounts of chemical pesticides are applied in cotton (37%) and paddys (20%), which together account for about 57% of the total pesticide consumption. In addition to these, wheat and pulses accounts for about 4%, plantation crops 7% and vegetables 9%. The pesticide-consuming states are Andhra Pradesh followed by Punjab and Maharastra (Bhardwaj and Sharma 2013).

Currently, pesticides are commonly found as contaminants everywhere—they are present in the biosphere, food and soil, causing problems for non-targeted organisms, plants, beneficial microorganisms, birds and animals. Overwhelming scientific evidence has reported major potential risks to human health and the environment we live in (Forget 1993; Igbedioh 1991). The indiscriminate use of pesticides has resulted in acute poisoning, causing diseases such as cancer, reproduction problems, malformation of the liver, impairment to the immune system, neurological impairment, etc. In some cases, diabetes and acute carcinogenic cases have also been reported due to the side effect of pesticides. Therefore, considering all these effects, biopesticides serve as alternative agents in managing crop pests and diseases. Their application method is similar to that of synthetic pesticides and is mostly harmless towards non-targeted pests. They are cheaper, effective and eco-friendly as compared to chemical pesticides. Therefore, the use of biopesticides can be done incorporating them as a vital component in Integrated Pest Management (IPM), thereby reducing the use of synthetic pesticides.

9.4 BIOPESTICIDE USAGE FOR SUSTAINABLE CROP PROTECTION

9.4.1 Micro-Bioagents as Sources of Biopesticides

There are currently several choices available in biopesticides based on various microorganisms, and they have been found to be effective in management of various pests and diseases. Biopesticides have now emerged as inevitable components

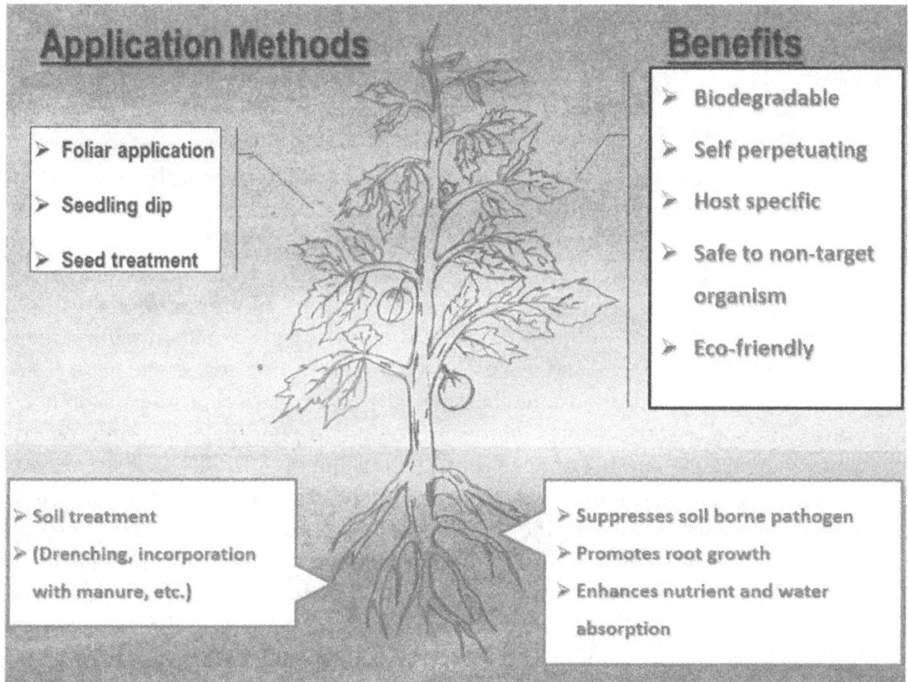

FIGURE 9.1 Application methods and benefits of biopesticide usage.

in agriculture. More importantly, organic farming has become of increasing importance. The bioagents are mostly naturally occurring and are beneficial. They have antagonistic activities and play a vital role in suppressing pest and pathogen populations and also in maintaining sustainable soil health and environmentally safe crops (Harman 2011; Junaid et al. 2013). The most important factors that govern their survival, development, conservation and multiplication are environmental factors (Figure 9.1). Biocontrol agents have long been recognized as vital components of Integrated Pest Management due to their specificity in targeting pests and pathogens, as well as being eco-friendly and leaving no residues on the crops, and being safe to all beneficial microorganisms. They are preferred over synthetic chemicals, as the former create problems like pest resurgence, pesticide resistance, pesticide residues, etc. Hazardous and phytotoxicity effects of the chemicals to the surrounding environment are now more obvious. On the other hand, biopesticides are more eco-friendly, efficient and economically viable, which makes them more attractive to researchers and users than in the past.

Several reports of biocontrol agents (BCA) have been documented and demonstrated to reduce pests and disease, either in laboratory, green house or field conditions. Many of these bioagents had very promising results in the experimental trial and are now mostly adapted and commercially available in the market. The following are the important and commercially available micro-bioagents that are widely in use.

9.4.1.1 Fungal Bioagents as Sources of Biopesticides

The important fungal bioagents that have shown promising effects for suppressing pests and diseases are as follows.

9.4.1.1.1 Trichoderma *Species*

Trichoderma are mostly soilborne and are also found in decomposing organic matter. It has been used as a fungal bioagent ever since it was first reported by C.H. Persoon almost 200 years ago. They are very rapid growers, have capacity to utilize diverse substrates and are easy to isolate and culture. Species of *Trichoderma* are ubiquitous in nature, free living and are found predominantly in agricultural or desert soil, salt marshes and forests of every climatic region. They mostly grow in their natural habitats such as on root surfaces and are effective in controlling roots, and even foliar diseases. One of the most important characteristic of *Trichoderma* sp. is its competing ability for space and nutrients, ability to solubilize phosphates and micronutrients; they increase the number of deep roots, increasing plant ability to resist various stresses and enhance germination and root and shoot growth, induce host resistance and, most importantly, increase the quality yield. The hydrolytic enzymes, peptaibols, antibiotics, competence in utilizing nutrients, the mechanisms of competition for nutrition and space, mycoparasitism, induction of host defense, etc. present in the *Trichoderma* spp. made them an efficient bioagent, helping in the management of various stresses (Pandey et al. 2021). To date, 89 species of *Trichoderma* genus and several strains have been reported as effective biocontrol agents against plant diseases. Some of the important species are *T. harzianum, T. viridae, T. hamatum, T. konigii*, etc., which live in close proximity to roots of plants and may even form endophytic associations with their hosts. They produce various types of antibiotics and toxins, such as trichothecin, sesquiterpine and trichodermin, which directly affect other competing organisms or mycoparasites. The hyphae of pathogens cause lysis by secreting different types of lytic enzymes such as chitinase, pectinase, glucanase, etc. Some of the important pathogens that are effectively inhibited by *Trichoderm*a are *Phytophthora* sp., *Fusarium* sp., *Sclerotium rolfsii, Alternaria* sp., *Rhizoctonia solani, Cladosporium fulvum, Ralstonia solanacearum,* etc. (Gangwar and Sharma 2013; Pandey and Pundhir 2013; Selvakumar 2008; Shahnaz et al. 2013; Vasanthakumari and Shivanna 2013; Yadav et al. 2013).

9.4.1.1.2 Beauveria bassiana

These are mostly entomopathogenic, which hamper the growth and development of various arthropod pests, ultimately leading to the death of the insect. Its entomopathogenic activity was first reported by Agostino Bassi, an Italian scientist, who discovered that the microbes were causing disease in silkworms. It produces white fluffy corpses resembling pastris in insect pests, which is known as white muscardine disease. *B. bassiana* are mostly non-pathogenic and non-toxic to the surrounding environment and are mostly found in undisturbed natural habitats, lesser in cultivated lands. Their mechanism of infestation in the insects begins with the landing of spores on the host surface and germinates through the insect's cuticle and causes severe infections (Keswani et al. 2013). The germinating tubes grow and destroy the exoskeleton due to enzymatic digestion or mechanical pressure creating a clean hole at the site of penetration (Keswani et al. 2013; Pinnamaneni and Potineni 2010). After that, on reaching hemolymph, the fungus produces blastospores and evades any innate immune response. The fungus then starts utilizing fat bodies and muscular tissue as a source of nutrients and causes damage to food tracts of epithelial cell walls, making insects weak. Thus, the fungus grows profusely, covering the whole body of the insect and causing death within a few days.

9.4.1.1.3 Metarhizium anisopliae

These are mostly soil-inhabiting entomopathogenic fungi causing green muscardine disease in insects. *M. anisopliae*, earlier known as *Entomphthora anisopliae*, was first reported from the larvae of the wheat cockchafer *Anisopliae austriaca* in Ukraine in 1879 by Metschnikoff. Sorokin was renamed as *Metarhizium anisopliae* in 1883 (Tulloch 1976). These are mostly cosmopolitan in nature and are found in the soil of both cultivated and non-cultivated habitats. They are mostly non-toxic to vertebrates but cause dermal or eye irritation to sensitive mammalian bodies or patients with compromised immune systems. The *M. anisopliae* mostly enter insects through pores and spiracles in the sense organ. After entry into the host, the hyphae grow laterally and proliferates causing damage to the internal system of the insect. The hyphae continue to grow inside the insects until all internal content is consumed. Then, the fungi break through the cuticle and sporulates, giving the fuzzy appearance of an insect (Aw and Hue 2017). Several reports have been documented to produce many secondary metabolites by this fungus such as destroxins, swainsinone and cytochalasin C, which has insecticidal properties. *M. anisopliae* have been found to be effective against various insect pests. The important orders that are affected by this fungus are Isoptera, Homoptera, Heteroptera, Dermaptera, Coleoptera, Orthoptera, Diptera, Lepidoptera, Hymenoptera and Siphanoptera (Sahayaraj and Borgio 2010). The major insects that are suppressed by these fungi include aphids, grasshoppers, whiteflies, nematodes, plant hopper, mosquitoes, etc.

9.4.1.1.4 Verticillium lecani

V. lecani, which is also known as white halo fungus due to its white mycelial growth on the edge of the infected scale insect, is mostly found in the soil of tropics and subtropic regions. They are mostly effective in killing their host by means of mechanical pressure or by releasing mycotoxins such as dipicolonic acid, cyclosporine, cyclodepsipeptide toxin, Bassianolide, etc. (Goettel et al. 2008; Ravindran et al. 2018). The fungus when comes in contact with insects germinates and enters the host body through the cuticle and proliferates extensively, killing the insect within 42–78 hrs. The insects that are suppressed by *V. lecani* include aphids, thrips, mealy bugs, whiteflies and other sucking pests (Jasim and Mohammed 2019; Patil et al. 2012).

9.4.1.1.5 Ampelomyces quisqualis

A. quisqualis are naturally occurring and found to be effective against several species of pathogens belonging to the family Erysiphaceae. They are eco-friendly and do not cause any negative effects on human health. They are most specific to powdery mildew fungi, where they hyperparsitize the pathogen and suppress its growth and development. However, their effects also have reported against more than 64 fungal species belonging to different genera such as *Leveillula, Microsphaera, Sphaerotheca, Uncinula* and *Podosphera* (Gilardi et al. 2008; Lee et al. 2007; Sundheim 2007). In addition, reports on hyperparasitism on anamorphic genera *Oidiopsis* and *Oidium* have also been reported. The mechanism of infection mostly includes direct penetration of hyphae and coniodophores through mechanical means or by secreting mechanical enzymatic activities (Kiss et al. 2004). Its effect in controlling pathogens can be seen within 8 to 10 days, provided the temperature is 20°C.

9.4.1.2 Bacterial Bioagents as Sources of Biopesticides

Unlike fungal bioagents, many bacterial bioagents have been reported by various researchers that are effective against various insects and pathogens. Some of the important bacterial bioagents that are commonly available in the market in various formulations are discussed here.

9.4.1.2.1 Pseudomonas fluorescens

P. fluorescens are the group of soil-colonizing microorganisms that are non-pathogenic having high saprophytic ability and are generally gram negative in nature. They are characterized by the presence of green florescent pigments known as pyoverdin, which are mostly induced when iron concentration is greater in the surrounding environment. The important characteristics which make them unique as bacterial bioagents are their faster rate of multiplication and high rhizosphere-colonizing activity, which acts as a systemic antagonist against various diseases of fungal and bacterial origin (Pandey and Chandel 2014; Prabhukarthikeyan and Thiruvengadam 2016). In addition, *P. fluorescens* can also produce growth-promoting hormones which makes them special not only as a bioagents but as a plant growth-promoting microorganism. They are very effective against a wide range of pathogens and restrict their growth with the production of antibiotics such as 2,4 diacetylphloroglucinol (DAPG), pyoluteorin, pyrrolnitrin, oomycin A and phenazine 1 carboxylic acid (PCA) (Fuente et al. 2008; Prabhukarthikeyan and Thiruvengadam 2016). In addition to these traits, they also induce systemic resistance (ISR) in many hosts, which help the plants to defend against pathogen attack and makes the plant resistant. The *P. fluorescens* are highly effective against *Rhizoctonia solani, Phytophthora, Pythium, Sclerotinia, Sclerotium*, etc.

9.4.1.2.2 Bacillus subtilitis

B. subtilis are mostly soilborne, gram-positive rod shape bacteria which generally forms endospores and helps the bacteria to withstand extreme environmental conditions. They help in inducing resistance and also inhibit the pathogen through the production of extracellular antibiotics, toxins and enzymes such as hydrolase and lipopeptidase (Mardanova et al. 2017; Wang et al. 2018). These lipopeptidases have not been found to be effective against many pathogens but are known to act as an effector molecule that helps in inducing resistance in the host. Some of the important pathogens controlled by them are *Xanthomonas, Rhizoctonia, Alternaria, Pythium, Phytophthora, Sclerotinia*, etc.

9.4.1.2.3 Bacillus thuringiensis

B. thuringiensis are gram-positive spore-forming bacterium that occur naturally in the soil and are very effective bioinsecticides. They produce a unique toxin which is found to be toxic against many insect species belonging to the orders Lepidoptera, Diptera, Hymenoptera and Coleopteran (Oguh et al. 2019; Sanchis and Bourguet 2008). They are also effective against invertebrates such as mites, nematodes and protozoans. The different types of biopesticide formulations have been developed utilizing *B. thuringiensis* strains. They contain the genes which codes for a unique crystal protein that are insecticidal in nature and have been successful in transferring it to many agricultural crops such as cotton, rice, soybean and corns and have proven to be a highly successful model in agriculture biotechnology. They are mostly used to control various lepidopteran pests such as American bollworm, looper, fruit borer, stem borer, leaf-feeding caterpillar, etc. During insect feeding the bacteria enters into the body of the insect, along with food materials. The crystal protein produced by the bacterial cells gets mixed in the midgut and gets solubilized (Chandler et al. 2011). After this, a crystal protein as delta endotoxin is formed which breaks down the lining of cells in the insect guts, causing the insect to stop feeding within 1–2 days. They also damage tissues and multiply within the insect body, causing the death of the insects (Bravo et al. 2005; Chandler et al. 2011).

9.4.2 Mechanisms Employed by Bioagents

In order to cause a disease in plants, three factors are most important—Host, Pathogen and the Environment. The biocontrol agents also interact with these factors in order to manage the disease. There are various mechanisms deployed by bioagents that inhibit the growth and development of pathogens. These mechanisms are broadly classified into two categories.

9.4.2.1 Direct Antagonism

This mechanism results from the high degree of selectivity for a particular pathogen and requires physical contact with each other. Direct antagonism includes the following.

9.4.2.1.1 Hyperparasitism

This is considered the most direct form of antagonism, which involves the growth of the bioagent over the targeted organism, resulting in coiling and dissolution of the cell wall and membrane by the secretion of various enzymes (Pal and Gardener 2006; Tiwari 1996). These kinds of mechanisms act against *Rhizoctonia solani*. Several cell wall-degrading enzymes such as chitinase and β-1, 3 glucanase are produced by *Trichoderma* for mediating biological control. These enzymes are produced as a result of gene expression, changes in which can lead to the production of enzymes that do not show active ability in cell wall destruction (Harman 2000). Such reduced ability to control the pathogen has been reported against *Botrytis cineria;* other examples of hyperparasitism are *Gliocladium virens, Acremonium altenatum,* etc., which parasitized and killed powdery mildew pathogen (Kiss 2003).

TABLE 9.1
List of Antibiotics against Plant Pathogens

Antibiotic	Source	Target Pathogen	References
Agrocin 84	*Agrobacterium radiobacter*	*Agrobacterium tumefaciens*	Murphy et al. 1981
2,4 Diacetylpholoroglucinol	*Pseudomonas fluorescence* F113	*Pythium*	De Souza et al. 2003; Shanahan et al. 1992;
Bacillomycin D	*Bacillus subtillus* AU195, *Bacillus* amylo liquefaciens strain FZB42	*Aspergillus flavus, Fussarium graminearum*	Gu et al. 2017; Moyne et al. 2001
Gliotoxin	*Trichoderma virens*	*Pythium, ultimum, Rhizoctonia solani*	Jones and Hancock 1988
Herbicolin	*Pantoea agglomerans* C91	*Erwinia amylovora*	Pusey et al. 2011

9.4.2.1.2 Antibiosis

The bioagents produce various low-molecular weight compounds or antibiotics that have direct effects on pathogen growth and development, known as antibiosis (Table 9.1). Several antibiotics are produced by different bioagents that are toxic toward the invading pathogen. Three types of antibiotics are reported to be produce by bioagents such as volatile/non-polar, non-volatile/polar and water soluble (Junaid et al. 2013). Among these antibiotics produced, volatile antibiotics are more important due to their effectiveness, and they have been found to act to all sides away from the site of production. Some of the lists of antibiotics produced by bioagents are as follows.

9.4.2.2 Indirect Antagonism

9.4.2.2.1 Competition

This mechanism involves active growing and colonizing the space in soil or living plant surfaces that are nutrient limited. The bioagents compete with the pathogen for nutrients and spaces for establishment in the given environment. The bioagents have more ability to utilize and uptake the nutrients and grow rapidly, occupying larger spaces and causing a shortage of nutrients and space for the competing pathogen.

9.4.2.2.2 Lytic Enzyme Secretion

Most of the bioagents are reported to produce enzymes that have the capability to hydrolyze various polymeric compounds present in the cell wall of the pathogens such as cellulose, hemicellulose, chitin, proteins and even the nucleic acid (Junaid et al. 2013). The secretions of enzymes are governed by various genes in the bioagent, a slight mutation of which can cause an inability to secrete the effective enzymes against the pathogen. Some of the important bioagents that produce enzymes are *Pseudomonas fluorescens*, which produce hydrogen cyanide (HCN) that are involved in suppression of root pathogens. The volatile compounds are produced by *Enterobacter cloacae*, which is effective in suppressing *Pythium altimum*, causing a damping off of cotton found to be ammonia (Howell et al. 1988).

9.4.2.2.3 *Plant Growth Promotion*

Many of the bioagents have been reported to promote the growth of the host by producing growth-promoting factors and help in escaping the incidence of disease. The best example of increasing growth during the early stages of plant growth, and thereby showing resistance, can be found in damping off of crops belonging to the Solanaceous family. Both the fungal and bacterial bioagents have been reported to produce growth-promoting agents and enhance growth by increasing nutrient solubilization, its uptake and root growth (Chaube et al. 2003). Some of the strains such as AN-27 of *Aspergillus niger* have been reported to produce growth-promoting compounds such as 2-carboxy-methyl-3-hexyl-maleic anhydride and 2 methylene -3 hexyl butanedioic acid, which are responsible for increasing root and shoot growth and also the biomass of crops. Woo et al. (2006) reported that when seeds of broad bean are coated with *T. viridae*, it helps in increasing the fresh and dry weight of roots and shoots and also the nodules.

9.4.2.2.4 *Induce Systemic Resistance*

This mechanism is one of the most important types of antagonism. The bioagents helps in inducing resistance either locally or systematically through expressions of various resistant genes. Out of various molecules, salicylic acid and non-expressor of pathogenesis-related genes (NPR1) are important agents in inducing systemic resistance. Such types of mechanisms can be seen in grapes to control *Botrytis cineria* when *T. harzianum* is inoculated on roots and leaves. In cotton, *Trichoderma virens* produces proteins that have hydrophobin and induces the development of roots and resistance to the disease (Desmukh et al. 2006; Dreuge et al. 2007). *Trichoderma* species secrete many fungal proteins such as cellulose, swollenins and xylanase to enhance the defense mechanism of the host (Martinez et al. 2001).

The effectiveness of the biopesticide depends upon the methods of application. Generally, they are applied in four ways such as foliar application, seedling dip, seed treatment and soil treatment.

9.5 BIOPESTICIDE PRODUCTION AND COMMERCIALIZATION: AN OVERVIEW

9.5.1 Types of Commercial Bioformulations

The bioformulation process involves mixing of microbial bioagents, along with carrier and adjuvants that are suitable for it to protect the actively living bioagents. It also helps them to improve durability, bioactivity, storability and efficiency, while contacting and interacting with the targeted pest and pathogens. The bioformulation should be prepared depending upon the convenience of the users. The process of formulation of biopesticides does not differ much from that of synthetic pesticides preparation. However, much care and attention have to be given to the process of bioformulation, since, as living organisms, their viability is of prime importance and must be present at required levels during the process of production, transportation and storage. Various factors such as method of application, time of application, presence of favorable conditions, etc., have a great influence on the establishment of bioagents and the effective control of the targeted pests and pathogens.

9.5.1.1 Categories of Biopesticide Formulation (Based on the Physical State)

9.5.1.1.1 Dry Formulation

This type of formulation is produced in large quantities as compared to other formulations. It is most commonly available and frequently produced. Mostly they are produced as wettable powders (WP), dust powders (D or DP), granules (GR), micro-granules (MG), water dispersible granules (WG), etc.

9.5.1.1.2 Liquid Formulation

This type of formulation is less produced as compared to dry formulations and are formulated either as suspension concentrates (SC) or occasionally as capsule suspensions (CS), which are prepared along with the presence of adjuvants such as stabilizers, osmoticant, stickers, surfactants, antifreeze compounds, additional nutrients, etc. They are readily available and are prepared using water, oil or their combinations as carriers. Liquid formulation meant for dilution includes suspension concentrates (SC), capsule suspensions (CS), emulsions, oil dispersions (OD), suspo-emulsions (SE) and ultra-low volume formulations (Knowles, 2005, 2006)

9.5.1.2 General Preparations of Biopesticides (Depending on the Types of Formulation)

9.5.1.2.1 Wettable Powder (WP)

They are mostly the inert fillers, which are based on dry and fine of a size less than 5 μ formulation. They are applied as suspension mostly by mixing with water. WPs are the most preferred formulations and are prepared in large quantities because of long duration storage stability, fine mixing ability with water and easy application. Teera-Arunsiri et al. (2003) formulated *Bacillus thuringiensis*-based biopesticides as wettable powders for increasing efficacy, longevity and easy handling of the product. *B. thuringiensis* subsp. *aizawai* was mixed with various adjuvants and spray dried. However, the negative effect lies in its dustiness, as it often raises human health issues not only during the process of preparation but also during its application in the field. The WPs are prepared depending on the bioagents taken. These are blended with the carrier base, along with dispersing, wetting, osmoticant and sticker agents.

9.5.1.2.2 Dust (DP)

The dust formulations are mostly based on mineral carrier formulations, which consist of active bioagents at 10% concentration, along with the anticaking agents. They are mostly clay- or talc-based and have a particle size of 50–100 μm and can be applied manually or by mechanical duster. Ultra violet protectants, anticaking agents and adhesive materials are used as formulants for enhancing adsorption. This formulation was used before granules were developed. Other dusts are manufactured very simply and they are still used today in many parts of the world. Due to their dustiness and application issues that cause problems to human health, they are nowadays restricted in use. However, this formulation is still being utilized in many countries (Knowles 2001).

9.5.1.2.3 Granule (GR)/Micro-Granule (MG)

This formulation is generally made from various mineral sources such as silica, kaolin, ground plant residue, starch, polymers, etc. (Tadros 2005). The bioagents that are actively growing are mixed in such minerals at a concentration of 5–20%. They differ from dust particles in size, as GR have a coarse particle size ranging between 100–1000 μ. They are mostly preferred due to the effectiveness, of the bioagents being spread or absorbed in the granules and easy application for management of various pests, pathogens, nematodes and weeds. Granules after application release their active ingredient slowly. Soil moisture is required in some granules for releasing their active ingredient (Knowles 2005; Tadros 2005)

9.5.1.2.4 Water Dispersible Granule (WG)

These are very similar to the wettable powders; however, the only difference is that the particles are present in granular form. This provides an additional advantage of WG over WP, particularly with all the disadvantages of WP dustiness. The storage stability is similar to that of WP and is usually produced following different processes such as spray drying, fluid bed drying, extrusion, etc. WGs are easily dispersed in water, with similar suspension as WP, which results in easy application. However, the major drawback of WG lies in the cost involved in the production techniques; consequently, they are not preferred by many users and producers.

9.5.1.2.5 Suspension Concentrate (SC)

This type of formulation is mostly prepared by dispersing solid active ingredients in a water or liquid phase. They also carry different ingredients such as wetting agents, dispersants, thickeners (viscosity modifiers), antifoaming agents, antifreezing agents, etc. They have a particle size distribution of 1–10 μm. They are produced by the wet grinding process where particle surfaces adsorbed inert ingredients preventing re-aggregation of small particles. Since they are water based, they are safe to the operator and also to the environment (Knowles 2005; Woods 2003). The only disadvantage of SC is its solid active ingredient, which needs to be suspended in a water base whenever needed and has to be agitated before each application.

9.5.1.2.6 Capsule Suspension (CS)

In this type of formulations the active agents are encapsulated in starch, gelatin or cellulose or any other polymer-based microcapsules of size 10^{-3} to 10^{-9} μm diameter and are suspended in stable aqueous phase for dilution with water before use. Microcapsule encapsulation have been used to provide a smaller size of higher efficiency to formulations of fungal biopesticide (Brar et al. 2006; Winder et al. 2005). The encapsulation adds an added advantage that facilitates the release of the bioagents slowly and also protects them from the various abiotic stresses. They are similar to SC in terms of use of adjuvants such as surfactants and thickeners, which help them to stabilize. However, despite these advantages, this type of CS formulation is less utilized and production is very low due to its high-cost involvement and its complex production technologies (Chen 2013; El-Sayed 2005).

9.5.2 Production Techniques of Bioagent-Based Pesticides

The production process of bioagent-based pesticides needs a strong knowledge about the bioagent, its identification, surviving ability, effectiveness against particular pest or pathogens, etc. In addition, the techniques used for production of biopesticides need to be well standardized for keeping in mind the viability of the bioagents and easy application. There are various steps involve in the production process of the bioformulation of the biopesticides such as:

- Conditioning of carrier material
- Preparation of microbial inoculants and adjuvants
- Blending of bioagents with carrier material and adjuvants
- Quantitative and qualitative evaluation of produced product
- Final packaging and storage

In the following section the production techniques of two important bioagents will be discussed, which are based on wettable powder (WP) formulations.

9.5.2.1 Production Techniques in *Trichoderma viride*-Based Biopesticides of 1% WP

For the production of *T. viride*-based biopesticides, talcum powder is most often used as carrier material. The steps generally involved in the production process are as follows.

9.5.2.1.1 Conditioning of Carrier Material

The well-sieved talc powder is used and approximately 960 g are filled in HDPE (high-density polyethylene) bags. The mouths of the bags are then closed with rubber bands or threads, which are mostly heat resistant and are sterilized at 121°C under 15 psi pressure for 15 minutes.

9.5.2.1.2 Preparation of T. viride *Inoculants*

The strain of *T. viride* from which the biopesticides has to be prepared is selected and re-isolated from the mother culture. It is then inoculated on the Potato Dextrose (PD) broth and shaken gently to lodge the spores. Then, it is inoculated in the shaker at 150 rpm and 28±1°C for 7 days. After the appearance of the dense mycelial growth in the flask they are used for further process.

9.5.2.1.3 Blending T. viride *with Carrier Material and Adjuvants*

The previously sterilized talc powder packets and broth culture of *T. viride* are taken, along with mannitol and carboxymethyl cellulose solution (CMC). The culture is well shaken and measured in a cylinder. The packets are then carefully opened in aseptic conditions and filled with the culture suspension of *T. viride*, along with 1% mannitol, and1% carboxymethyl cellulose solution (CMC) and are made airtight. The inoculated packet is then placed in an incubator at 27±1°C for 7 days to allow the bioagents to multiply. Only after a 24-hour incubation period, the packets are shaken using a mechanical shaker in order to achieve uniform spread of the bioagents within the formulation.

The process is repeated alternately for seven days. After that, they are subjected to evaluation for the presence of the desired population of *T. viride* (Bora 2017).

9.5.2.1.4 Quantitative and Qualitative Evaluation of Produced Products

For evaluation of the product, 12–15 days old, bioformulation packets are selected randomly and re-isolation is done on the PDA media by keeping them on incubator at 28±1°C for 72 h. The colonies developed are examined and counted on the colony counter and cfu/g of bioformulation are calculated taking following equations:

$$\text{Viable cell/g (cfu/g)} = \text{Mean plate count} \times \text{dilution factor Aliquot taken}$$

One must ensure the colonies are 2×10^6 cfu/gm formulation (as per CIB guidelines) before it is released in the market given to the users. If the count is less than the required population count, then re-inoculation has to be done repeatedly (Bora 2017).

9.5.2.1.5 Final Packing and Storage

After the bioformulation examination, the bioagents are ready for final packaging. They must be packed in white-colored LDPE bags of which thickness must not be less than 0.062 mm. The packets are then sealed to avoid any kind of leakage. They are labeled with certain information such as manufacturer's name, license number, batch number, manufacturing date and expiry date before storage. They are then kept in the storage room, avoiding direct sunlight and chemicals (Bora 2017).

9.5.2.2 Production Techniques of *Pseudomonas*-Based Biopesticides of 1% WP

The production process involved in the preparation of *Pseudomonas*-based biopesticides is similar to that of *T. viride*, such as conditioning of carrier material, blending of bioagents with carrier material and adjuvants, quantitative and qualitative evaluation of the formulated product and final packing/storage. The only difference is in the preparation of inoculants of *P.* fluorescens, which involves using King's B media for growth and multiplication instead of PD broth or Nutrient broth.

9.5.3 Commercially Available Bioproducts for Pest Management

The use of indiscriminate synthetic pesticides at present has created huge problems, not only to the soil health and environment but also has resulted in the development of resistance problems from insects and pathogens. For this reason, the use of biopesticides has been implemented in an IPM concept to overcome all the challenges and issues created by chemical pesticides. Because of their practical, economical and eco-friendly approach, they have become a powerful tool to manage pests and diseases and for the production of sustainable agriculture. They have now proven to be an alternative to most problematic synthetic pesticides without creating any harmful effects on the surrounding environment and public health. Research is being carried out today on identifying microbial pathogens against insects and pests. Many strains of beneficial microbes are now commercially available in the market under various trade names (Tables 9.2 and 9.3). On average, microbial pesticides alone

TABLE 9.2
List of Commercially Available Microbial Pesticides That Are Effective against Insect Pests

Microbial Agents	Commercial Name	Target Pests	References
Bt kurstaki	Dipel, Halt, Biobit, Thuricide	Tea looper caterpillar, Bunch caterpillar, *Andraca bipunctata*, *Buzura suppressaria*, etc.	Nair et al. 2011; Vanlaldiki et al. 2013
Bt tenebrionis	Trident, Novovdor	Colorado potato beetle, alder leaf beetle, small hive beetle	Buchholz et al. 2006; Eski et al. 2017
Bt aizawai	Agree, Certan	Nettle caterpillar	Plata-Rueda et al. 2020
Bt galleriae	Spicturin	Lepidopteran pest	
Bt sandiego	Foil, Novodor, Bt New leaf	Colorado potato beetle	Kalushkov and Batchvarova 2014
Bt israelensis	Vectobac, Arobe, Skeetal	Dipteran pests such as black flies and mosquitoes	Molloy 1990; Pistrang and Burger 2012
Bacillus popillae	Doom	Japanese beetle, European chaffer	Rippere et al. 1998
Beauveria basssiana	Boverin, Conidia, Naturalis	Pod borer, tobacco caterpillar, rice hispa, tea looper, white grub, whitefly, leafhopper	Chavan et al. 2008; Karthikeyan and Selvanarayanan 2011; Sahayaraj and Namachivayam 2011
B. brongniartii	Melocont, Betel	*Holotrichia serrata*	
Verticillum lecanii	Mycotal, Vetalec	Aphids, coffee green scale, citrus green scale, mealy bugs	Balfour and Khan 2012; Chavan et al. 2008; Kulkarni and Patil 2013
Metarrhizium anisopliae	Bio-Blast, BIO 1020	*Orycetes rhinoceros*, *Holotrichia* sp., *Dysdercus cingulate*, *Oxycarenus hyalinipennis*, termites, *Tribolium castaneum*	Batta and Safieh 2005; Sahayaraj and Borgio 2010
Nomuraea rileyi	-	*Spodoptera litura*, *H. armigera* and *Thysonoplusia orichalcea*	Bhargavi et al. 2018; Devi et al. 2003
Granulosis virus (GV)	Madmax, Carpo	Sugarcane shoot borer, *Chilo infuscatellus*, *cydia pomonella*, *Pieris rapae* and Cassava horne worm	Pavan et al. 1983; Ramanujam et al. 2014
Nuclear Polyhedrosis Virus (NVP)	Ha-NPV, SI-NPV, Spodopterian	*Spodoptera* spp. and *Heliothus* spp.	Rosell et al. 2008
Heterorhabditis sp.	Nema top, Nema green, Grubstake	*Corcyra cephalonica*, *Galleria mellonella*	Dhirta et al. 2019
Steinerneme sp.	X-Gnat, Ecomast, Green commandos, Megnet	Squash vine borer, Larvae of soil dwelling insect, *Tribolium confusum*, *Ephesia kuehniella*	Athanassiou et al. 2008; Canhilal and Carner, 2006

TABLE 9.3
List of Commercially Available Microbial Pesticides That Are Effective against Plant Pathogens

Microbial Agents	Commercial Name	Target Pathogen	References
Agrobacterium radiobacter	Galtrol, Nagol	*Agrobacterium tumefaciens*	Moore and Warren 1979; Vicedo et al. 1993
P. fluorescence	Frostban	Bunch rot, fire blight pathogen, gray mold	Cabrefiga et al. 2007; Mikani et al. 2008
Bascillus subtillus	Kodiac companion, Biopro	*Aspergillus* sp., *Rhizoctonia* sp., *Erwinia amylovora, Phytophthora cactorum*	Broggini et al. 2005; Utkhede et al. 2001
Streptomycine griseoviridis	Mycostop	Soilborne pathogens	
Trichoderma harzianum	Trichodex	*Botrytis cinerea*	Batta 2004
Trichoderma sp.	Plant shield, Root shield	*P. cactorum, Armillaria mellea*	Elkins et al. 1998; Smith et al. 1990
Gliocladium virens	GL-21	Nematodes	Ashraf and Khan 2007
G. catenulatum strain JI446	Prima stop soil guard	Soilborne pathogens	Fravel et al., 1998; McQuilken et al. 2001
Ampelomyces quisquallis isolate M-10	AQ10	Powdery mildew	Kiss 2003

occupy around 1.3% of the world's total pesticide market of which, 90% of them are used as insecticides (Hajek and St. Leger 1994). They constitute the largest group of broad-spectrum biopesticides, covering about 1500 naturally occurring insect-specific microorganisms (Table 9.2). Over 200 microbial biopesticides are available worldwide, out of which 53 numbers were registered in the United States, but the products registered for use in Asia are variable (Thakore 2006). The total production of biopesticides is over 3000 tons/yr worldwide, sharing about 1.4% to 2.5% of the US$28 billion global pesticide market (Gupta and Dikshit 2010). Approximately, 90% of the world biopesticide market is dominated by *Bacillus thuringiensis*, followed by neem-based products and formulations (Glare and Callaghan 2000; Kandpal 2014; Yang and Wang 1998).

9.6 ISSUES AND CHALLENGES OF BIOPESTICIDE USAGE

Even though biopesticides comprise a small share (5–6%) of the total global crop protection market, their value lies between $3 and $4 billion (Dunham and Trimmer 2018). Some of the important issues and challenges of biopesticide production and usages are as follows:

- The categorization, stringency of biopesticides and regulations differ widely, with countries such like United States Environmental Protection Agency (US-EPA) recognizing plant-incorporated protectants (PIPs) such as *Bt* genes as biopesticides. However, the European Union does not recognize PIPs as biopesticides and uses the term biological control agent. Such type of divergences make the regulation process difficult for production companies (Balog et al. 2017)

- Due to the prolonged process in registering bioactive substances in many countries, manufacturing gets delayed even though the substances have been isolated and formulated. However, there are some exceptions with countries such as the United States, which follows a similar model to register biopesticide active substances as chemical pesticides
- One of the important factors in the lengthy process for the registration of biopesticides is the risk assessment factor, which needs to be determined
- The high cost involvement in registering new agents is another important factor that hinders the registration of biopesticides (Damalas and Koutrobas 2018)
- For proper identification and formulation of biopesticides, there needs to be an innovative chain starting from research to developer and finally to delivery. For all of these purposes, skilled human resources, sufficient facilities to carry out the work and proper infrastructure is needed for the startup and to ensure that the developed products have potential for commercialization (Glare et al. 2012)
- The main drawback associated with the use of biopesticides is the requirement of intensified management practices, which is slower and very less efficient in comparison to the process for chemical pesticides (Ash 2009; Glare et al. 2016)
- Several viruses have been reported that can infect insects belonging to the family Baculoviridae and can control them. But due to their slow killing action and different technical problems, their uses are, however, limited
- Botanical pesticides, even though having potentialities in controlling pests through various mechanisms such as anti-feeding, anti-juvenile, hormonal activities, hatching or oviposition deterrent, repellent and growth disruption their uses are limited due to the presence of chemical molecules in the crude extract of plants (Table 9.4). In addition, botanical pesticides are quickly degradable and degrade within few hours or days. This rapid degradation makes them more expensive in production and application. Due to various difficulties such as large-scale production, commercialization, standardization of chemical compounds, approval from regulatory agencies, slow action and cost, only a handful of botanical pesticides have been commercialized (Khater 2012)
- Bioherbicides also play a vital role in overcoming the challenge of chemical pesticides. However, there are many attributes which make them limited in availability, such as less host specificity and lack of persistence. Therefore, incorporation of advance techniques is needed to upgrade and improve some of their important characteristics, such as extending host ranges and improving formulation and field persistence to make it a significant bioherbicide (Radhakrishnan et al. 2018)
- The most important myco-herbicide agents are the fungal species, followed by bacteria and viruses. Even though many species of fungi have

TABLE 9.4
Lists of Botanical Pesticides That Are Effective against Insect Pests

Plants	Products	Target Pests	References
Azadirachta indica	Azadirachtin	Desert locust, borers of various crops, sucking pests, flies and beetles	Ahmad et al. 2015; Chandramohan et al. 2016; Senthil Nathan et al. 2009
Nicotiana tabaccum	Nicotine	Aphids, leafhoppers, bugs, mites, thrips	Oguh et al. 2019; Shivkumara et al. 2019
Chrysanthemum cinerariaefolium	Pyrethrum/ Pyrethrins	Aerosol bombs for mosquitos	Shivkumara et al. 2019; Oguh et al. 2019
Derris elliptica	Rotenone	Leaf-eating caterpillars, beetles and fish poisoning	Oguh et al. 2019; Shivkumara et al. 2019
Ryania speciosa	Ryania	Caterpillars (European corn borer, Corn earworm, and others), Codling moth and thrips	Oguh et al. 2019; Shivkumara et al. 2019
Schoenocaulon officinale	Sabadilla	Squash bugs and Citrus thrips	Oguh et al. 2019; Shivkumara et al. 2019

proved their pathogenicity against various weeds, very few have been commercialized

- One of the important issues of the bioherbicides is its host specificity and host range. For identification and development one must be very specific in choosing the species and should have a proper knowledge in relationship between host phylogeny and specificity of pathogen
- Host range is a complex issue in the development of a bioherbicide as most of these products are host specific. The selection of microbial species has to be done effectively in a host range test because some of the microbes create difficulties as they colonize in the host, creating risk. Some also colonize latently, which interferes with the viability of the commercial product (Casella et al. 2010)
- The viability of bioagents are important aspects as far as effectiveness of biopesticides are concerned. In many cases it has been found that bioagents have much less ability to compete with other microorganism for nutrients and space and that reduces the antagonistic activities. Also, the toxic substance produced by the plant leachate possesses negative impacts on effectiveness of bioagents
- Commercial development of biopesticides depends upon feasibility of large-scale production of bioagents which are pathogenic and have genetically stable propagules (Boyette and Hoagland 2015). Therefore, more technologies that are specific and policy-driven, are needed for making biopesticides more economical and popular among farmers. Also, in order to give biopesticides more potential as commercial products, additional research must be carried out that considers the importance of environmental and biological factors on the pathogenesis of biopesticides (Aneja et al. 2017)

- Most of the virus-based insecticides are very effective, highly specific and virulent. They are also non-toxic to non-humans and possess significant potentialities of biopesticides (Barrera et al. 2011) but their efficiency depends upon various factors such as prevailing climatic conditions, humidity, temperature, formulation, method and time of spray, etc., which mostly influence the pace of kill (Cisneros et al. 2002)
- Entomopathogenic nematodes (EPNs) also serve as an alternative to chemical pesticides and are widely used in horticultural production for management of soilborne pests and diseases and are found to be eco-friendly and safe to humans. However, the problem lies with the application method and timing. For example, for management of fall army worm they need to be applied during early morning or at late night when they are active. During this period nematodes are also less exposed to UV light which is harmful to nematodes (Shapiro-Ilan et al. 2006)
- One of the important factors in biopesticide usage is the profitability to the users and the need to expose uninformed farmers of the importance and benefits. The significance of biopesticides should be realized and should be used as sustainable solutions to overcome or mitigate the drawbacks of synthetic pesticides
- A major challenging task is the development of biopesticides that will perform better than the chemical pesticides, having a longer shelf life, showing broad-spectrum activity, and also its stability toward environmental variables should be well considered (de la Cruz et al. 2019)

9.7 CONCLUSION AND FUTURE PROSPECTS

New research achievements in the field of microbial pesticides can be achieved with effective, safe and eco-friendly integrated pest management strategies. However, in order to utilize the biopesticides effectively and efficiently, several gaps in knowledge need to be identified and addressed properly for their sustainable use. The effect of bioagent application on non-target pests and organisms needs to be carried out properly in order to determine the risk potential. Formations of simpler guidelines for biopesticide production and marketing are indeed the need of the hour so that small and medium scale enterprises can boost their production. The most important drawback in biopesticide use is its marketing, which needs time for production as well as for selling. An understanding of various mechanisms of bioagents is needed in order to determine the pest resistance to synthetic insecticides. Further studies on the molecular level in order to determine biochemical and cellular defense needs is to be investigated thoroughly. In order to achieve and included potential bioagents in future pest management programs and to achieve sustainable production of agriculture, the involvement of all the experts are needed.

With more advancement in technologies, it has now been become easier to identify the bioagents and to determine the potentialities to combat the challenges of pests and pathogens. The only drawback in biopesticide use is its narrow range, slow working ability and dependency on environmental factors which makes it less effective than chemical pesticides. Therefore, to overcome these obstacles more research in identifying the bioagents, and formulation of new products needs to be done. In addition, the dissemination of knowledge and commercialization of the final product is of great concern, as it is less preferred by the user due to its slow effectiveness.

REFERENCES

Agrow. 2013. Biopesticides. http://www.agrow.com

Ahmad, S., Ansari, M.S., Muslim, M. 2015. Toxic effects of neem based insecticides on the fitness of *Helicoverpa armigera* (Hubner). *Crop Protection* 68:72–78.

Ananthi, M. 2019. Review on generation types of botanical pesticides and its mode of action. *International Journal of Chemistry Studies* 3:43–48.

Aneja, K.R., Khan, S.A., Aneja, A. 2017. Bioherbicides: Strategies, challenges and prospects. In *Developments in Fungal Biology and Applied Mycology*. Springer: Singapore, pp. 449–470.

Ash, G.J. 2009. The science, art and business of successful bioherbicides. *Biological Control* 52: 230–240.

Ashraf, M.S. and Khan, T.A. 2007. Efficacy of *Gliocladium virens* and *Talaromyces flavus* with and without organic amendments against *Meloidogyne javanica* infecting egg-plant. *Asian Journal of Plant Pathology* 1:18–21.

Athanassiou, C. G., Palyvos, N. E., Duarte, T. K. 2008. Insecticidal effect of *Steinernema feltiae* (Filipjev) against *Tribolium confusum* du Val and *Ephestia kueniella* (Zeller) in stored wheat. *Journal of Stored Products Research* 44:52–57.

Attia, S., Grissa, K.L., Lognay, G., Bitume, E., Hance, T., Mailleux, A.C. 2013. A review of the major biological approaches to control the worldwide pest *Tetranychus urticae* (Acari: Tetranychidae) with special reference to natural pesticides. *Journal of Pest Science* 86:361–386.

Aw, K.M.S., Hue, S.M. (2017). Mode of infection of *Metarhizium* spp. fungus and their potential as biological control agents. *Journal of Fungi* 3:30.

Bailey, A., Chandler, D., Grant, W.P., Greaves, J., Prince, G., Tatchell, M. 2010. Pest management with biopesticides. In: Bailey, A., Chandler, D., Grant, W.P., Greaves, J., Prince, G., Tatchell, M. (Eds.) *Biopesticides: Pest Management and Regulation*. CABI: Wallingford, UK, pp. 71–130.

Balfour, A., Khan A. 2012. Effects of *Verticillium lecanii* (Zimm.) Viegas on *Toxoptera citricida* Kirkaldy (Homoptera: Aphididae) and its parasitoid *Lysiphlebus testaceipes* Cresson (Hymenoptera: Braconidae). *Plant Protection Science* 48:123–130.

Balog, A., Hartel, T., Loxdale, H.D., Wilson, K. 2017. Differences in the progress of the biopesticide revolution between the EU and other major crop-growing regions. *Pest Management Science* 73:2203–2208.

Barrera, G., Simón, O., Villamizar, L., Williams, T., Caballero, P. 2011. *Spodoptera frugiperda* multiple nucleopolyhedrovirus as a potential biological insecticide: Genetic and phenotypic comparison of field isolates from Colombia. *Biological Control* 58:113–120.

Batta, Y.A. 2004. Effect of treatment with *Trichoderma harzianum* Rifai formulated in invert emulsion on postharvest decay of apple blue mold. *International Journal of Food Microbiology* 96:281–288.

Batta, Y.A., Safieh, D.I.A. 2005. A study of treatment effect with *Metarhizium anisopliae* and four types of dusts on wheat grain infestation with red flour beetles (*Tribolium casteneum* Herbs). *Journal of Islamic University Gaza* 13:11–22.

BCC Research. 2012. Global markets for biopesticides. http://www.bccresearch.com/pressroom/report/code/CHM029D>

Bhardwaj, T., Sharma, J.P. 2013. Impact of pesticides application in agricultural industry: An Indian scenario *International Journal of Agriculture and Food Science Technology* 4: 817–822.

Bhargavi, G.B., Manjula, K., Rao, A.R., Reddy, B.R. 2018. Efficacy of oil based formulations of *Nomuraea rileyi* (Farlow) Samson against *Spodoptera litura* in vitro. *International Journal of Current Microbiology and Applied Sciences* 7:3413–3422.

Bora, L.C. 2017. Major steps in production process of microbial bioagents based bio-pesticides. *Compendium of Lectures, ICAR sponsored training on Cultural Techniques for Bio Fertilizers and Bio-Pesticides*, DBT-AAU Centre, Assam Agricultural University, Jorhat, 138–142.

Boyette, C.D., Hoagland, R.E. 2015. Bioherbicidal potential of *Xanthomonas campestris* for controlling *Conyza canadensis*. *Biocontrol Science and Technology* 25:229–237.

Brar, S.K., Verma, M., Tyagi, R.D., Valero, J.R. 2006. Recent advances in downstream processing and formulations of *Bacillus thuringiensis* based biopesticides. *Process Biochemistry* 41:323–342.

Bravo, A., Gill, S.S., Soberón, M. 2005. Bacillus thuringiensis mechanisms and use. *Comprehensive Molecular Insect Science* 7: 175–205

Broggini, G.A.L., Duffy, B., Holliger, E., Scharer, H.J., Gessler, C., Patocchi, A. 2005. Detection of the fire blight biocontrol agent *Bacillus subtilis* BD170 (Biopro®) in a Swiss apple orchard. *European Journal of Plant Pathology* 111:93–100.

Buchholz, S., Neumann, P., Merkel, K., Hepburn, H.R. 2006. Evaluation of *Bacillus thuringiensis* Berliner as an alternative control of small hive beetles, *Aethina tumida* Murray (Coleoptera: Nitidulidae). *Journal of Pest Science* 79:251.

Cabrefiga J, Bonaterra A, Montesinos E. 2007. Mechanisms of antagonism of *Pseudomonas fluorescens* EPS62e against *Erwinia amylovor*a, the causal agent of fire blight. *International Journal of Microbiology* 10:123–132.

Canhilal, R., Carner, G.R. 2006. Efficacy of entomopathogenic nematodes (Rhabditida: Steinernematidae and Heterorhabditidae) against the squash vine borer, *Melittia cucurbitae* (Lepidoptera: Sesiidae) in South Carolina. *Journal of Agricultural and Urban Entomology* 3:27–39.

Casella, F., Charudattan, R., Vurro, M. 2010. Effectiveness and technological feasibility of bioherbicide candidates for biocontrol of green foxtail (*Setaria viridis*). *Biocontrol Science and Technology* 20:1027–1045.

Chandler, D., Bailey, A., Tatchell, G.M., Davidson, G., Greaves, J., Grant, W.P. 2011. The development, regulation and use of biopesticides for integrated pest management. *Philosophical Transaction of the Royal Society Bulletin* 386:2–13.

Chandramohan, B., Murugan, K., Madhiyazhagan, P., Kovendan, K., Kumar, P.M., Panneerselvam, C. 2016. Neem by products in the fight against mosquito – borne diseases: Biotoxicity of neem cake fractions towards the rural malaria vector *Anopheles culicifacies* (Diptera: Culicidae). *Asian Pacific Journal of Tropical Biomedicine* 6:472–476.

Chaube H.S., Mishra, D.S., Varshney, S., Singh, U.S. 2003. Biocontrol of plant pathogens by fungal antagonists: A historical background, present status and future prospects. *Annual Review of Plant Pathology* 2:1–42.

Chavan, B.P., Kadamand, J.R., Saindane, Y.S. 2008. Bioefficacy of liquid formulation of *Verticillium lecanii* against aphid (*Aphis gossypii*). *International Journal of Plant Protection* 1:69–72.

Chen, K.N., Chen, C.Y., Lyn, Y.C., Chen, M.J. 2013. Formulation of a novel antagonistic bacterium based biopesticide for fungal disease control using microencapsulation techniques. *Journal of Agricultural Science* 5:153–163.

Cisneros, J., Pérez, J.A., Penagos, D.I., Ruiz V.J., Goulson, D., Caballero, P., Williams, T. 2002. Formulation of a Nucleopolyhedrovirus with boric acid for control of S*podoptera frugiperda* (Lepidoptera: Noctuidae) in maize. *Biological Control* 23:87–95.

Damalas, C.A., Koutroubas, S.D. 2018. Current status and recent developments in biopesticide use. *Agriculture* 8:13–18.

de la Cruz, R., Cruz Maldonado, J.J., Rostro Alanis Md, J., Torres, J.A., Parra Saldívar, R. 2019. Fungi-based biopesticides: Shelf-life preservation technologies used in commercial products. *Journal of Pest Science* 92:1003–1015.

de Souza, J.T., Arnould, C., Deulvot, C., Lemanceau, P., Gianinazzi-Pearson, V., Raaijmakers, J.M. 2003. Effect of 2,4-diacetylphloroglucinol on pythium: Cellular responses and variation in sensitivity among propagules and species. *Phytopathology* 93:966–975.

Deshmukh, S., Hueckelhoven, R., Schaefer, P., Imani, J., Sharma, M. 2006. The root endophytic fungus *Piriformospora indica* requires host cell death for proliferation during mutualistic symbiosis with barley. *Proceedings of National Academic Sciences USA* 103:18450–18457.

Devi, P.S., Prasad, Y.G., Chowdary, D.A., Rao, L.M., Balakrishnan, K. 2003. Identification of virulent isolates of the entomopathogenic fungus *Nomuraea rileyi* (F) Samson for the management of *Helicoverpa armigera* and *Spodoptera litura* (identification of virulent isolates of *N. rileyi*). *Mycopathologia* 156:365–373.

Dhirta, B., Khanna, A.S., Sharma, P.L., Singh, M. 2019. Bioassay of *Heterorhabditis bacteriophora* against various insects. *Journal of Entomology and Zoology Studies* 7:318–323.

Druege, U., Baltruschat, H., Franken, P. 2007. *Piriformospora indica* promotes adventitious root formation in cuttings. *Scientia Horticulturae* 112:422–426.

Dunham, W., Trimmer, M. 2018. Biological products around the world. Bioproducts Industry Alliance Spring Meeting & International Symposium (www.bpia.org). Accessed on 20 August 2019.

Elkins, R.B., Rizzo, D.M., Whiting, E.C. 1998. Biology and management of Armillaria root disease in pear in California. *Acta Horticulture* 475:453–458.

El-Sayed, W. 2005. Biological control of weeds with pathogens: Current status and future trends. *Journal of Plant Disease and Protection* 112:209–221.

El-Wakeil, N.E. 2013. Botanical pesticides and their mode of action. *Gesunde Pflanzen* 65:125–149.

Eski, A., Demir, İ., Sezen, K. 2017. A new biopesticide from a local *Bacillus thuringiensis* var. *tenebrionis* (Xd3) against alder leaf beetle (Coleoptera: Chrysomelidae). *World Journal of Microbiology and Biotechnology* 33:95.

Forget, G.1993. Balancing the need for pesticides with the risk to human health. In Forget, G., Goodman, T., de Villiers, A. (Eds.), *Impact of Pesticide Use on Health in Developing countries*. IDRC: Ottawa, p. 2.

Fravel, D.R., Connick, W.J., Lewis Jr, J.A. 1998. Formulation of microorganisms to control plant diseases. In Burgess, H.D. (Ed.), *Formulation of Microbial Biopesticides*. Kluwer Academic Publishers: Dordrecht, The Netherlands, pp.186–202.

Fuente, L.D.L., Mavrodi, O., BAjsa, N., Mavrodi, D. 2008. Antibiotics produced by fluorescent *Pseudomonas*. In Sorvari, S., Pirtilla, A.M. (Eds.), *Prospects and Applications for Plant-Associated Microbes*. BioBien Innovations: Finland, pp. 249–255.

Gangwar, O.P., Sharma, P. 2013. A growth and survival of *Trichoderma harzianum* and *Pseudomonas fluorescens* on different substrates and their temporal and spatial population dynamics in irrigated rice ecosystem. *Annals of Plant Protection Sciences* 21:104–108.

Gilardi, G., Manker, D. C., Garibaldi, A., Gullino, M.L. 2008. Efficacy of the biocontrol agents *Bacillus subtilis* and *Ampelomyces quisqualis* applied in combination with fungicides against powdery mildew of zucchini. *Journal of Plant Diseases and Protection* 115:208–213.

Glare, T.R., Callaghan, M. 2000. *Bacillus thuringiensis: Biology, Ecology and Safety*. J. Wiley & Sons: Chichester, pp. 350.

Glare, T.R., Caradus, J., Gelernter, W.D., Jackson, T.A., Keyhani, N.O., Köhl, J., Marrone, P.G., Morin, L., Stewart, A. 2012. Have biopesticides come of age? *Trends in Biotechnology* 30:250–258.

Glare, T.R., Gwynn, R.L., Moran-Diez, M.E. 2016. Development of biopesticides and future opportunities. In Glare, T., Moran-Diez, M. (Ed.), *Microbial-Based Biopesticides. Methods in Molecular Biology*. Humana Press: New York, Vol 1477, pp. 211–221.

Goettel, M.S., Koike, M., Kim, J.J., Brodeur, J. 2008. Potential of *Lecanicillium* spp. for management of insects, nematodes and plant diseases *Journal of Invertebrate Pathology* 98:256–261

Gu, Q., Yang, Y., Yuan, Q., Shi, G., Wu, L., Lou, Z., Huo, R., Wu, H., Borriss, R., Gao, X. 2017. Bacillomycin D Produced by *Bacillus amyloliquefaciens* is involved in the antagonistic interaction with the plant-pathogenic fungus *Fusarium graminearum*. *Applied and Environmental Microbiology* 83:1075–1017.

Gupta, S., Dikshit, A. 2010. Biopesticides: An eco-friendly approach for pest control. *Journal of Biopesticides* 3:186–188.

Hajek, A.E., St. Leger, R.J. 1994. Interactions between fungal pathogens and insect hosts. *Annual Review of Entomology* 39:293–322.

Harman, G.E. 2000. Myths and dogmas of Bio control: Changes in the perceptions derived from research on *Trichoderma harzianum* T-22. *Plant Disease* 84:377–393.

Harman, G.E. 2011. Trichoderma-not just for biocontrol anymore. *Phytoparasitica* 39:103–108

Hashimi, M.H., Hashimi, R., Ryan, Q. 2020. Toxic effects of pesticides on humans, plants, animals, pollinators and beneficial organisms. *Asian Plant Research Journal* 5:37–47.

Howell, C.R., Beier, R.C., Stipanovic, R.D. 1988. Production of ammonia by *Enterobacter cloacae* and its possible role in the biological control of Pythium pre-emergence damping-off by the bacterium. *Phytopathology* 78:1075–1078.

Igbedioh, S.O. 1991. Effect of agricultural pesticides on human, animal and higher plants in developing countries. *Archives of Environmental and Occupational Health* 46:218–224.

Isman, M.B, Machial, C.M. 2006. Pesticides based on plant essential oils: From traditional practice to commercialization. *Advances in Phytomedicine* 3:29–44.

Isman, M.B. 2006. Botanical insecticides, deterrents, and repellents in modern agriculture and an increasingly regulated world. *Annual Review of Entomology* 51:45–66.

Jan, S., Singh, R., Bhardwaj, R., Ahmad, P., Kapoor, D. 2020. Plant growth regulators: A sustainable approach to combat pesticide toxicity *Biotechnology* 10:466.

Jasim, W.A., Mohammed, A.A. 2019. Efficacy of entomopathogenic fungi *Verticillium lecanii* and *Isaria fumosorosea* against *Myzus persicae* under laboratory conditions *Plant Archives* 19:1416–1419.

Jones, R., Hancock J.G. 1988. Mechanism of gliotoxin action and factors mediating gliotoxin sensitivity. *Microbiology* 134:2067–2075.

Jones, R.S. 2010. Biochemical pesticides: Green chemistry designs by nature. In Anastas (Ed.) *Handbook of Green Chemistry*. Wiley-VCH Verlag GmbH & Co. Online pp. 329–347. https://doi.org/10.1002/9783527628698.hgc106

Junaid, J.M., Dar, N.A., Bhat, T.A., Bhat, A.H., Bhat, M.A. 2013. Commercial biocontrol agents and their mechanism of action in the management of plant pathogen. *International Journal of Modern Plant & Animal Sciences* 1:39–57.

Kalushkov, P., Batchvarova R. 2014. Effectiveness of Bt Newleaf® Potato to control *Leptinotarsa decemlineata* (Say) (Coleoptera: Chrysomelidae) in Bulgaria. *Biotechnology and Biotechnological Equipment* 19: 28–34.

Kandpal, V. 2014. Biopesticides. *International Journal of Environmental Research* 4:191–196.

Karthikeyan, A., Selvanarayanan, V. 2011. In vitro efficacy of *Beauveria bassiana* (Bals.) Vuill. and *Verticillium lecanii* (Zimm.) viegas against selected insect pests of cotton. *Recent Research in Science and Technology* 3:142–143.

Keswani, C., Singh, S.P., Singh, H.B. 2013. *Beauveria bassiana*: Status, mode of action, applications and safety issues. *Biotech Today* 3:16–20.

Khater, H.F. 2012. Prospects of botanical biopesticides in insect pest management. *Pharmacologia* 3:641–656.

Kiss, L. 2003. A review of fungal antagonists of powdery mildews and their potential as bio agents. *Pest Management Science* 59: 475–483.

Kiss, L., Russell, J.C., Szentiva, O., Nyi, Xu, X., Jeffries, P. 2004. Biology and biocontrol potential of *Ampelomyces mycoparasites*, natural antagonists of powdery mildew fungi. *Biocontrol Science and Technology* 14:635–651.

Knowles, A. 2001. *Trends in Pesticide Formulations*. Agrow Reports, UK, PJB Publications Limited, pp.89–92.

Knowles, A. 2005. *New Developments in Crop Protection Product Formulation*. Agrow Reports UK, T&F Informa UK Limited, pp.153–156.

Knowles, A. 2006. *Adjuvants and additives*. Agrow Reports, T&F Informa UK Limited, pp. 126–129.

Kulkarni, S.R., Patil S.K. 2013. Efficacy of different biopesticides and insecticides against mealy bugs on custard apple. *Pest Management in Horticultural Ecosystem* 19:113–115.

Lee S.Y., Kim Y.K., Kim H.G., Shin H.D. 2007. New hosts of *Ampelomyces quisqualis* hyper-parasite to powdery mildew in Korea. *Research in Plant Disease* 13:127–207.

Leng, P., Zhang, Pan, G., Zhao, M. 2011. Applications and development trends in biopesticides. *African Journal of Biotechnology* 10:19864–19873.

Lyn, M.E., Burnett, D., Garcia, A.R., Gray, R. 2010. Interaction of water with three granular biopesticide formulations. *Journal of Agricultural and Food Chemistry* 58:1804–1814.

Mardanova, A.M., Hadieva, G.F., Lutfullin, M.T., Khilyas, I.V., Minnullina, L.F., Gilyazeva, A.G., Bogomolnaya, L.M., Sharipova, M.R. 2017. *Bacillus subtilis* strains with antifungal activity against the phytopathogenic fungi. *Agricultural Sciences* 8:1–20.

Martinez, C., Blanc, F., Le, C.E., Besnard, O., Nicole, M., Baccou, J.C. 2001. Salicylic acid and ethylene pathways are differentially activated in melon cotyledons by active or heat-denatured cellulase from *Trichoderma longibrachiatum*. *Plant Physiology* 127: 334–344.

Mathur, S.C. 1999. Future of Indian pesticides industry in next millennium. *Pesticide Information* 24:9–23.

McQuilken, M.P., Gemmell, J., Ahdenperae, M-L.L. 2001. *Gliocladium catenulatum* as a potential biological control agent of damping-off in bedding plants. *Journal of Phytopathology* 149:171–178.

Mikani, A., Etebarian, H.R., Sholberg, P.L., Gorman, D.T., Stokes, S., Alizadeh, A. 2008. Biological control of apple gray mold caused by *Botrytis mali* with *Pseudomonas fluorescens* strains. *Postharvest Biology and Technology* 48:107–112.

Molloy, D.P. 1990. Progress in the biological control of black flies with *Bacillus thuringiensis israelensis*, with emphasis on temperate climates. In de Barjac H., Sutherland D.J. (Eds), *Bacterial Control of Mosquitoes & Black Flies*. Springer, Dordrecht.

Moore, L.W., Warren, G. 1979. *Agrobacterium radiobacter* strain 84 and biological control of crown gall. *Annual Review of Phytopathology* 17:163–179.

Moyne, A.L., Shelby, R., Cleveland, T.E., Tuzun, S. 2001. Bacillomycin D: An iturin with antifungal activity against *Aspergillus flavus*. *Journal of Applied Microbiology* 90:622–629.

Murphy, P.J., Tate, M.E., Kerr, A. 1981. Substituents at N6 and C-5' control selective uptake and toxicity of the adenine-nucleotide bacteriocin, agrocin 84, in agrobacteria. *European Journal of Biochemistry* 115:539–543.

Nair, N., Tudu, B., Debnath, M.R., Dey, P.K., Somchoudhury, A.K. 2011. Efficacy of *Bacillus thuringiensis* var. Kurstaki and NPV against tea looper, *Hyposidra infixaria* Walk. (lepidoptera: Geometridae). *Journal of Entomological Research* 35:15–18.

Oguh, C.E., Okpaka, C.O., Ubani, C.S., Okekeaji U., Joseph P.S., Amadi, E.U. 2019. Natural pesticides (biopesticides) and uses in pest management—A critical review. *Asian Journal of Biotechnology and Genetic Engineering* 2:1–18.

Pal, K.K., Gardener, B.M. 2006. *Biological Control of Plant Pathogens. The Plant Health Instructor*, 1–25. doi : 10.1094/PHI-A-2006-1117-02.

Pandey, R.N., Jaisani, P., Yadav, D.L. 2021. *Trichoderma* spp. in the management of stresses in plants and rural prosperity. *Indian Phytopathology* 74:453–467.

Pandey, S., Pundhir, V.S. 2013. Mycoparasitism of potato black scurf pathogen (*Rhizoctonia solani* Kuhn) by biological control agents to sustain production. *Indian Journal of Horticulture* 70:71–75.

Pandey, S.K., Chandel S.C.R. 2014. Efficacy of *Pseudomonas* as biocontrol agent against plant pathogenic fungi. *International Journal Current Microbiology and Applied Sciences* 3:493–500.

Patil, S.B., Udikeri, S.S., Vandal, N.B. 2012. Bioefficacy of *Verticillium lecanii* (1.150/0 WP) against sucking pest complex on transgenic Bt cotton. *Journal of Cotton Research Development* 26:222–226.

Pavan, O.H.O., Boucias, D.G., Almeida, L.C., Gaspar, J.O., Botelho, P.S.M., Degaspari, N. 1983. Granulosis Virus of *Diatraea saccharalis* (DsGV): Pathogenicity, replication and ultrastructure," in *Proceedings of the International Congress of the International Society of Sugar Cane Technologists (ISSCT '83)*, La Havana, Cuba, vol. 2, pp.644–659.

Pinnamaneni, R., Potineni, K. 2010. Mechanisms involved in the entomopathogenesis of *Beauveria bassiana*. *Asian Journal of Environmental Science* 5:65–74.

Pistrang, L., Burger, J. 1984. Effect of *Bacillus thuringiensis* var. *israelensis* on a genetically-defined population of black flies (Diptera: Simuliidae) and associated insects in a montane New Hampshire stream. *Canadian Entomologist* 116:975–981.

Plata-Rueda, A., Quintero, H.A., Serrão, J.E., Martínez, L.C. 2020. Insecticidal activity of *Bacillus thuringiensis* strains on the Nettle caterpillar, *Euprosterna elaeasa* (Lepidoptera: Limacodidae). *Insects* 11:310.

Prabhukarthikeyan, S.R., Thiruvengadam, R. 2016. Antifungal metabolites of *Pseudomonas fluorescens* against *Pythium aphanidermatum*. *Journal of Pure and Applied Microbiology* 10:579–584.

Pusey, P.L., Stockwell, V.O., Reardon, C.L., Smits, T.H.M., Duffy, B. 2011. Antibiosis activity of *Pantoea agglomerans* biocontrol strain E325 against *Erwinia amylovora* on apple flower stigmas. *Phytopathology* 101:1234–1241.

Radhakrishnan, R., Alqarawi, A.A., Abd_Allah, E.F. 2018. Bioherbicides: Current knowledge on weed control mechanism. *Ecotoxicology and Environmental Safety* 158:131–138.

Ramanujam, B., Rangeshwaran, R., Sivakmar, G., Mohanand, M., Syandigeri, M. 2014. Management of insect pests by microorganisms. *Proceedings of Indian National Science Academy* 80:455–471.

Ravindran, K., Sivaramakrishnan, S., Hussain, M., Dash, C.K., Bamisile, B., Qasim, M., Wang, L. 2018. Investigation and molecular docking studies of Bassianolide from *Lecanicillium lecanii* against *Plutella xylostella* (Lepidoptera: Plutellidae) *Comparative Biochemistry and Physiology* 206–207:65–72.

Regnault-Roger, C. 2012. Trends for commercialization of biocontrol agent (biopesticide) products. In Merillon, J.M., & Ramawat, K.G. (Eds), *Plant Defence: Biological Control*, Springer: Dordrecht, Heidelberg, London, New York, pp. 139–160.

Rippere, K.E., Tran, M.T., Yousten, A.A., Hilu, K.H., Klein, M.G. 1998. *Bacillus popilliae* and *Bacillus lentimorbus*, bacteria causing milky disease in Japanese beetles and related scarab larvae. *International Journal of Systematic Bacteriology* 48:395–402.

Rosell, G., Quero, C., Coll, J. 2008. Biorational insecticides in pest management. *Journal of Pesticide Science* 33:103–121.

Sahayaraj, K., Borgio, J.F. 2010. Virulence of entomopathogenic fungus metarhizium anisopliae (metsch.) Sorokin on seven insect pests *Indian Journal of Agricultural Research* 44:195–200.

Sahayaraj, K., Borgio, J.F. 2010. Virulence of entomopathogenic fungus *Metarhizium anisopliae* (METSCH.) sorokin on seven insect pests. *Indian Journal of Agricultural Research* 44:195–200.

Sahayaraj, K., Namachivayam, S.K.R. 2011. Field evaluation of three entomopathogenic fungi on groundnut pests. *Tropicultura* 29:143–147.

Sanchis, V., Bourguet, D. 2008. *Bacillus thuringiensis*: Applications in agriculture and insect resistance management. A Review. *Agronomy for Sustainable Development* 28:11–20.

Selvakumar, R. 2008. *Bioformulations for management of late light of potato in North eastern India.* Third International Late Blight Conference, Beijing.

Senthil–Nathan, S., Choi, M.Y., Seo, H.Y., Paik, C.H., Kalaivani, K. 2009. Toxicity and behavioural effect of 3β, 24, 25 – trihydroxycycloartane and beddomei lactone on the rice leaf folder *Cnaphalocrocis medinalis* (Guenee) (Lepidoptera: Pyralidae). *Ecotoxicology and Environmental Safety* 72:1156–1162.

Shahnaz, E., Razdan, V.K., Rizvi, S.E.H., Rather, T.R., Gupta, S., Andrabi, M. 2013. Integrated disease management of foliar blight disease of onion: A case study of application of confounded factorials. *Journal of Agricultural Sciences* 5:17–22.

Shanahan, P., O'Sullivan, D.J., Simpson, P., Glennon, J.D., O'Gara, F. 1992. Isolation of 2,4-Diacetylphloroglucinol from a fluorescent pseudomonad and investigation of physiological parameters influencing its production. *Applied Environmental Microbiology* 58:353–358.

Shapiro-Ilan, D.I., Gouge, D.H., Piggott, S.J., Fife, J.P. 2006. Application technology and environmental considerations for use of entomopathogenic nematodes in biological control. *Biological Control* 38:124–133.

Shivkumara, K.T., Manjesh, G.N., Roy, S., Manivel, P. 2019. Botanical insecticides; prospects and way forward in India: A Review. *Journal of Entomology and Zoology Studies* 7:206–211.

Smith, V.L., Wilcox, W.F., Harman, G.E.1990. Potential for biological control of Phytophthora root crown rots of apple by *Trichoderma* and *Gliocladium* spp. *Phytopathology* 80:880–885.

Sundheim, L. 2007. Control of cucumber powdery mildew by hyperparasite *Ampelomyces quisqualis* and fungicides. *Plant Pathology* 31:209–214

Tadros, F. (2005). *Applied Surfactants: Principles and Applications.* Wiley-VCH Verlag GmbH & Co. KGaA, Weinheim, pp. 187–256.

Teera-Arunsiri, A., Suphantharika, M., Ketunuti, U. 2003. Preparation of spray-dried wettable powder formulations of *Bacillus thuringiensis*-based biopesticides. *Journal of Economic Entomology* 96:292–299.

Thakore, Y. 2006. The biopesticide market for global agricultural use. *Industrial Biotechnology* 2:194–208.

Tiwari, A.K. 1996. *Biological control of chick pea wilt complex using different formulations of Gliocladium virens through seed treatment.* Ph.D. thesis submitted to G. B. Pant University of Agriculture and Technology, Pantnagar India, pp.167.

Tulloch, M. 1976. The genus *Metarhizium. Transactions of the British Mycological Society* 66:407–411.

Utkhede, R.S., Sholberg, P.L., Smirle, M.J. 2001. Effects of chemical and biological treatments on growth and yield of apple trees planted in *Phytophthora cactorum* infested soils. *Canadian Journal of Plant Pathology* 23:163–167.

Vanlaldiki, H., Singh, M.P., Sarkar, P.K. 2013. Efficacy of eco-friendly insecticides on the management of diamondback moth (*Plutella xylostella* Linn.) on cabbage. *The Bioscan* 8:1225–1230.

Vasanthakumari, M.M., Shivanna, M.B. 2013. Biological control of anthracnose of chilli with rhizosphere and rhizoplane fungal isolates from grasses. *Archives of Phytopathology and Plant Protection* 46:1641–1666.

Vicedo, B., Penalver, R., Asins, M.J., Lopez M.M.1993. Biological control of *Agrobacterium tumefaciens*, colonization, and pAgK84 transfer with *A. radiobacter* K84 and the tra-mutant strain K1026. *Applied and Environmental Microbiology* 59:309–315.

Wang, X.Q., Zhao, D.L., Shen, L.L., Jing, C.L., Zhang, C.S. 2018. Application and mechanisms of bacillus subtilis in biological control of plant disease. *Role of Rhizospheric Microbes in Soil,* 225–250.

Winder, R.S., Wheeler, J.J., Conder, N., Otvos, S.S., Nevill, R., Duan, L. 2005. Microcapsulation: A strategy for formulation of inoculation. *Biocontrol Science and Technology* 13:15–169.

Woo, S.L., Scala, F., Ruocco, M., Lorito, M. 2006. The molecular biology of the interaction between *Trichoderma* spp, phytopathogenic fungi and plants. *Phytopathology* 6:181–185.

Woods, T.S. 2003. Pesticide Formulations. In Plimmer, J.R., Gammon, D. W., & Ragsdale, N.R. (Eds.) *Encyclopedia of Agrochemicals*, John Wiley & Sons, New York, pp.1–11.

Yadav, S.K., Dave, A., Sarkar, A., Singh, H.B., Sarma, B.K. 2013. Coinoculated biopriming with *Trichoderma, Pseudomonas* and *Rhizobium* improves crop growth in *Cicer arietinum* and *Phaseolus vulgaris. International Journal of Agriculture Environment and Biotechnology* 6:255.

Yang, X.M., Wang, S.S. 1998. Development of *Bacillus thuringiensis* fermentation and process control from a practical perspective. *Biotechnology and Applied Biochemistry* 28:95–98.

Yüzbaşıoglu, E., Dalyan, E. 2019. Salicylic acid alleviates thiram toxicity by modulating antioxidant enzyme capacity and pesticide detoxifcation systems in the tomato (*Solanum lycopersicum* Mill.). *Plant Physiology and Biochemistry* 135:322–330.

Zhang, W., Jiang, F., Ou, J. 2011. Global pesticide consumption and pollution: With China as a focus. *Proceedings of the International Academy of Ecology and Environmental Sciences* 1:125–144.

Zhou, Y., Xia, X., Yu, G., Wang, J., Wu, J., Wang, M., Yang, Y., Shi, K., Yu, Y., Chen, Z., Gan, J. 2015. Brassinosteroids play a critical role in the regulation of pesticide metabolism in crop plants. *Science Report* 5:9018.

10 Plastic Degrading Microorganisms

Eco-Friendly Tool for Combating Plastic Pollution in Soil

Rashmi Rawat and Meena
Hemvati Nandan Bahuguna Garhwal University, Srinagar Uttarakhand, India

CONTENTS

DOI: 10.1201/9781003147091-10

10.1 INTRODUCTION

The term plastic has its origins in the Greek word "Plastikos", which indicates that the material can be given shape (Lecture 4.1: Thermoplastics and Thermosets, NPTEL). Plastics are manufactured from a vast variety of organic polymers such as polyethylene, PVC, polypropylene etc. An extensive variety of plastics can be broadly categorized into three types based on their reaction to temperature/heat treatment: (1) thermoplastics, (2) thermosets and (3) elastomers (Klein 2011). Since their discovery, plastics have acquired considerable importance in our lifestyle and living without using plastic seems almost impossible. A plethora of useful qualities such as being lightweight, flexible, moisture resistant, durable and low cost makes plastics a preferable component in numerous sectors, including infrastructure, construction, agriculture, consumer goods, packaging and telecommunication, etc. of which the packaging industry consumes the largest share of plastic produced and is a major contributor in plastic pollution (Ritchie and Roser 2018). High rates of production and thoughtless overconsumption of plastic has resulted in the accumulation of plastic waste in the environment at an alarmingly increasing rate. First, although synthetic plastic was produced in 1907, large-scale plastic production was observed in 1950s and it has been growing rapidly ever since (Geyer et al. 2017). The world has witnessed a steep increase in the plastic production rate after 1990. In 1950 global plastic production was 1.5 million metric tonnes but reached 367 million metric tonnes in 2020. This represents a roughly 244-fold increase in plastic production. In recent decades global plastic production has seen 135.93% growth, from 270 million metric tonnes in 2010 to 367 million metric tonnes in 2020 (Ritchie and Roser 2018) (Tiseo 2021). According to an estimate of all the plastics produced from 1950 to 2015, 30% are still in use and 70% plastics contributed to the plastic waste generated amounting 6300 million metric tonnes. Approximately 21% of the total waste generated was treated by incinerating (12%) and recycling (9%). A major portion of the total waste, around 60%, was discarded in landfills or natural environment (Geyer et al. 2017). While most of the plastic waste ends up in terrestrial systems, around 3% of global plastic waste enters ocean. Approximately 80% of ocean waste comes from terrestrial sources and 20% comes from marine sources. "Great Pacific Garbage Patch" (GPGP) is the most well-known example of large plastic accumulation in surface water (Ritchie and Roser 2018). The situation has become aggravated with a sharp increase in single use plastic consumption as no efficient systems are in place to safely dispose or recycle such large amounts of plastic waste. The problem has become a global menace and is intensifying day by day, with large amounts of plastic being continuously produced and consumed and a meager amount of the waste being responsibly consumed and managed.

Plastic pollution has devastating effects on aquatic and terrestrial species. About 700 marine species and more than 50 freshwater species have been reported to have ingested or become entangled in plastic debris. A variety of terrestrial organisms have also been reported to have lost their lives or been harmed due to ingestion of plastic waste (Lau. et al. 2020). The COVID-19 pandemic has been a big blow to the fight against plastic pollution, as huge amounts of waste are being generated daily in the form of discarded masks, gloves, protective gear and biomedical waste (Plastics and the Environment 2021). An estimated 1.6 million metric tonnes of plastic waste have been generated since the outbreak, and 3.4 billion single use face masks or face shields are discarded daily due to the COVID pandemic (Benson et al. 2021).

Durability and resistance to degradation makes plastics virtually impossible for nature to degenerate and assimilate. The situation desperately calls for well-designed management strategies to effectively handle the behemoth of plastic waste. Reuse, recycle, responsible use, dematerialization and conversion technologies must be considered and employed carefully to design an efficient system (Geyer et al. 2017). A variety of microorganisms have the capability to degrade different kinds of plastics to some degree. Biodegradation of plastics by microorganisms should be considered and researched seriously as it has the potential to be a viable tool in the fight against plastic pollution.

The present chapter will walk us through an introduction to plastics, the biodegradation process and the long road of the biodegradation journey that we have traveled so far.

10.2 TYPES OF PLASTICS

Plastics can be categorized into the following types.

10.2.1 Thermoplastics

The thermoplastics' structure consists of molecular chains that lack crosslinkage between them, or in other words, chains are independent of each other. In the case of thermoplastics chain assembly is done by "polyaddition reaction" (Figure 10.1). Weak interactions, such as van der Waals forces, play an important role in connecting the chains. The reason for the ability of thermoplastics to be reversibly melted and reshaped by cooling is the lack of crosslinking. These have relatively low melting points. The optical properties of thermoplastics are determined by type and structure of the material and may vary from transparent to opaque. The macromolecular chains can contain statistical oriented side chains, or statistical distributed crystalline phases can be formed. Thermoplastics can be differentiated into: (a) Amorphous type and (b) Semicrystalline type, based on the type of organization of macromolecular chains they are comprised. Some examples and uses of thermoplastics are listed in Table 10.1.

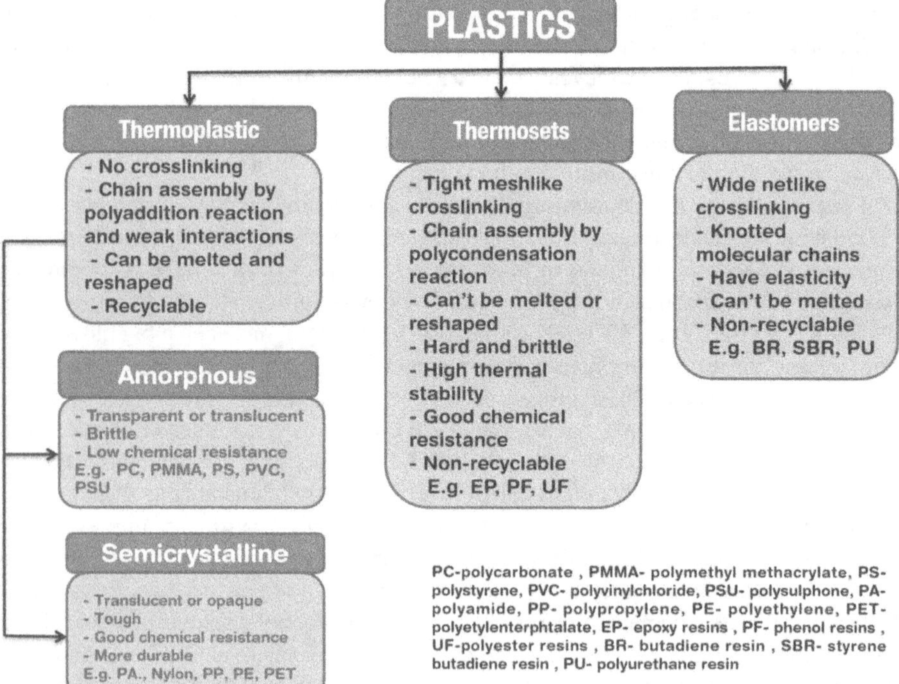

FIGURE 10.1 Types and properties of plastics.

TABLE 10.1
Examples of Thermoplastics and Their Uses

Plastic	Properties	Uses
PET (Polyethylene-terepthalate)	• Impermeable to gases, clear, hard, solvent resistant	Mineral water bottles, carbonated drink bottles, boil-in-meal pouches, polyester fibers, thermoformed sheets, carpeting, building insulation (Millet et al. 2018; Alabi et al. 2019)
HDPE (High-density polyethylene)	• Hard to semi flexible, permeable to gas, strong	Washing liquid bottles, shampoo and conditioner bottles, snack food boxes, milk bottles, non-carbonated bottles, buckets, agricultural pipes, plant pots, crates (Alabi et al. 2019; Oliveira et al. 2020)
LDPE (Low-density polyethylene)	• Good chemical resistance, flexible, limited U.V resistance	Packaging films and bags, squeezable (dispensing) bottles, wash bottles, cling wraps, mulch films, garbage bags, crates (Alabi et al. 2019) (Oliveira et al. 2020)
PP (Polypropylene)	• Tough, moderate heat resistance, sensitive to U.V.	Auto parts, industrial fibers, food containers, dishware, plastic furniture, microwaveable meal trays, vehicle upholstery, syringes, straws, crates (Alabi et al. 2019)
PS (Polystyrene)	Lightweight, transparent, good water resistance	Packaging foam, take out cartons, disposable plates and cups, cafeteria trays, insulation boards, plastic cutlery, yogurt pots, CD cases, low-cost brittle toys, crates (Alabi et al. 2019)
Polycarbonate	Can absorb impacts	Eye protection, machine guards, strings, ropes, kitchen equipment, reusable bottles, crates (Alabi et al. 2019; Oliveira et al. 2020)
PVC (Poly-vinyl chloride)	Stiff, tough, lightweight	Used for insulating cables, plastic windows, plumbing pipes, roof sheets, shoe soles, cosmetic container, crates (Alabi et al. 2019)
Acrylic (Polymethylmetacrylate)	Weather resistant, good electrical insulator, sunlight resistant	Signs, covers for car lights, used in cameras, air craft windows and canopies (Rhodes 2018)
Polyamide (Nylon)	Strong, slippery, sensitive to sunlight, good resistance to chemicals	Washers, spacers, brushes, cloth fibers, gear wheels, bearings, casing for power tools (Rhodes 2018)

10.2.1.1 Amorphous Thermoplastics

These are comprised of statistically oriented macromolecular chains without any near order. These plastics retain their normal characteristics below glass temperature (Tg). Above that, they become soft elastic. Amorphous thermoplastics possess the following properties:

- Are optically transparent or translucent
- Have low propensity to creep
- Have good structural stability

- Have low propensity to get deformed
- Are brittle
- Have low chemical resistance
- Can break down due to environmental stress

Examples of amorphous thermoplastics include: polycarbonate (PC), polymethyl methacrylate (PMMA), polystyrene (PS), polyvinylchloride (PVC) and polysulfone (PSU). These are suitable for use in housing components (*Thermoplastic classification* [n.d.]; NPTEL Lecture 4.1: Thermoplastics and Thermosets [n.d.]; Klein 2011).

10.2.1.2 Semicrystalline Thermoplastics

Semicrystalline thermoplastics are comprised of the amorphous phase with integrated crystalline phases, constructed by near order forces. These are generally opaque and retain their normal characteristics above glass temperature also and have tough elastic to hard form. Semicrystalline thermoplastics have the following characteristics:

- Are translucent or opaque
- Have good fatigue resistance
- Are tough
- Have good chemical resistance
- Have good sliding characteristics
- Are wear resistant
- Exhibit glass transition
- Have some degree of crystallinity

Examples of semicrystalline thermoplastics include: polyamide (PA), nylon, polypropylene (PP), polyethylene (PE), polyetylenterphtalate (PET). They are tougher and more durable than amorphous thermoplastics. These are suitable for making the components that are subjected to mechanical wear and for sliding materials.

10.2.2 Thermosets

Thermosets are comprised of molecular chains with tight mesh like crosslinkage. Chain assembly of thermosets is achieved by "polycondensation" reaction. Because of strong crosslinking they cannot be reshaped or melted and are hard and brittle. Extensive crosslinking also imparts desirable qualities such as increased mechanical strength (Figure 10.1), good chemical resistance and high level of thermal stability. Examples of thermosets include epoxy resins (EP), phenol resins (PF) and polyester resins (UF). Some types of thermosets and their uses are mentioned in Table 10.2.

10.2.3 Elastomers

Elastomers have wide netlike crosslinking of "knotted" molecular chains. This type of crosslinking provides a high level of dimensional stability. These are elastic––that means by applying a load (for e.g. tensile load) the chains become entangled, but

TABLE 10.2
Examples of Thermosets and Their Uses

Plastic	Properties	Uses
Epoxy resin	Good chemical resistance, good electrical insulator	Used as adhesive (araldite) or can be reinforced with carbon fiber to produce composite material used in aerospace (Rhodes 2018)
Melamine formaldehyde	Stiff, hard, strong, resistant to some chemicals	Electric insulation, laminates for work surfaces (Rhodes 2018)
Polyester resin	Stiff, hard, good chemical resistance	Casting and encapsulation, bonding of other materials (Rhodes 2018)
Urea formaldehyde	Stiff, strong, brittle, good electric insulator	Electrical fittings, handles and control knobs, adhesives (Rhodes 2018)

after removal of the load they relax again; they can be stretched to over twice their length and then immediately return to their original length. Like thermoset plastics elastomers cannot be melted. Examples of elastomers include butadiene resin (BR), styrene butadiene resin (SBR) and polyurethane resin (PU) (Figure 10.1). Some examples of elastomers and their applications are listed in Table 10.3.

Thermoplastics are recyclable while thermosets and elastomers are non-recyclable.

10.3 BIOPLASTICS

Bioplastics is a broad term that includes bio-based plastics and biodegradable plastics. These are not necessarily biodegradable (Folino et al. 2020; Jia 2020). Bioplastics can be categorized into three groups (Figure 10.2): (1) Bio-based plastics that are biodegradable, (2) Bio-based plastics that are not biodegradable, (3) Petroleum origin biodegradable plastics (Tokiwa et al. 2009; Bartolo et al. 2021).

10.3.1 Bio-Based and Biodegradable Plastics

These are manufactured fully or partially from biomass or renewable sources, such as cellulose, and are biodegradable. Important examples of these are PLA (polylacatic

TABLE 10.3
Examples of Elastomers and Their Uses

Plastic	Uses
PUR Polyurethane	Low-density foam used in upholstery, bedding and automotive seating, footwear, straps and bands, mattress padding, waterproofing of outerwear, shower curtains (Rhodes 2018)
BR Butadiene-elastomer	Tire manufacturing, golf ball core, used as fuel in combination with an oxidizer in various solid rocket boosters (Rhodes 2018)
SBR Styrene-butadiene elastomer	Shoe heels and soles, gaskets, coated paper, binder in Li-ion battery electrodes, used as part of cement-based basement waterproofing systems

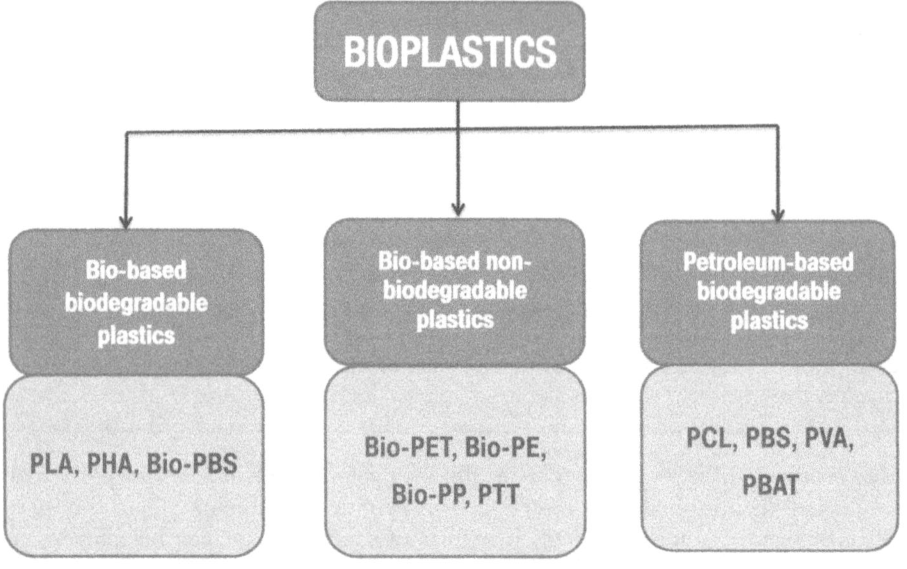

PLA- polylacatic acid, **PHA**- polyhydroxyalkanoates, **PBS**- poly butylene succinate, **PET**- polyethylene, **PE**- polyetylenterphtalate, **PP**- polypropylene, **PTT**- polytrimethylene terephthalate, **PCL**- polycaprolactone, **PVA**- polyvinyl alcohol, **PBAT**- polybutylene adipate terephthalate

FIGURE 10.2 Types of bioplastics.

acid), PHA (polyhydroxyalkanoates) and bio-PBS (bio-based poly butylenes succinate) (Bartolo et al. 2021).

10.3.2 Bio-Based Non-Biodegradable Plastics

These are non-biodegradable despite being produced from biological resources. Their chemical structure and properties are the same as petroleum-based plastics and are named after their petroleum-based counterparts by adding the term "bio" to the name. Examples of such bioplastics include drop-ins such as bio-PET, bio-PE, bio-PP and technical performance polymers such as PTT, TPS-ET (European Bioplastics 2018; McGuire 2019; Bartolo et al. 2021).

10.3.3 Petroleum-Based Biodegradable Plastics

These plastics are fossil-based and not made from biological or renewable sources but are biodegradable (Tokiwa et al. 2009*)*. Examples of petroleum-based biodegradable plastics are PCL (polycaprolactone), PBS (poly butylene succinate), PVA (polyvinyl alcohol) and PBAT (polybutylene adipate terephthalate) (Tokiwa et al. 2009; European Bioplastics 2018; Bartolo et al. 2021).

10.4 DEGRADABLE PLASTICS

A plastic is considered degradable if it can break down into simpler subunits and no longer retain its original properties. Plastics may be degradable or biodegradable (Kjeldsen et al. 2018). Two main mechanisms of degradation of plastics are (1) photo and oxo-degradation and (2) biodegradation (Jia 2020).

10.4.1 Photo and Oxo-Degradable Plastic (Additive-Based Plastics)

A minute amount of a polyolefin-based additive is added to regular fossil-based plastics during manufacturing of photo and oxo-degradable plastics, which helps to degrade plastic in the presence of sunlight and oxygen. Additive formulation controls the programmed service life, after which the degradation starts. These are degraded by the process of oxidative degradation and after that they are susceptible to degradation under compostable conditions. The idea behind this was that the plastic would be degraded by microbes into carbon and water but it was found that photo and oxo-degradable plastics are not biodegraded efficiently and in a reasonable time duration, leading to accumulation of incompletely degraded plastic fragments in the environment. Therefore, these cannot be considered biodegradable (CIPET 2009; Jia 2020).

10.4.2 Biodegradable Plastics

Biodegradable plastics can be converted into biomass, CO_2 water by enzymatic actions of microorganisms (Lau et al. 2020). The process of biodegradation occurring at higher rates in controlled conditions is called composting (McGuire 2019). Biodegradable plastic consists of a variety of polymers from different origins and of different chemical structures. More than 20 types of biodegradable plastics have been developed; of these, only three groups are produced on a commercial scale: (1) starch blends, (2) PLA and (3) polybutylene-based polymers (including PBS and PBAT). Both starch blends and PLA production depend heavily on plant feedstocks such as potato, cassava, corn and sugar cane (Jia 2020).

1. *Starch blend*: Starch is extensively used to manufacture and develop biodegradable plastics because of its good availability and low cost. It has low strength and water resistance; therefore, it is blended with other polymers to achieve desired properties. In early blends starch was added to conventional plastics; these types of starch blends are only partially degradable. TPS (thermoplastic starch) is a fully degradable starch blend. The Italian company Nova Mont is a lead producer of starch blends
2. *PLA (polylactic acid)*: PLA is the most well-known biodegradable plastic. It has a relatively low cost and various attractive mechanical properties, as compared to other polymers. The American company Nature Works leads PLA production
3. *Polybutylene-based polymers (PBS and PBAT)*: PBS and PBAT are fossil-based biodegradable plastics. Development of bio-based equivalent for these is still under development. The German company BASF is leading in the production of PBS/PBAT (Jia 2020; Bartolo et al. 2021)

10.5 APPLICATIONS OF BIOPLASTICS

The major application of bioplastics is for packaging companies to produce snack packaging, plastic bags and other products. Another application is in agriculture and horticulture industries for mulching films (Jia 2020). Bio-PET is being used in the manufacturing of plant-based beverage bottles (McGuire 2019). PHA is widely used in the food packaging industry due to its good tensile strength, printability, flavor and odor barriers, resistance to grease and oil and high stability to temperatures. It can also be used in making compostable personal hygiene products. PHA also finds many uses in the medical industry in sutures and fasteners, meniscus repair and regeneration devices, rivets, tacks, staples and screws, surgical mesh, cardiovascular patches, vein valves, bone marrow scaffolds, ligament and tendon grafts, etc.

10.6 BIODEGRADATION OF PLASTICS BY MICROBES

Biodegradation: This is transformation of a polymer into basic elemental components (water, CO_2/CH_4), energy and biomass by actions of microbes. Microorganisms use the polymer as substrate for growth. The process of biodegradation is a complex process involving several steps that can be categorized into three steps (Figure 10.3): (1) Biodeterioration, (2) Biofragmentation, (3) Mineralization and Assimilation (Kjeldsen et al. 2018).

Biodegradability: This is the ability of matter to break down completely into harmless products by the action of microorganisms. In the case of plastics, they can be

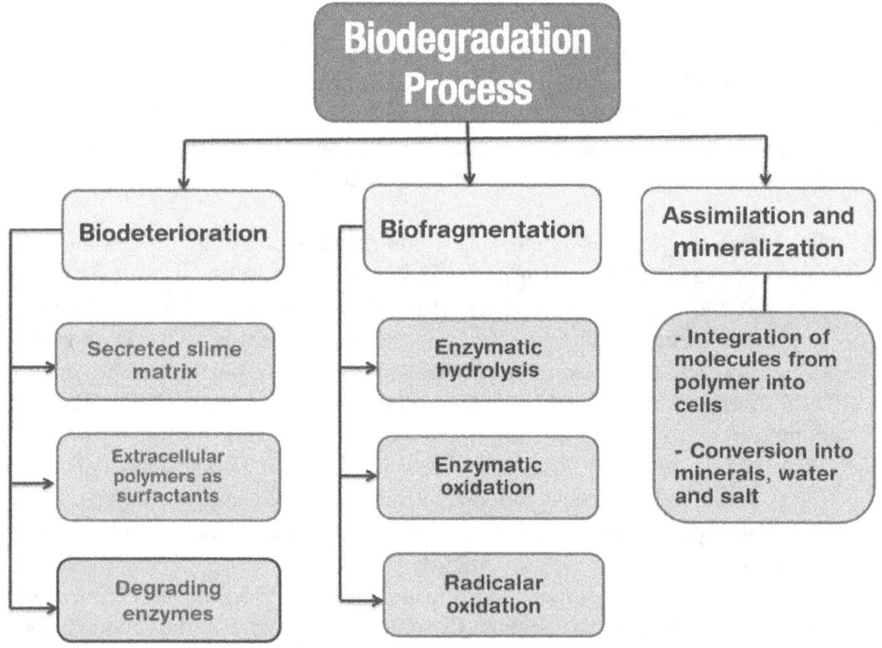

FIGURE 10.3 Process of biodegradation of plastic.

rendered biodegradable only when they are completely mineralized into harmless products in a defined time period, according to different guidelines set by different standards viz. ASTM, OECD, EN and ISO.

Biodegradibility depends on various factors, including composition of plastic, formulation and additives, surface area and environmental conditions (soil, water, temperature, microbial load) (Kershaw 2015; Van der Oever et al. 2017; Kjeldsen et al. 2018).

10.6.1 The Process of Biodegradation

The biodegradation process of plastics can be summarized in the following steps.

10.6.1.1 Biodeterioration

This is the initial stage of biodegradation in which superficial damage results in modification of physical and chemical characteristics of the plastic. Both abiotic and biotic factors play an important role at this stage (Urbanek et al. 2018). Abiotic factors such as mechanical forces, light, temperature and chemicals contribute greatly to weakening the polymer structure. They can either initiate the biodegradation process or act as synergistic components in the process (Kjeldsen et al. 2018).

Attachment of microorganisms to plastic surface initiates the process and microbial attachment lead to the development of biofilm. Biofilm formation depends on structure and composition of plastic, as well as environmental conditions (Urbanek et al. 2018). Biodeterioration renders the plastic more susceptible for microbial attack. Important factors that aid the process are: a) secreted slime matrix, b) extracellular polymers acting as surfactants, and c) production of degrading enzymes (Figures 10.3 and 10.4). Various studies have implemented pre-treatment of plastics such as U.V. exposure (Oet al. 2004; Singh et al. 2015), thermal treatment (Singh et al. 2015), gamma-irradiation and chemical treatment to facilitate degradation by microorganisms (Lucas et al. 2008, Arutchelvi et al. 2008). The rate of initial breakdown depends on various factors, including polymer chain length, size and shape, crystallinity, molecular weight distribution, surface porosity, pore size and water diffusion in polymer matrix (Kjeldsen et al. 2018).

10.6.1.1.1 Secreted Slime Matrix

Microbial adhesion to various surfaces is aided by specific secreted extracellular compounds. These compounds are polysaccharide or polypeptide in nature and form a slimy matrix that protects microbes against desiccation. The slimy matrix can enter the pores of substrates and alter the distribution of pores and change moisture degrees and thermal transfers. Mechanical pressure of expansion caused by penetrating matrix enlarges the pores and induces cracks, which consequently weakens the resistance and longevity of plastic. Development of mycelia of filamentous microorganisms into the matrix can assist in biodegradation.

10.6.1.1.2 Extracellular Polymers Acting as Surfactants

Extracellular polymers secreted by microbes can act as surfactants by making interactions between hydrophilic and hydrophobic phases easier; hence enabling microbes to enter the pores of plastics. Various active chemicals secreted by bacteria help in

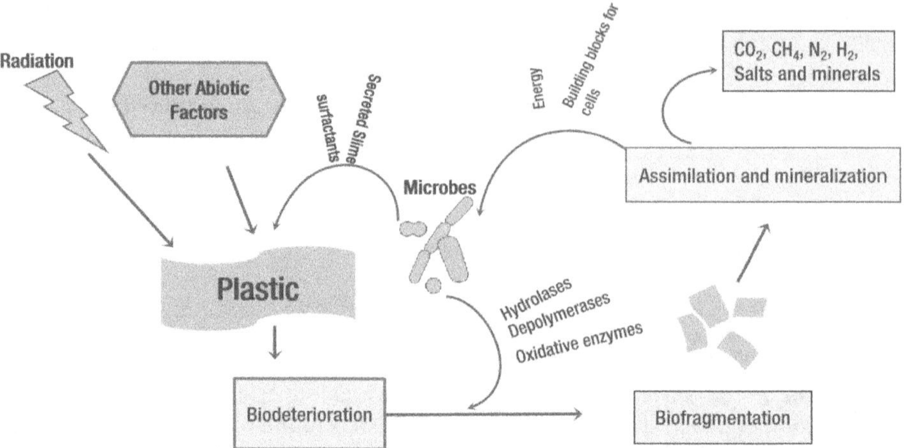

FIGURE 10.4 Process of biodegradation of plastic by microbes.

withering of plastics, including nitric acid (*Nitrobactor*), nitrous acid (*Nitrosomonas*), sulfuric acid (sulfur oxidizing bacteria), oxalic acid, citric acid, gluconic acid, oxaloacetic acid and lactic acid. Withering facilitates entry of water in the polymer matrix causing hydrolysis of polymer into oligomers and monomers (Figure 10.4). As the degradation progresses, the formation of pores changes the microstructure of the polymer, leading to the release of oligomers and monomers (Ghatge et al. 2020).

10.6.1.1.3 Production of Degrading Enzymes

Some microbes produce lipases, esterases, ureases and proteases that help in the degradation of plastic polymers.

10.6.1.2 Biofragmentation

As plastic polymers are broken down into smaller fragments, they become more susceptible to microbial attack. These fragments are further transformed into oligomers and/or monomers by the process of biofragmentation and are taken up by microbial cells to be used as a substrate for growth. Microorganisms employ different mechanisms to break down the polymers including: (a) enzymatic hydrolysis (Figure 10.4), (b) enzymatic oxidation, and (c) radicalar oxidation (Figure 10.3) (Mohan and Srivastava 2010).

10.6.1.2.1 Enzymatic Hydrolysis

Some microorganisms break down certain polymers by producing depolymerases. The first hydrolytic depolymerase discovered was "PHB depolymerases." Since then, a few other enzymes have been discovered such as polyurethanase, PHA depolymerase and PHBV depolymerase (Shah et al. 2008; Gewert et al. 2015). These enzymes can be lipases, esterases or endopeptidases. Lipases and esterases attack carboxylic linkages, while endopeptidases attack amide groups. Some microorganisms might be able to breakdown the polymers but not utilize the process products. These breakdown products can be used by some other microbes; thus making complete biodegradation of plastic a synergistic process.

10.6.1.2.2 Enzymatic Oxidation

Enzymatic oxidation is useful where cleavage by specific enzymes is not feasible. The oxidation process incorporates additional groups to polymer structure by the action of enzymes belonging to class oxidoreductases. Oxygenase enzymes add oxygen atoms, forming alcohol or peroxyl groups, making the polymer susceptible for fragmentation. Other enzymes of this class are peroxidases and oxidases. Peroxidases catalyze reaction between a peroxyl molecule and an electron acceptor group. Oxidases catalyze hydroxylation and oxidation reactions.

10.6.1.2.3 Oxidation by Free Radicals

Some oxidation reactions produce free radicals that can accelerate the process of biodegradation (Lucas et al. 2008).

10.6.1.3 Assimilation and Mineralization

Monomers and oligomers resulting from depolymerization by biodeterioration and the biofragmentation process are transported through microbial cell membranes into cells and utilized as carbon and energy sources (Mohan and Srivastava 2010). Assimilation and mineralization processes go hand in hand (Figure 10.3). Conversion of monomers into energy and building blocks of cells involve reactions that result in mineralization, forming gases (CO_2, CH_4, N_2, and H_2), water, salts and minerals as byproducts, which are harmless when released into environment (Zee 2005; Eskander and Saleh 2017).

10.7 ESTIMATION OF BIODEGRADATION

Polymers are considered biodegradable only when they are degraded completely into harmless products which can be recycled in biogeochemical cycles. If not degraded completely, the hydrophobic, high surface residues cause damage to environment. Hydrophobic compounds such as PCB and DDT can be attracted by degraded plastic residues and be held up to one million times background levels. When these chemicals are at background levels in the environment and diluted out, they do not present a serious risk, but when these are concentrated on degraded plastic residues they become a danger to the health of the environment and the lives inhabiting contaminated areas (Narayan 2009).

To test the biodegradability of a plastic polymer under various conditions, different protocols are adopted. The efficiency and extent of biodegradation for particular types of polymers are estimated by the following guidelines set by different standards such as ASTM, OECD, EN and ISO (Figure 10.5). The methods for studying biodegradation under different conditions include:

1. Field tests
2. Simulation tests
3. Laboratory tests
4. Biodegradation under anaerobic conditions (Muller 2004)

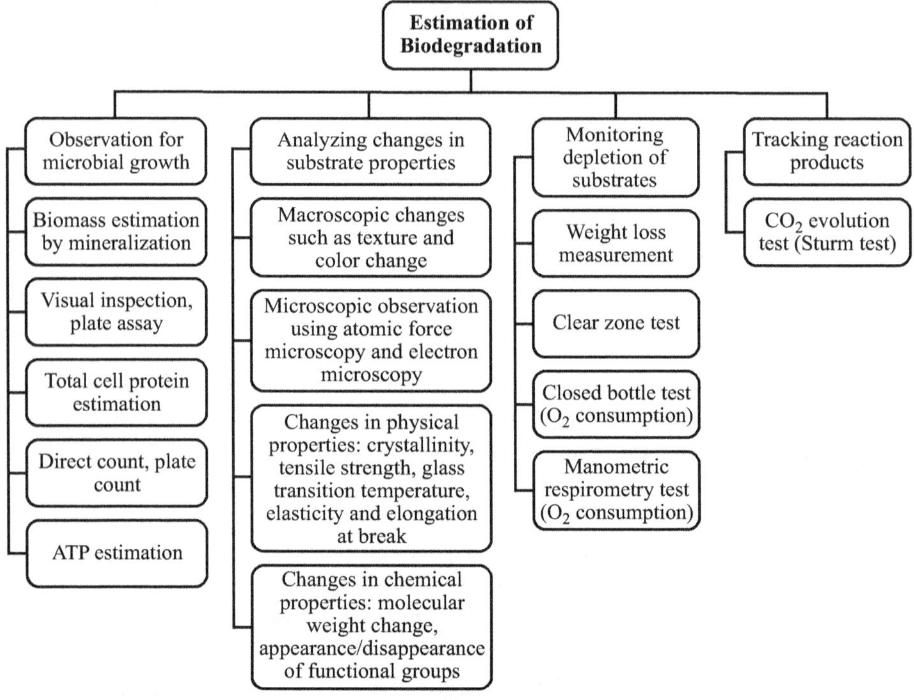

FIGURE 10.5 Methods for estimation of biodegradation.

After putting the polymer through one of these methods, the extent and efficacy of the biodegradation process can be analyzed using four approaches (Andrady 1994; Zee 2005; Mohan and Srivastava 2010; Harrison et al. 2018; Ruggero et al. 2019):

i. Observation for microbial growth
ii. Analysis of changes in substrate properties
iii. By monitoring depletion of substrates
iv. By tracking reaction products

10.8 METHODS FOR STUDYING BIODEGRADATION

10.8.1 Field Tests

Field tests provide the nearest estimation of degree of plastic biodegradation in a particular environment (aquatic/terrestrial) because the plastic is exposed to natural conditions. In this test, plastic samples of defined weight and dimensions are exposed to the conditions that we want to study biodegradation in; i.e., the sample is buried in soil or immersed in water for different intervals of time. Once the incubation period is over, the sample is studied for different aspects of biodegradation; i.e., microbial growth estimation, weight loss estimation and change in structure of substrate. The downside to this method is that these are not reproducible and little information can be obtained about the biodegradation process, factors such as pH, temperature, humidity, microbial community composition, the inability of the supply of oxygen to be regulated and its time-consuming nature (Figure 10.5). However, this method holds significance for extrapolating results obtained in laboratory conditions to results of field conditions (Muller 2004; Zee 2005; Singh and Sharma 2008; Mohan and Srivastava 2010).

10.8.2 Laboratory Scale Simulation Tests

These tests provide data for biodegradation in a specific environment. This is accomplished by designing complex media to mimic natural conditions and using native microbial flora (mixed culture from specific environment) for inducing biodegradation (OECD guideline for testing of chemicals 2005). Different factors such as pH, temperature, moisture content, microbial community and supply of oxygen can be regulated in simulation tests. Media for simulation tests include compost, soil and seawater or substrate mimicking the environment under study.

Simulation tests are reproducible and analytical methods used to determine biodegradation in field tests can be used for assessing biodegradation. Additional tests that are applicable for simulation studies include intermediate products of degradation process, determining CO_2 evolution or O_2 consumption, thus giving better insight into the degradation process (Muller 2004; Zee 2005; Singh and Sharma 2008; Ruggero et al. 2019).

10.8.3 Laboratory Tests

Laboratory tests are reproducible and useful for studying the basic mechanism of degradation. In lab tests often a higher degree of degradation is achieved than observed under natural conditions. Pre-treatment as UV exposure, chemical treatment and exposure to gamma-radiation usually precedes the process of treatment with microorganisms. This makes the process of biodegradation easy and rapid to follow. After pre-treatment, pre-weighed disinfected films are aseptically placed in sterilized defined media and inoculated with either a mixed population or pure microbial strains; this process is done in multiple replicates. An inoculated medium is incubated at specific temperature for a specific period of time. after the specified time the polymer is removed from the culture and tested for visual changes and other parameters to estimate the degree of biodegradation ((Muller 2004; Hadad et al. 2005; Singh and Sharma 2008).

10.8.4 Biodegradation under Anaerobic Conditions (in Anaerobic Sludge Treatment)

Anaerobic biodegradation can be estimated by measurement of gas production. In this method the test substrate is exposed to diluted anaerobically digested sludge. Biodegradability is assessed by monitoring the increase in headspace pressure resulting from the formation of CO_2, CH_4 and total inorganic carbon (Figure 10.5). These tests are implied for assessment of potential anaerobic biodegradability in an anaerobic digester See OECD guidelines for testing of chemicals 2005 (OECD 2005).

10.9 TESTING METHODS FOR ESTIMATION OF BIODEGRADATION

10.9.1 Methods for Monitoring Microbial Growth

10.9.1.1 Visual Inspection/Plate Assay

This method is used to determine whether test plastic supports microbial growth or not. In this test the plastic is placed on the surface of a mineral salt agar containing no other carbon source. The test plastic and agar surface is sprayed with inoculum. Plates are incubated at constant temperature and examined after specified incubation period at for amount of growth. A rating of 0–4 is given to quantify the extent of growth based on the rating system defined in ASTM G21 and G29 (Andrady 1994; Zee 2005).

10.9.1.2 Plate Counts, Direct Counts and Turbidity Measurements

The medium in which biodegradation is carried out can be examined directly by microscopy, by standard dilution plate technique or turbidity (Singh et al. 2015; Atuanya et al. 2016) can be measured to estimate the amount of growth.

10.9.1.3 Mineralization of Biomass

Microbial growth can be exposed to suitable vapor to kill the cells and then biomass can be estimated by mineralization (Andrady 1994).

10.9.1.4 Total Cell Protein Estimation

Biomass can be quantified by determining the concentration of extractable protein (Andrady 1994; Orr et al. 2004; Hadad et al. 2005).

10.9.1.5 Total ATP Estimation

ATP from biomass can be extracted and quantified to estimate biomass (Andrady 1994).

10.9.2 Analysis of Changes in Substrate Properties

Biodegradation is usually followed by changes in physical, chemical and mechanical characteristics of plastic. The measure of these changes can be considered as indicators of biodegradation.

10.9.2.1 Evaluation of Macroscopic Modifications

Primary screening for assessment of biodegradability includes observation for changes in texture, color, appearance of holes or cracks and formation of biofilm, but these tests do not affirm that test plastic is biodegraded (Shah et al. 2008; Atuanya et al. 2016).

10.9.2.2 Microscopic Observations

Test plastic is analyzed using different microscopic techniques such as photonic microscopy, atomic force microscopy and electron microscopy (Orr et al. 2004; Pramila and Ramesh, 2011; Kavitha et al. 2014) for detailed study of surface changes that have occurred during the test period.

10.9.2.3 Measuring Changes in Physical Properties

Changes in physical properties such as crystallinity, tensile strength, glass transition temperature, elongation at break and elasticity are measured to estimate biodegradability. Crystallinity is measured using X-ray diffraction crystallography (XRD), tensile strength by tensile tester elongation at break by mechanical tester and elasticity by dynamic mechanical thermal analysis. Differential scanning calorimeter (DSC) measures glass transition temperature (Tg), cold crystalline temperature (Tcc) and melting temperature (Tm). Density, viscosity, molecular weight distribution and contact angles are determined.

10.9.2.4 Identifying Changes in Chemical Properties

Appearance of low molecular weight fragments and the formation of new or disappearance of functional groups signifies biodegradation. These changes can be figured out using a number of techniques. GPC (gel permeation chromatography) is used to separate oligomers of different molecular weights. Oligomers and monomers can be separated by HPLC (high-performance liquid chromatography) and GC (gas chromatography). MS (mass spectroscopy) can be employed for identifying intermediate products after purification by the above techniques and with the help of NMR (nuclear magnetic resonance spectroscopy) monomer structures are determined. Detection of changes in functional groups is achieved by FTIR (Fourier transform infrared spectroscopy) (Orr et al. 2004; Kavitha et al. 2014; Atuanya et al. 2016; Ruggero et al. 2019).

10.9.3 Monitoring Depletion of Substrates

10.9.3.1 Weight Loss Observations

Weight of the sample before and after investigation is measured, and the loss in weight is taken as indicative of biodegradation but this criterion does not confirm biodegradation (Singh et al. 2015; Atuanya et al. 2016).

10.9.3.2 Clear Zone Test

The clear zone test is employed to screen the ability of a microbe to degrade a particular plastic (Figure 10.5). Fine powder of a plastic is dispersed in a hot defined medium as the sole carbon source. The medium has opaque appearance on solidifying. The plates are inoculated with test microbial strains. If the plastic is degraded by the microbe, a clear zone is observed around the colony (Singh et al. 2015).

10.9.3.3 Closed Bottle Test (O_2 Consumption)

This test determines consumption of oxygen during the test period. The test polymer is added to liquid mineral medium in fine powder form as the only source of organic carbon and inoculated with test culture. The medium is contained in completely filled, closed bottles in the dark at constant temperature. Degradation is followed by analyzing dissolved oxygen over a period of 28 days. Biodegradation is determined by the ratio between the oxygen consumption (Figure 10.5) caused by degradation (corrected against blank values of inoculum) and the ThOD (theoretical oxygen demand) or COD (chemical oxygen demand). If >60% ThOD is obtained over the 28 day test period with a 10 day window period after the lag phase, test polymer is considered biodegradable. Different guidelines viz. OECD 301D, Regulations (EC) no 440/2008, C.4-E, DIN EN ISO 10707 include closed bottle test as standard test (Hydrotox, *Closed Bottle Test*, Laboratory of Ecotoxicology and Water Protection GmbH–Freiburg, 2017; OECD guideline for testing of chemicals, Ready Biodegradability, 1992).

10.9.3.4 Manometric Respirometry Test (O_2 Consumption)

This method determines oxygen consumption in closed respirometers (e.g. Oxitop apparatus). Known concentration of test plastic is added as sole organic carbon source in fine powder form to a specific volume of liquid mineral medium containing inoculum. The test vessel is stirred at a constant temperature for 28 days. Consumption of oxygen is followed by measuring the change in volume and/or pressure in the respirometer flask. Released CO_2 is trapped in a solution of potassium hydroxide or sodium hydroxide. The amount of oxygen consumed by the microbial population during the degradation process (corrected against blank inoculum) is expressed preferably as % ThOD or %COD. A degradation value of > 60% ThOD (or less preferred, COD) within 28 days and a window period of 10 days after the end of the lag phase is set as pass level to consider test plastic as biodegradable. (OECD guideline for testing of chemicals, Ready Biodegradability, 1992; Hydrotox, *Manometric Respirometry Test*, Laboratory of Ecology and Water Protection GmbH–Freiburg, 2017; Chemex, 2017).

10.9.4 By Tracking Reaction Products

10.9.4.1 CO_2 Evolution Test (Sturm Test)

Microorganisms oxidize carbon sources to derive energy, a process that results in production of CO_2. Thus, the measure of CO_2 evolved can be used to assess the biodegradation of plastic (Narayan 2009). Test polymer is added to a liquid mineral medium containing inoculum, in a known concentration as sole source of organic carbon (Figure 10.6). Aeration of the medium is achieved by passing CO_2 free air at a controlled rate in the dark or in diffuse light. Estimation of CO_2 production is done to monitor degradation over 28 days. The CO_2 is trapped in barium hydroxide or sodium hydroxide and measured by titration of the residual hydroxide or as dissolved inorganic carbon, (DIC), respectively. CO_2 produces from test polymer (corrected against the CO_2 produced from blank i.e. inoculated medium without test polymer) is expressed as percentage of $ThCO_2$. Pass level of >60% $ThCO_2$ (theoretical CO_2 production) within 28 days, with a window period of 10 days after lag phase has been set to categorize a substance as readily biodegradable. This method is included as a standard test for biodegradability in various guidelines viz. OECD 301 B, Regulation (EC) no. 440/2008, C.4-C, DIN EN ISO 9439, ISO 14852 and ASTM D5864–11 (Hydrotox *CO_2 Evolution Test*, Laboratory of Ecotoxicology and Water Protection GmbH, 2017; OECD guideline for testing of chemicals, Ready Biodegradability, 1992; Pramila and Ramesh 2011; Singh et al. 2015). In a study conducted by CIPET, Chennai plastic films collected from several commercial manufacturers were subjected to tests for conforming their biodegradability, according to Biodegradability

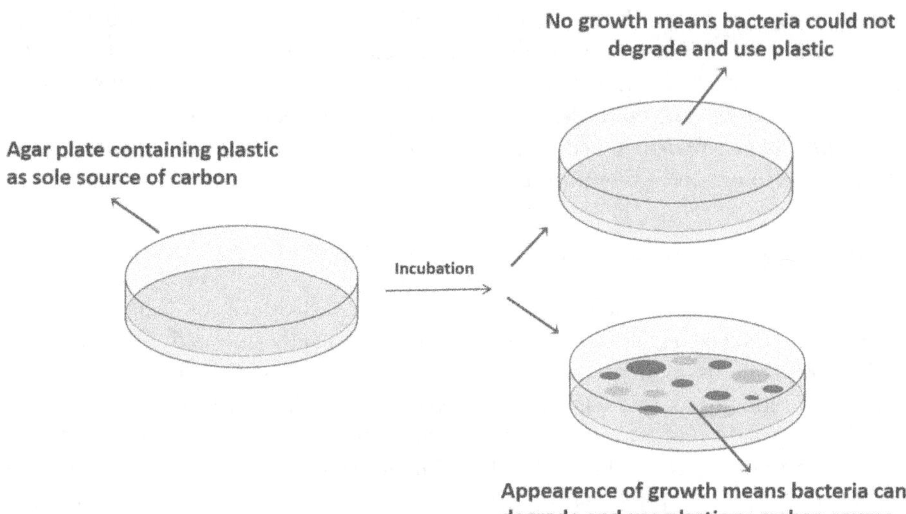

FIGURE 10.6 Plate assay for monitoring microbial growth.

standards ASTM D-6400 (degradation under composting conditions) and ASTM D-5338 (degradation in aerobic conditions) (CIPET, 2009).

10.10 DEGRADATION OF PLASTIC BY MICROBES

10.10.1 Degradation of PE (Polyethylene)

Due to its high hydrophobicity and molecular weight PE is almost fully resistant to biodegradation (Ghatge et al. 2020). For making it susceptible to biodegradation, some kind of modification such as UV, heat and chemical treatment is required (Shah et al. 2008; Ghatge et al. 2020). Extensive research has shown that microbes possess ability to degrade PE that needs to be explored. A number of bacteria, including *Pseudomonas putida*, *Pseudomonas aeruginosa*, *Bacillus amyloliquefaciens* (Ibiene et al. 2013; Singh et al. 2015, Shahreza et al. 2019), *Bacillus mycoides*, *Streptomyces* sp. (Atuanya et al. 2016), *Rhodococcusruber* (Orr et al. 2004), *Klebsiella pneumonia* and *Brevibacillus borstelensis* have been reported to degrade PE. Fungal species capable of degrading PE include *Aspergillus* spp. (Pramila and Ramesh 2011) and *Aspergillus clavatus* strain JASK1 (Gajendiran et al. 2016), *Cladosporium*, *Fusarium*, *Penicillium*, *Phanerochaete* (Muhonja et al. 2018; Ghatge et al. 2020).

10.10.2 Degradation of PET (Polyethyleneterephthalate)

The presence of aromatic groups makes PET difficult to be utilized by microorganisms but microbial communities can utilize its subunit DTP (diethylene glycol terephthalate) as a carbon source (Webb et al. 2013). Biodegradation of PET to some extent was achieved by *Aspergillus niger* and *Bacillus subtilis* (Asmita et al. 2015). A novel bacterium *Ideonella sakaiensis* 201-F6can utilizes PET as energy and a carbon source causing almost complete degradation of PET films. *Ideonella* produces two enzymes which hydrolyze PET into terephthalic acid and ethylene glycol (Yoshida et al. 2016). Enzymes that can hydrolyze PET include carboxylic ester hydrolase enzymes such as lipases, esterases and cutinases. These enzymes have been isolated from both bacterial (*Ideonella sakaiensis* and *Thermobifida fusca*) and fungal (*Fusarium solani, Humicola insolens* and *Aspergillus oryzae*) sources (Carr et al. 2020).

10.10.3 Degradation of PUR (Polyurethane)

Polyurethanes are susceptible to microbial attack, particularly by fungi. Polyester type polyurethanes are more susceptible, whereas polyether polyurethanes are still highly resistant to biodegradation. *Comamonas acidovorans* can use polyester type polyurethanes as sole carbon, nitrogen source (Nakajima-Kambe et al. 1995). *Bacillus safensis* and *Bacillus subtilis* demonstrated good degradation activity for PU Impranil™ (Nakkabi et al. 2015). Fungal species capable of degrading Pu include *Psetalotiopsis microspora, Fusariumsolani, Spicaria* spp., *Alternaria solani* and *Aspergillus flavus* (Ibrahim et al. 2011; Russell et al. 2011). The PU degrading enzyme has been purified and characterized (Eubeler 2010).

10.11 BIODEGRADATION OF BIODEGRADABLE PLASTICS

Biodegradable plastics vary in their susceptibility to degradation depending on the composition and environmental conditions of degradation process. Table 10.4 lists some of the degradation methods used for assessing degree of biodegradation different types of plastics and plastic degrading microorganisms studied.

10.11.1 Degradation of PCL (Polycaprolactone)

PCL is a synthetic biodegradable polyester that is partially crystalline and is formed by ring-opening polymerization of caprolactone. It can be biodegraded aerobically, as well as anaerobically. Both bacteria and fungi have been shown to degrade PCL including *Fusarium, Penicillium, Aspergillus* and *Clostridium.* PCL is degraded by lipases, esterases, PCL depolymerases and cutinase. Studies have shown that cutinase acts as PCL depolymerase (Shimao 2001; Tokiwa et al. 2009).

10.11.2 Degradation of PPL (Poly Beta-Propiolactone)

PPL can be degraded by PHB depolymerases and lipases. A number of microbes can degrade PPL, the majority of which belong to the genus *Bacillus,* other degraders include: *Acidovorax*, *Variovorax paradoxus*, *Sphingomonas paucimobilis* and *Streptomyces* (Tokiwa et al. 2009).

10.11.3 Degradation of PLA (Polylactic Acid)

PLA is synthesized either by condensation polymerization of lactic acid or by ring opening polymerization of lactide. It is extensively used as a polymer in medicine

TABLE 10.4
Results Summary of Some Research Work Done on Microbial Plastic Degradation

S.N.	Tested Plastic	Method and Techniques Used	Microorganism	Incubation Period	Observations	References
1	PET and modified PET with polyester "Bionolle"	Weight loss analysis, SEM, FTIR and XPS	*Penicillium funiculosum*	84 days	Slight weight loss of 0.07%, emergence of small holes, chemical changes indicating enzymatic hydrolysis and oxidation	Nowak et al. 2011
2	PE (HDPE, LDPE)	Morphological changes, FE-SEM, AFM, FTIR	*Penicillium chrysogenum, Penicillium oxalicum*	90 days	Structural changes like grooves, cracks, damaged layer, fragileness, pits and roughening of the surface were observed	Ojha et al. 2017
3	PET	Morphological changes, CO_2 evolution, H-NMR spectroscopy, Gel permeation chromatography, SEM, HPLC, Enzyme assays	*Ideonella sakaiensis* 201-F6	42 days	The PET film was extensively and almost completely damaged Two enzymes capable of hydrolyzing PET and the reaction intermediate, mono(2-hydroxyethyl) terephthalic acid (MHET) are produced by this strain	Yoshida et al. 2016
4	LDPE	Weight loss analysis, CO_2 evolution, SEM, AFM, FTIR	*Aspergillus clavatus* strain JASK1	90 days	30% weight loss of LDPE films	Gajendiran et al. 2016
5	PE	Weight loss analysis, FTIR	*Kocuria palustris* M16, *Bacillus pumilus* M27 and *Bacillus subtilis* H1584	30 days	The weight loss observed was 1%, 1.5% and 1.75% with M16, M27 and H1584 isolates, respectively	Harshvardhana and Jhaa 2013
6	LDPE and Pro-oxidant containing LDPE-L0235	Photo-oxidative pre-treatment, weight loss analysis, FTIR	*Brevibaccillus borstelensis* strain 707	30 days	Weight loss observed was 6.2% and 7.8% for LDPE and LDPE-L0235, respectively	Hadad et al. 2005

(*Continued*)

TABLE 10.4 (*Continued*)
Results Summary of Some Research Work Done on Microbial Plastic Degradation

S.N.	Tested Plastic	Method and Techniques Used	Microorganism	Incubation Period	Observations	References
7	LDPE	Weight loss analysis, FTIR, GC-MS	*Aspergillus oryzae* strain A5, *Aspergillus fumigates* strain B2, *Aspergillus nidulans* E1, *Bacillus cereus* strain A5, *Brevibacillus borstelensis* strain B2	112 days	The highest weight reduction by fungi was 36.4±5.53% by *Aspergillus oryzae* strain A5. The highest weight reduction by bacteria was 35.72± 4.01% by *Bacillus cereus* strain A5	Muhonja et al. 2018
8	PE	Chemical-alkali pre-treatment, Weight loss analysis, SEM	*Sphingomonas, Pseudomonas aeruginosa, Aspergillus niger, Aspergillus flavus*	30 days	In treated plastic surface erosions, folding, cracks and extensive roughening of the surface with pit formation was observed Highest weight loss of 56% was observed for plastic treated with *Sphingomonas*	Padmanabhan et al. 2020
9	PE	SEM, GC-MS, FTIR, XRD	*Bacillus licheniformis, Lysinibacillus fusiformis*	30 days	Maximum weight loss achieved was 2.97 ± 0.5% in plastic treated with bio-surfactant produced by *B. licheniformis* and degradation by *L. fusiformis*	Mukherjee et al. 2016
10	PS	Morphological changes, SEM, XPS, FT-NMR	*Pseudomonas* sp. DSM 50071	60 days	Edge smoothing, holes formation was observed on treated PS surface Oxidative degradation was indicated	Kim et al. 2020
11	LDPE	Weight loss analysis, AFM, SEM, FTIR	*Stenotrophomonas* sp. P2, *Achromobacter* sp. DF22	100 days	~8% weight reduction, structural deformity, chemical stability change	Dey et al. 2020
12	LDPE	SEM, FTIR	*Microbulbifer hydrolyticus* IRE-31		Morphological changes in the polymer surface and formation of additional carbonyl groups in the polymer chains were observed	Li et al. 2020

as it is absorbed in animals and humans; degradation is probably by non-enzymatic hydrolysis (Shimao, 2001). PLA is relatively resistant to microbial attack; therefore, not many microbes capable of PLA degradation are reported. PLA degrading bacterial genera include *Amycolatopsis, Lentzea, Kibdelosporangium, Streptoalloteichus* and *Saccharotrix* (Tokiwa et al. 2009; Sukkhum and Kitpreechavanich 2011).

10.11.4 Degradation of PVA (Polyvinyl Alcohol)

Polyvinyl alcohol is a biodegradable and water soluble polymer. PVA degraders have been reported from bacterial genera *Pseudomonas, Sphingomonas* and fungal genus *Penicillium* (Shimao 2001; Kim et al. 2003; Kawai and Hu 2009).

10.12 CONCLUSION AND FUTURE PROSPECTS

As our understanding of the mechanisms behind microbial degradation of plastic has improved, more attempts can be made to identify the enzymes involved and test their efficacy of degradation and possibilities to use those enzymes on a larger scale. Table 10.5 provides a quick view of the potential of developing biodegradation as a tool for tackling plastic waste and some difficulties that need to be addressed. The search for the microbes capable of diverse types also needs to be continued. In the wake of the plastic menace a lot of research has been done and much is going

TABLE 10.5
SWOT Analysis for Feasibility of the Use of Microbial Degradation of Plastics

Strengths	Weaknesses
• Microbial degradation process does not harm the environment • It does not release toxic products that are harmful for health • Plastic degrading capabilities have been identified in many microbes • Some are able to degrade plastic almost completely; e.g. *Ideonella sakaiensis* 201-F6 degrades PET efficiently • Advanced techniques like gene sequencing and recombinant DNA technology can greatly help in designing an efficient degradation process	• It is a slow process so it will take longer time • Not all the plastics are readily degraded by microbes • More research is needed to design a feasible and efficient process • Well-structured research schemes need to be created to give direction and for the research work to be fruitful
Opportunities	**Threats**
• Databases are being developed (Gan and Zhang 2019), which can help in accelerating the research • Efficient biodegraders can be further studied, and genes can be explored to advance the research into a practical solution • Biodegradable and compostable plastics are gaining a place in our lives. If we can replace the major share of plastic with these, it will be easier to develop a process for rapid degradation and manage the plastic waste	• There is a lack of co-ordination in research that might delay finding a solution • Effective research schemes are not employed to devise a strategy for development of effective solution

on to find a feasible and eco-friendly solution to the problem. It is evident from the observations of numerous works of research that a number of microbes—notably fungi and bacteria—are capable of degrading plastics. Different types of plastics are degraded efficiently by different types of microbes that possess mechanisms fit for the purpose. By improving their degradation capability through recombinant technology and gene transfer it is possible to attain higher degree of degradation; in fact, a novel bacterium *Ideonella sakaiensis* 201-F6, discovered by Yoshida et al. is capable of using PET as an energy source and achieves nearly 100% degradation. Therefore, it can be concluded that with some more work a working solution for plastic pollution can be found in the form of microbes.

Significant work has been done in the direction of understanding microbial degradation of plastics. Yet, much remains to be done to be able to employ this method on a larger scale. Many microbes have been shown to efficiently degrade plastic. Further studies might focus on improving the efficacy of these microbes with the use of recombinant DNA technology.

REFERENCES

Alabi, O., Ologbonjaye, K., Awosolu, O., Alalade, O. 2019. Public and environmental health effects of plastic wastes disposal: A review. *Journal of Toxicology and Risk Assessment* 5: 021.

Andrady, A. 1994. Assessment of environmental biodegradation of synthetic polymers. *Journal of Macromolecular Science, Part C: Polymer Reviews* 34(1): 25–76.

Arutchelvi, J., Sudhakar, M., Arkatkar, A., Doble, M., Bhaduri, S., Uppara, P. 2008. Biodegradation of polyethylene and polypropylene. *Indian Journal of Biotechnology* 7: 9–22.

Asmita, K., Tanwar, S., Shanbhag, T. 2015. Isolation of plastic degrading micro-organisms from soil samples collected at various locations in Mumbai, India. *International Research Journal of Environment Sciences* 4(3): 77–85.

Atuanya, E., Omoregbee, N.N., Udochukwu, U. 2016. Biodegradation of polyethylene by *Streptomyces* sp. isolated from plastic enriched composting soil. *Open Journal of Biomedical & Life Sciences* 2(4): 10–20.

Bartolo, A.D., Infurna, G., Dintcheva, N.T. 2021. A review of bioplastics and their adoption in the circular economy. *Polymers* 13: 1229.

Benson, N., Bassey, D.E., Palanisami, T. 2021. COVID pollution: Impact of COVID-19 pandemic on global plastic waste footprint. *Heliyon* 7: e06343.

Carr, C., Clarke, D., Dobson, A. 2020. Microbial polyethylene terephthalate hydrolases: Current and future perspectives. *Frontiers in Microbiology* 11: 571265.

Chemex. 2017. Manometric respirometry (OECD 301F) Available at: http://www.chemex.co.uk/products/manometric-respirometry-oecd-301f.

CIPET, Chennai 2009. *Bio-degradable Plastics- Impact on Environment.* Central Pollution Control Board, Ministry of Environment & Forests, Government of India. Chennai.

Dey, A., Bose, H., Mohapatra, B., Sar, P. 2020. Biodegradation of unpretreated low-density polyethylene (LDPE) by *Stenotrophomonas* sp. and *Achromobacter* sp., isolated from waste dumpsite and drilling fluid. *Frontiers in Microbiology*, 11: 603210.

Eskander, S., Saleh, H.E.-D. Biodegradation: Process Mechanism. In: *Environmental Science and Engineering.* Studium Press LLC: Houston, TX, pp. 1–31.

Eubeler, J.P. 2010. *Biodegradation of synthetic polymers in the aquatic environment.* Available at: http://elib.subb.uni-bremen.de/edocs/00101809–1.pdf.

European Bioplastics. 2018. *Fact Sheet European Bioplastics*. European Bioplastics. Berlin, Available at: https://www.european-bioplastics.org/news/publications/#FactSheets4.

Folino, A., Karageorgiou, A., Calabrò, P.S., Komilis, D. (2020). Biodegradation of wasted bioplastics in natural and industrial environments: A review. *Sustainability* 12: 6030.

Gajendiran, A., Krishnamoorthy, S., Abraham, J. 2016. Microbial degradation of low-density polyethylene (LDPE) by *Aspergillus clavatus* strain JASK1 isolated from landfill soil. *3 Biotech* 6: 52.

Gan, Z., Zhang, H. 2019. PMBD: A comprehensive plastics microbial biodegradation database. *Database: The Journal of Biological Databases and Curation* 2019: baz119.

Gewert, B., Plassmann, M., MacLeod, M. 2015. Pathways for degradation of plastic polymers floating in the marine environment. *Environmental Science Processes and Impact* 17: 1513–1521.

Geyer, R., Jambeck, R.J., Law, K.L. 2017. Production, use, and fate of all plastics ever made. *Science Advances* 3: e1700782.

Ghatge, S., Yang, Y., Ahn, J.-H., Hur, H.-G. 2020. Biodegradation of polyethylene: A brief review. *Appl Biol Chem* 63: 27.

Hadad, D., Geresh, S., Sivan, A. 2005. Biodegradation of polyethylene by the thermophilic bacterium *Brevibacillus borstelensis. Journal of Applied Microbiology* 98: 1093–1100.

Harrison, J., Boardman, C., O'Callaghan, K., Delort, A.-M., Song, J. 2018. Biodegradability standards for carrier bags and plastic films in aquatic environments: A critical review. *Royal Society Open Science* 5: 171792., http://dx.doi.org/10.1098/rsos.171792.

Harshvardhana, K., Jhaa, B. 2013. Biodegradation of low-density polyethylene by marine bacteria from pelagic waters, Arabian Sea, India. *Marine Pollution Bulletin* 77(1–2): 100–106.

Hydrotox—Laboratory of Ecotoxicology and Water Protection GmbH—Freiburg 2017. *Closed bottle test*. Available at: http://www.hydrotox.de/en/services/laboratory-services/biological-degradation/ready-biological-degradation/closed-bottle-test.html

Hydrotox—Laboratory of Ecotoxicology and Water Protection Gmbh - Freiburg. 2017. Manometric respirometry test. Available at: http://www.hydrotox.de/en/services/laboratory-services/biological-degradation/ready-biological-degradation/manometric-respirometry-test.html.

Hydrotox—laboratory of Ecotoxicology and Water Protection Gmbh—Freiburg. 2017. *CO_2-evolution test*. Available at: http://www.hydrotox.de/en/services/laboratory-services/biological-degradation/ready-biological-degradation/co2-evolution-test.html.

Ibiene, A.A., Stanley, H.O., Immanuel, O.M. 2013. Biodegradation of polyethylene by *Bacillus* sp. indigenous to the Niger Delta Mangrove Swamp. *Nigerian Journal of Biotechnology* 26: 68–79.

Jia, M.Z. 2020. *Biodegradable Plastics: Breaking Down the Facts*. Greenpeace East Asia, Beijing.

Kavitha, R., Anju, K.M., Bhuvaneswari, V. 2014. Biodegradation of low density polyethylene by bacteria isolated from oil contaminated soil. *International Journal of Plant, Animal and Environmental Sciences* 4(3): 601–610.

Kawai, F., Hu, X. 2009. Biochemistry of microbial polyvinyl alcohol degradation. *Applied Microbiology and Biotechnology* 84: 227–237.

Kershaw, P.J. 2015. *Biodegradable Plastics and Marine Litter. Misconceptions, Concerns and Impacts on Marine Environments*. Nairobi: United Nations Environment Programme (UNEP).

Kim, B., Sohn, C., Lim, C., Lee, J.P. 2003. Degradation of polyvinyl alcohol by *Sphingomonas* sp. SA3 and its symbiote. *Journal of Industrial Microbiology and Biotechnology* 30: 70–74.

Kim, H., Lee, H., Yu, H., Jeon, E., Lee, S., Li, J., Kim, D.H. 2020. Biodegradation of polystyrene by *Pseudomonas* sp. isolated from the gut of superworms (larvae of *Zophobas atratus*). *Environmental Science & Technology* 54: 6987–6996.

Kjeldsen, A., Price, M., Lilley, C., Guzniczak, E. 2018. *A Review of Standards for Biodegradable Plastics*. Industrial Biotechnology Innovation Centre.

Klein, R. 2011. Material properties of plastics. In: Klein, R. (Ed.) *Laser Welding of Plastics, First Edition*. Wiley-VCH Verlag GmbH & Co. KGaA. Weinheim, Germany, pp. 3–69.

Lau., W.W.Y., Shiran, Y., Bailey, R.M., Cook, E., Stuchtey, M.R., Koskella, J., Velis, C.A., Godfrey, L., Boucher, J., Murphy, M.B., Thompson, R.C., Jankowska, E., Castillo, A.C., Pilditch, T.D., Dixon, B., Koerselman, L., Kosior, E., Favoino, E., Gutberlet, J., Baulch, S., Atreya, M.E., Fischer, D., He, K.K., Petit, M.M., Sumaila, U.R., Neil, E., Bernhofen, M.V., Lawrence, K., Palardy, J.E. 2020. Evaluating scenarios toward zero plastic pollution. *Science* 369 (6510): 1455–1461.

Li, Z., Wei, R., Gao, M., Ren, Y., Yu, B., Nie, K., Xu, H., Liu, L. 2020. Biodegradation of low-density polyethylene by *Microbulbifer hydrolyticus* IRE-31. *Journal of Environmental Management* 263: 110402.

Lucas, N., Bienaime, C., Belloy, C., Queneudec, M., Silvestre, F., Nava-Saucedo, J. 2008. Polymer biodegradation: Mechanisms and estimation techniques. *Chemosphere* 73: 429–442.

McGuire, M. 2019. *Bioplastics—A better option for the environment?* Available at: https://edis.ifas.ufl.edu/publication/FR418.

Millet, H., Vangheluwe, P., Block, C., Sevenster, A., Garcia, L., Antonopoulos, R. 2018. The nature of plastics and their societal usage. In: Harrison, R., Hester, R.E. (Eds.), *Plastics and the Environment*. Royal Society of Chemistry, London, pp. 1–20.

Mohan, S.K., Srivastava, T. 2010. Microbial deterioration and degradation of polymeric materials. *J Biochem Tech* 2(4): 210–215.

Muhonja, C.N., Makonde, H., Magoma, G., Imbuga, M. 2018. Biodegradability of polyethylene by bacteria and fungi from Dandora dumpsite Nairobi-Kenya. *PLoS ONE* 13(7): e0198446.

Mukherjee, S., Chaudhuri, U., Kundu, P. 2016. Bio-degradation of polyethylene waste by simultaneous use of two bacteria: *Bacillus licheniformis* for production of bio-surfactant and *Lysinibacillus fusiformis* for biodegradation. *RCS Advances* 6: 2982–2992.

Muller, R.J. 2004. *Biodegradability of polymers: Regulations and methods for testing*. Available at: https://www.onlinelibrary.wiley.com/doi/abs/10.1002/3527600035.bpola012.

Nakajima-Kambe, T., Onuma, F., Kimpara, N., Nakahara, T. 1995. Isolation and characterization of a bacterium which utilizes polyester polyurethane as a sole carbon and nitrogen source. *FEMS Microbiology Letters* 129: 39–42.

Nakkabi, A., Sadiki, M., Fahim, M., Ittobane, N., Ibnsouda, K.S., Barkai, H., El abed, S. 2015. Biodegradation of poly(ester urethane) by *Bacillus subtilis*. *International Journal of Environmental Research* 9(1): 157–162.

Narayan, R. 2009. Fundamental principles and concepts of biodegradability—Sorting through the facts, hypes, and claims of biodegradable plastics in the marketplace. *BioPlastics Magazine* 4: 28–31.

Nowak, B., Pajak, J., Labuzek, S., Rymarz, G., Talik, E. 2011. Biodegradation of poly(ethylene terephthalate) modified with polyester "Bionolle" by *Penicillium funiculosum*. *Polimery* 56(nr 1): 35–44.

NPTEL. Lecture 4.1: Thermoplastics and Thermosets. (n.d.). Available at: http://nptel.ac.in/courses/112107085/module4/lecture1/lecture1.pdf (accessed March 01, 2019).

OECD 1992. Test No. 301: Ready Biodegradability, OECD guidelines for the testing of chemicals, Section 3, OECD Publishing, Paris.

OECD 2005. OECD guideline for testing of chemicals. Proposal for revised introduction to the OECD guidelines for testing of chemicals, section 3, part 1: principles and strategies related to the testing of degradation of organic chemicals. ENV/JM.TG(2005)5/REV1.

Ojha, N., Pradhan, N., Singh, S., Barla, A., Shrivastava, A. 2017. Evaluation of HDPE and LDPE degradation by fungus, implemented by statistical optimization. *Scientific Reports* 7: 39515.

Oliveira, J., Belchior, A., da Silva, V., Rotter, A., Petrovski, Ž., Almeida, P.L., Lourenço, N.D., Gaudêncio, S.P. 2020. Marine environmental plastic pollution: Mitigation by microorganism degradation and recycling valorization. *Frontiers in Marine Science* 7: 567126.

Orr, I.G., Hadar, Y., Sivan, A. (2004). Colonization, biofilm formation and biodegradation of polyethylene by a strain of *Rhodococcus ruber. Applied Microbiology and Biotechnology* 65: 97–104.

Padmanabhan, L., Varghese, S., Patil, R., Rajath, H., Krishnasree, R., Shareef, M. 2020. Ecofriendly degradation of polyethylene plastics using oil degrading microbes. *Recent Innovations in Chemical Engineering* 13: 29–40.

Plastics and the Environment. 2021. Geneva Environment Network. Available at: https://www.genevaenvironmentnetwork.org/resource/updates/plastics-and-the-environment/

Pramila, R., Ramesh, K.V. 2011. Biodegradation of low density polyethylene (LDPE) by fungi isolated from marine water–A SEM analysis. *African Journal of Microbiology Research* 5(28): 5013–5018.

Rhodes, C.J. 2018. Plastic pollution and potential solutions. *Science Progress* 101(3): 207–260.

Ritchie, H., Roser, M. 2018. *Plastic Pollution.* Our World In Data. Available at: https://ourworldindata.org/plastic-pollution

Ruggero, F., Gori, R., Lubello, C. 2019. Methodologies to assess biodegradation of bioplastics during aerobic composting and anaerobic digestion: A review. *Waste Management & Research* 37(10): 959–975.

Russell, J., Huang, J., Anand, P., Kucera, K., Sandoval, A., Dantzler, K.W., Hickman, D., Jee, J., Kimovec, F.M., Koppstein, D., Marks, D.H., Mittermiller, P.A., Nunez, S.J., Santiago, M., Townes, M.A., Vishnevetsky, M., Williams, N.E., Vargas, M.P.N., Boulanger, L.-A., Bascom-Slack, C, Strobel, S.A. 2011. Biodegradation of polyester polyurethane by endophytic fungi. *Applied and Environmental Microbiology* 77(17): 6076–6084.

Shah, A., Hasan, F., Hameed, A., Ahmed, S. 2008. Biological degradation of plastics: A comprehensive review. *Biotechnology Advances* 26: 246–265.

Shahreza, H., Akhavan Sepahy, A., Hosseini, F., Khavari Nejad, R. 2019. Molecular identification of *Pseudomonas* strains with polyethylene degradation ability from soil and cloning of alkB gene. *Archives of Pharmacy Practice* 10(4): 43–48.

Shimao, M. 2001. Biodegradation of plastics. *Current Opinion in Biotechnology* 12: 242–247.

Singh, B., Sharma, N. 2008. Mechanistic implications of plastic degradation. *Polymer Degradation and Stability* 93: 561–584.

Singh, J., Gupta, K., Shrivastava, A. 2015. Isolation and identification of low density polyethylene (LDPE) degrading bacterial strains from polythene polluted sites around Gwalior City (M.P.). *Journal of Global Biosciences* 4(8): 3220–3228.

Sukkhum, S., Kitpreechavanich, V. 2011. New insight into biodegradation of poly (l-lactide), enzyme production and characterization. In: Carpi, A. (Ed.), *Progress in Molecular and Environmental Bioengineering—From Analysis and Modeling to Technology Applications.* InTechOpen, London.

Thermoplastic classification. (n.d.). Retrieved March 01, 2017, from Ensinger: www.ensinger-online.com.

Tiseo, I. 2021. *Global plastic production 1950–2020.* Available at: https://www.statista.com/statistics/282732/global-production-of-plastics-since-1950/.

Tokiwa, Y., Calabia, B.P., Ugwu, C.U., Aiba, S. 2009. Biodegradability of plastics. *International Journal of Molecular Sciences* 10: 3722–3742.

Urbanek, A.K., Rymowicz, W., Mirończuk, A.M. 2018. Degradation of plastics and plastic-degrading bacteria in cold marine habitats. *Applied Microbiology and Biotechnology* 102: 7669–7678.

Van der Oever, M., Molenveld, K., Van der Zee, M., Bos, H. 2017. *Bio-based and Biodegradable Plastics – Facts and Figures.* Wageningen: Wageningen Food & Biobased Research.

Webb, H.K., Arnott, J., Crawford, R.J., Ivanova, E.P. 2013. Plastic degradation and its environmental implications with special reference to poly(ethylene terephthalate). *Polymers* 5: 1–18.

Yoshida, S., Hiraga, K., Toshihiko, T., Taniguchi, I., Yamaji, H., Maeda, Y., Toyohara, K., Miyamoto, K., Kimura, Y., Oda, K. 2016. A bacterium that degrades and assimilates poly (ethylene terephthalate). *Science* 351(6278): 1196–1199.

Zee, M.V. 2005. Biodegradability of polymers – Mechanisms and evaluation methods. In: Bastioli, C. (Ed.), *Handbook of Biodegradable Polymers*, Shawbury, UK: Rapra Publishing, 1–33.

11 An Integrated Approach for the Rehabilitation of Mine Spoil Based on Plant-Microbe Interactions

Ekta Mishra, Shilpi Jain, and Shahenaz Jadeja
The Maharaja Sayajirao University of Baroda
Baroda, India,

Disha Mishra
CSIR, Lucknow, India

CONTENTS

DOI: 10.1201/9781003147091-11

11.1 INTRODUCTION

Mining has an enormous influence on the country's socio-economic conditions by generating employment and revenue. With the increase in electricity demand, especially in developing countries, coal has been the primary source for electricity production and meets nearly 40% of the world's electricity demand and about 63% in India (Tripathi et al. 2014). Other significant sectors that depend on coal include iron and steel, cement, and chemical sub-sectors (WEO 2019). Coal is mainly extracted through surface or opencast mining and underground mining (WCA 2019). India is one of the top five coal-producing countries and had generated 480.04 million metric tons (MT) of coal during 2019–2020 with a coal demand of 695.49 MT (MOC 2019). Although coal production helps to meet the country's economic growth, it damages the environment involving habitat and biodiversity destabilization due to deforestation, nutrient loss, pollution of soil, air, surface water, and groundwater. The risks associated with mining also include disasters such as landslides, mud floods, and ground subsidence. The mining operations, especially during surface mining, involves removal of vast quantities of topsoil, subsoil, and overburdens, and backfilling of the spoil in the excavated region leads to the stockpiling of overburdens and mine rejects, resulting in building up of anesthetic landscapes (Tripathi et al. 2014; Juwarkar and Jambulkar 2008). The resulted mine spoil is characterized by a lack of soil organic carbon (SOC) and other essential nutrients, acidic pH, low cation exchange capacity, heavy metal pollution, and other low physicochemical and biological properties as a result of nutrient leaching from soil erosion makes it difficult to support plant growth. Therefore, the reclamation of coal mine spoil by a suitable scientific approach, which is more advantageous than natural restoration, will facilitate revegetation and stabilization of mine spoil. Various methods have been studied to reclaim coal mine spoil, such as agroforestry, phytoremediation with native and exotic plant species, microbe-assisted phytoremediation, organic soil amendment (manure, sludge, biochar) aided plantations (Ghosh and Maiti 2020). Biological remediation technologies like microbe-assisted phytoremediation are more suitable than other physicochemical remediation technologies based on their cost-effective approach and long-term effectiveness to achieve a productive rhizosphere and self-sustaining ecosystem. They take into account several factors that affect these interactions, such as contamination, prevailing climate, nature of spoil, restoration practices, selection of plant species, and microbial communities. The role of microbe-assisted phytoremediation in reclaiming the coal mine spoil is illustrated in Figure 11.1.

Soil properties affect growth of plant and community development. Hence, inoculation of beneficial soil microbial strains of plant growth-promoting rhizobacteria (PGPR) and arbuscular mycorrhizal fungi (AMF), along with selected plant species in mine spoils have been done to support plant growth and improve soil functions. The selection of plant species is essential to restore the rhizospheric functions and

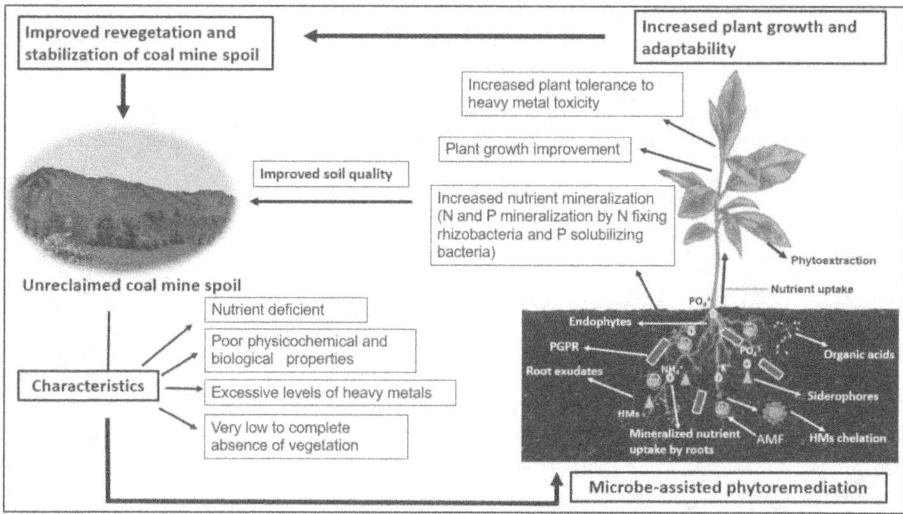

FIGURE 11.1 Role of microbe-assisted phytoremediation of coal mine spoil.

soil fertility in nutrient storage and cycling (Kumari and Maiti 2019). Therefore, this chapter deals with the source and nature of toxic pollutants released from the coal mining industry, and benefits of integrated plant-microbe application for rehabilitation of affected land, and ends with a discussion on challenges and recommendations for future research.

11.2 PHYSICOCHEMICAL AND BIOLOGICAL CHARACTERISTICS OF THE AFFECTED LAND

Due to its low physicochemical and biological properties, coal mine spoils' ability to sustain plant growth is limited. It must be characterized before planting, which is the most important initial soil growth step during ecosystem redevelopment (Jha and Singh 1991).

11.2.1 Physicochemical Properties

Several studies concerning coal mine spoil reclamation have involved initial physicochemical characterization of coal mine spoils. The significant physicochemical properties were identified as low Electrical Conductivity (EC) and water-holding capacity (WHC), high bulk density, sand and silt content, low levels of nutrient content, and excessive levels of toxic elements. Some of these have been depicted in Table 11.1.

11.2.1.1 Soil Texture

Soil texture signifies relative quantities of sand (2.0–0.5 mm), silt (0.05–0.002 mm), and clay (<0.002 mm) in a particular soil form (Mukhopadhyay et al. 2019). It is also a significant physical feature since silt and clay can enhance organic-mineral complexes' creation (Tripathi et al. 2016). Mishra and Shukla (2018) recorded sand, silt,

TABLE 11.1
Account of Physicochemical Properties of Coal Mine Spoils Found across the World

Location	pH	EC	Bulk Density	Lithology/ Mineralogy	TN (mg/kg)	TC (%)	SOC (%)	CEC	References
Jharia coalfields, Dhanbad, Jharkhand, India	6.07	–	0.56 mg m^{-3}	–	0.06	–	0.19	4.22 cmol Kg^{-1}	(Ahirwal and Maiti 2018)
Loess Plateau, China	8.21–8.28	–	1.64–1.83 g cm^{-1}	–	190–200	–	–	–	(Liu et al. 2017)
Hunter Valley, New South Wales, Australia	5.9–8.8	0.94 dS m^{-1}	–	–	1120	–	–	–	(Spargo and Doley et al. 2016)
Okpara coal mines, Enugu, Southeastern Nigeria	–	–	–	Medium to fine grained sandstones	–	–	–	–	(Nganje et al. 2011)
Jharia coalfields, Dhanbad, Jharkhand, India	4.61	0.66 dS m^{-1}	1.71 mg m^{-3}	–	–	–	1.82	4.63 cmol 100 g^{-1}	(Mukhopadhyay and Maiti 2011)
Douro coalfield, Portugal	3.3–5.4	48–393 µS cm^{-1}	–	Carbonaceous shales and lithic arenites	–	8.3–19.5	–	–	(Ribeiro et al. 2010)
Lajkura coal mine, Orissa, India	5.26	1.33 mS cm^{-1}	1.59 g cm^{-1}	–	0.14%	–	0.73	–	(Juwarkar and Jambhulkar 2008)

Abbreviations: EC, electrical conductivity; TN, total nitrogen; TC, total carbon; SOC, soil organic carbon; CEC, cation exchange capacity.

and clay levels of up to 85%, 12.6%, and 2.6%, respectively. Ahirwal and Maiti (2018) have recorded a high level of sand and silt in unreclaimed mine spoil, which can be attributed to poor reclamation conditions. Kumar and Maiti (2019) had reported the concentrations of sand, silt, and clay in the level of 82.2%, 14.7%, and 3.12%, which shows a high level of sand content in unreclaimed coal mines spoil.

11.2.1.2 Bulk Density

Bulk density is expressed as dry soil mass per unit volume of bulk, including soil solids and pore space (Mukhopadhyay et al. 2019). The spoil of the coal mine is characterized by compacted and high bulk density than other soils, leading to a decrease in soil porosity and aeration. The bulk density of the active natural soil usually varies between 1.1–1.5 mg m^{-3}, while the bulk density range of the mine soils reaches 1.6 mg m^{-3} (Ahirwal et al. 2016). Kumari and Maiti (2019) reported coal mine spoil's bulk density at a level of 2.3 mg m^{-3}. At the level of 1.52 g cm^{-3}, bulk density was reported by Mishra and Shukla (2018).

11.2.1.3 pH

Soil pH has a vital role concerning the facilitation of nutrient availability for plant uptake, the effect on biological processes, and elemental release in quantities which are toxic for plant growth (Mukhopadhyay and Maiti 2011). Moreover, both acidity and alkalinity are harmful for successful revegetation (Mukhopadhyay and Maiti 2011). A wide range of pH values have been reported. pH values from acidic (4.61, Kumari and Maiti 2019; 5.94, Jambhulkar and Kumar 2019; 4.2–4.7, Santos et al. 2016; 3.14–3.30, Bes et al. 2014; 2.5, Conesa et al. 2006), neutral to mildly alkaline (6.85, Ranđelović et al. 2016; 7.1–7.5, Parraga-Aguado et al. 2014; 6.6–7.3, Conesa et al. 2006), to slightly alkaline values (8.52–8.66, Fernandez et al. 2017; 8.2–8.4, Gomez-Ros et al. 2013).

11.2.1.4 Electrical Conductivity (EC)

The value of EC is considered as an index of soluble salt levels, which must be lower than 1.2 dS m^{-1} for successful vegetation growth, as the increase in soluble salt levels can lead to a decline in plant growth, site productivity, and soil-water balance (Mukhopadhyay et al. 2019). High electrical conductivity values (EC) (0.352 dS/m, Gajic et al. 2016; 150 µS/cm, Pandey and Singh 2014) reported indicate a large number of soluble salts in mine spoil. Many scientists have reported EC values in the range of 0.66 to 0.9 dS/m (Mukhopadhyay and Maiti 2011; Spargo and Doley et al. 2016; Jambhulkar and Kumar 2019). Lastly, the EC range is also varied (2.49–15.5 dS/m, Bes et al. 2014; 1.7–3.4 dS/m, Parraga-Aguado et al. 2014; 6.4–18 dS/m, Conesa et al. 2006).

11.2.1.5 Nutritional Attributes

The coal mine soils are deficient in nutrients including, soil organic carbon (SOC), as it is the crucial indicator for the activation of soil biological processes (Mukhopadhyay et al. 2019). Kumari and Maiti (2019) had reported the lower levels of available N, available P, extractable K, and SOC as 18.33 ppm, 0.016 ppm, 147.3 ppm, and 2.59% in mine spoil instead of the higher level of SOC due to coal carbon, which later oxidizes (Kumari and Maiti 2019). Jambhulkar and Kumar (2019) reported the available N, P, and K as 0.0025%, 0.013%, and 0.0027%, and low organic carbon (OC) as 0.21%. Ahirwal and Maiti (2018) had reported a significant reduction in available N, available P, and exchangeable K and showed a reduction in SOC and TN by 87 and 71%, which can be attributed to the loss of vegetation and fertile soil during surface coal mining (Ahirwal and Maiti 2018). Jain et al. (2014) had reported a higher level of sulfur (3.32%), carbon (22%), along with pyrite and other mineral matter in mine rejects. The wide range of OC (15.1–33.3%, Santos et al. 2016; 0.17–0.48%, Bes et al. 2014; 1,55%, Ranđelović et al. 2016; 3.14–6.65%, Parraga-Aguado et al. 2014; 0.5%, Conesa et al. 2006), total N (0.04%, Bes et al. 2014; 0.29–0.62%, Parraga-Aguado et al. 2014; 0.5–1.2%, Santos et al. 2016) as well as available content of K_2O (42.3–129.0 g/kg, Santos et al. 2016; 7.67–37.0 mg/100 g, Ranđelović et al. 2016) and P_2O_5 (0.9–74.3 mg/kg, Santos et al. 2016; 0.05–24.0 mg/100 g, Ranđelović et al. 2016) also reported in various studies.

11.2.2 Biological Properties

11.2.2.1 Microbiological Properties

To promote the growth of the rhizosphere, the existence of critical microbes in the soil is important. Rhizosphere microorganisms have the capability to degrade, transform and volatilize organic and inorganic pollutants. Different heavy metal resistance bacterial strains have been isolated and reported from the Bokaro Coal mines, which include *Enterobacter ludwigii*, *Klebsiella pneumonia*, *Klebsiella oxytoca*, *Enterobacter cloacae*, *Acinetobacter gyllenbergii*, and *Enterobacter cloacae*. (Gandhi et al. 2015). Singh and Narzary (2021) had reported the presence of *Acinetobacter, Exiguobacterium, Bacillus, Leclercia, Lysinibacillus, Pseudarthrobacter, Microbacterium*, and *Pseudomonas* in overburden samples from the Assam coal mine area, which had a high degree of resistance to heavy metals. These strains showed resistance to multiple heavy metals (Cd, As, Cr, Ni, and Cu). An abandoned coal mine in southeast Kansas showed the presence of 13 acid-resistant bacteria. Most bacterial species were Phylum *Firmicutes*, followed by *Actinobacteria* and *Proteobacteria*. Recently, the existence of *Pseudomonas syringae* as main phosphate solubilizers was reported. This result was in contrast with *Bacillus subtillis*, followed by *Pantoea dispersa, Bacillus circulans* in soils of coal mine landfills in Chhattisgarh.

Some studies, such as Jambhulkar and Kumar (2019), have recorded that at Western Coalfields Limited (WCL), Durgapur, Maharashtra, India, microbial groups such as Actinomycetes, N fixer strains such as *Azotobacter* and *Rhizobium*, and Vesicular-arbuscular mycorrhiza (VAM) have been found to be negligible. The existence of bacteria, fungi, and actinomycetes was stated by Juwarkar and Jambhulkar (2008) as 4.1×10^1 CFU/g, 9.0×10^1 CFU/g, and 6.0×10^1 CFU/g. In addition, in the coal mine spoil dump at Lajkura coal mine, Orrisa, India, the strains of critical N fixer microbes such as *Azotobacter*, *Rhizobium* and VAM are completely absent. Mine spoil with unfordable mechanical composition caused a reduction in the number of N fixing microorganisms, and a decrease in the occurrence of mycorrhizae has been reported (Adriano et al. 1980; Pavlovic et al. 2004; Mitrovi et al. 2008; Haynes 2009; Pandey et al., 2013; Pandey and Singh 2014; Gajic et al. 2016; Gajic and Pavlovic 2018).

11.2.2.2 Enzyme Activity

Soil enzymes mediate the degradation, transformation, and mineralization of soil organic matter and nutrients. Enzymatic activity is associated with soil fertility, cycling of nutrients, microbial activity, and degree of pollution (Mukhopadhyay et al. 2019). The enzymes are primarily released intracellularly or extracellularly from the cells of bacteria, plants, and animals, which metabolize organic matter and release nutrients in a simpler form (Maiti and Ahirwal 2019). Enzymatic activities such as cellulase, phosphatase, and other hydrolases may be considered to signify any soil disturbance and the stoichiometric relationship between the soil organic and mineral matter. The biochemical reactions of C, N, and sulfur in the soil are considered to catalyze β-glucosidase, urease, and aryl sulfatase, whereas alkaline phosphatase and acidic phosphatase are responsible for phosphorus mineralization. Dehydrogenases (DHA) are crucial enzymes in soil because they are indicative of the overall microbial activity. Mukhopadhyay and Masto (2016) recorded a growing trend in DHA levels with the reclamation age attributed to higher coverage of vegetation and organic C content. However, Jain et al. (2016) reported that coal overburden had a negative impact on enzymatic properties such as DHA, alkaline phosphatase, urease, and β-glucosidase.

11.3 HEAVY METALS: SOURCE AND TOXICITY

The excessive number of heavy metals has been found in the mine spoils such as Zn, Fe, Cr, Cu, Cd, Ni, and Pb (Li et al. 2014). However, the few metals as trace elements are essential for efficient plant growth. Their presence beyond the threshold limit becomes toxic for plant growth and the surrounding environment. Kumari and Maiti (2019) had reported high levels of Zn, Cd, Cr, Fe, Ni, Cu, and Pb in mine samples. As well, Mishra and Shukla (2018) had reported the presence of high metal contents and their relative abundance in the order; Cd<Ni<Cu<Cr<Zn in the mine spoil. Metal-rich sulfide bearing minerals such as pyrite (FeS_2), sphalerite (ZnS), and chalcopyrite ($CuFeS_2$) are easily weathered and oxidized to release these metals (Ribeiro et al. 2010; Li et al. 2014). When sulfide-rich minerals (pyrites, pyrrhotite, arsenopyrite, and marcasite) are exposed to water and air, they oxidize, resulting in massive AMD formation. Further acceleration from acidophilic bacteria (*Thiobacillus ferroxidans)* naturally persists in sites that encounter acid wash, when coming in contact with mine spoil, catalyzes the secondary reaction for the dissolution of metal cations such as Al, Zn, Mn, Cu, Cd, Cr, Pb, etc. (Choudhury et al. 2020). Therefore, the generation of heavy metals in available form takes place in the mine spoil. The weathering and oxidation of minerals found in the waste dump that generates metal ions like sulfate and iron can also affect the surrounding arable land and lead to a decline in fertility (Choudhury et al. 2020). The siltation of close by water bodies due to blown out and runoff particles from waste dumps via wind and water can also cause water quality deterioration (Choudhury et al. 2020).

In short, mine spoil showed multi-element contamination with high concentrations of metals such as Pb, As, Sd, Cd, Mn, Cu, Hg, Fe, Zn and, Ni that exceeded local background values and the maximum allowable limits for usage in agriculture, residential, commercial or industrial usage based mainly on geochemical partitioning (Gomez-Ros et al. 2013; Olivares et al. 2013; Bes et al. 2014; Parraga-Aguado et al. 2014; Ranđelović et al. 2016; Santos et al. 2016).

11.4 PHYTOREMEDIATION WITH DIFFERENT AMENDMENTS FOR COAL MINE SPOIL RECLAMATION

"Phytoremediation," as published by Oxford University, is described as "the use of various plant species capable of growing on polluted sites and removing high concentrations of heavy metals (loids) or other toxic substances to remediate contaminated soil, water or sediments." The principle of phytoremediation is that plant roots either degrade the contaminant down in the soil or absorb the contaminant, storing it in the plant's roots, stems and leaves (Rungwa et al. 2013). In comparison with traditional techniques, revegetation and phytostabilization are sustainable approaches for remediation of derelict mine spoils. Unfortunately, the extreme pH, high salinity/acidity, low water-holding capacity, high heavy metals content, and poor soil organic matter and fertility (Wong and Ho 1993; Wong and Ho 1994; Asensio, et al. 2013) characteristic of mine tailings do not favor natural plant growth. Hence, soil amelioration techniques should be functional to improve the physical, chemical and biological properties of mine tailings to improve plant setup (Santibañez et al. 2012; Courtney et al. 2009). Thus, to tackle these issue soil amendments on mine tailings has been profusely opted using organic, inorganic and microbial amendments for optimal plant growth. Different possible amendments are given in Figure 11.2. Inorganic amendments improved the physical characteristics of mine tailings, such as drainage, as well as limited chemical properties, such as pH and high metal ion solubilization (Pourrut et al. 2011; Janoš et al. 2010). Contrastingly, organic amendment has been proven to buffer soil pH, thereby indirectly affecting the metals adsorption and complexation in mine tailings (Teixeira et al. 2011; Santibáñez et al. 2008; Sopper 1993). The application of beneficial microbes which would support the plant growth and improve the soil quality is suggested as a promising alternative approach, which would meet the need of sustainable and economical restoration of mine spoil tailings (Guo et al. 2014).

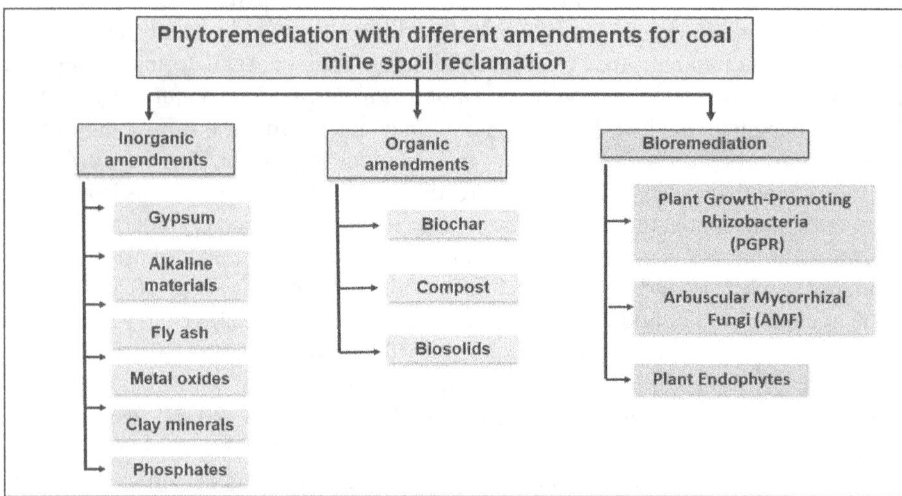

FIGURE 11.2 Different types of amendments for coal mine spoil reclamation.

The soil microbial communities play a vital role in maintaining the ecological balance in soil and sustaining the various types of ecosystems. The useful rhizospheric microbe-plant interactions have beneficial effects on both, soil quality and plant growth (Bi et al. 2018). Thus, the amended phytostabilization is proposed as the most potential phytoremediation technique to reduce the mobilization of toxicants from mine tailing.

11.4.1 Phytoremediation with Inorganic Amendments

The types of inorganic amendments used for reclamation of degraded lands include lime, gypsum, apatite, zeolites, iron and manganese oxides, clay minerals, fly ash, and various other materials (Pinto et al. 2018). Immobilizations of cationic trace elements by using the alkaline materials and some clay minerals have been reported, whereas oxyanion immobilization is achieved by using elemental metal oxides (Pinto et al. 2018). The inorganic amendment can enhance the soil's physical and chemical properties but not the soil's biological properties.

11.4.2 Phytoremediation with Organic Amendments

11.4.2.1 Biochar

The application of biochar to coal mine spoils adds benefits based on its sizeable porous surface and functional groups that include improving the soil's physical, chemical, and biological properties and increasing the bioavailability of essential elements in coal mine spoil (Arif et al. 2017; Ghosh and Maiti 2020). It improves the metal phytoextraction ability of plants, as well as mining waste's phytostabilization. (Ghosh and Maiti 2020). However, biochar's application involves deactivation in the long term that can occur or decrease sorption capability, and alteration of soil conditions could release potentially toxic elements from the surface of biochar (Ghosh and Maiti 2020).

11.4.2.2 Compost

The addition of compost to the soil can help plant growth, increase the nutrient availability for plant uptake, facilitate nutrient recycling, increase soil microbial activity, and enhance the soil-forming process (Chirakkara and Reddy 2015; Vuppaladadiyam et al. 2019). Although the compost making process can be influenced by weather conditions and generate greenhouse gases, the presence of heavy metals in the compost beyond the normal range can lead to its uptake in the plants, and excessive application can lead to soil salinization (Manciulea et al. 2017; Ning et al. 2017).

11.4.2.3 Biosolids

Nutrient rich organic materials generated during wastewater treatment process are called biosolids (Palansooriya et al. 2020). They also play a vital role in improving the physicochemical and biological characteristics of mine spoils, and hence, help add value to the organic by-products (Pinto et al. 2018). However, the drawbacks of biosolids for mine spoil restoration can lead to health and environmental risks. Biosolids, particularly sludge from both industrial and home effluents, typically contain higher levels of heavy metals than those found in natural soils (Pinto et al. 2018). As a result, prior to its application to soil, a suitable ecotoxicological characterization is recommended to ensure that human health and the environment are protected (Pinto et al. 2018).

11.4.3 Mine Spoil Rehabilitation through Integrated Plant-Microbe Interactions

Another approach to enhancing plant efficiency for phytoremediation is the use of plant-associated microorganisms (rhizospheric microorganisms). The rhizosphere microbial community will directly enhance root proliferation, stimulate plant growth, increase heavy metal sufferance and plant health (Fasani et al. 2018; Gupta et al. 2013). The use of various microbes to promote phytoremediation for the rehabilitation of coal mine spoils has been documented in various studies in recent years. A few of the following microorganisms are listed below.

11.4.3.1 Role of Arbuscular Mycorrhizal Fungi (AMF)

AMF (colonize plant roots) can form mutualistic potentially symbiotic relationships with 80% of all terrestrial plant species (Bi et al. 2019a). AMF facilitates assimilation and mobilization of soil nutrients in exchange for plant sugars (Ezeokoli et al. 2020), and thus, the mycorrhizal symbiosis can enhance plant establishment by improving nutrient uptake (Guo et al. 2014). Besides, AM fungi play roles in plant fitness and development, plant-pathogen control, soil structure improvement (soil aggregation) via glomalin production, and plant community succession (Ezeokoli et al. 2020). AMF can also reduce plant stresses caused by biotic and abiotic factors (Miransari 2010; Pankaj et al. 2019). Furthermore, the AMF has been shown to protect their host plants against heavy metal toxicity by mobilizing heavy metals from the soil and assisting in phytoremediation (Singh et al. 2019; Deb et al. 2020). The various possible mechanisms, such as immobilization by chelation, sequestration of metals into vacuoles, binding of metals to biopolymers in the cell wall, can help protect host plants by AMF through metal mobilization from soil (Deb et al. 2020). They have also been reported to exhibit resistance against drought, which can be through direct water transport and uptake through fungal hyphae to the host plants (Zhao et al. 2015), tolerance to hazardous metal concentrations and excessive salt content (Khan et al. 2014). Therefore, these advantageous properties of AMF can be of great benefit to climate change, specifically for water shortage and for the revegetation of degraded ecosystems, together with coal mine spoil banks (Zhao et al. 2015). There are various types of AMF strains used, along with phytoremediation for coal mine spoil reclamation. *Glomus* sp. and *Gigaspora* sp. are common AMF used to improve plant growth, nutritional conditions, and soil conditions of coal mine spoils. The beneficial effects of other AMF strains have been tabulated in Table 11.2.

11.4.3.2 Role of Plant Growth-Promoting Rhizobacteria (PGPR)

PGPR consists of bacteria that can populate the rhizosphere by triggering immune responses, modulating plant hormone balance, protecting against plant pathogens,

TABLE 11.2
Microbes Used for Ecological Restoration of Coal Mine Spoil and Coal Mine-Affected Areas

Type of Experiment	Microbial Strains	Plant Species	Beneficial Effects	References
Greenhouse pot experiment with 3 substrates: • S1: Recent discharged • S2: Weathered • S3: Spontaneously combusted mine spoils	*Glomus aggregatum*, *Rhizophagus intraradices* and *Funneliformis mosseae* (AMF strains)	Maize (*Zea mays* L.)	• Increased plant tolerance to heavy metal toxicity in S1, S2, and S3 • Increased uptake of N, P, and K • *F. mosseae* showed more suitability for revegetation of recently discharged coal mine spoil	(Guo et al. 2014)
Greenhouse pot experiment (with 2 substrates of coal mine spoil, S1: Weathered and S2: Combusted mine spoils)	*Rhizophagus intraradices* (AMF strain)	Maize (*Zea mays* L.)	• The AMF inoculation enhanced plant drought tolerance by increasing nutrient uptake, adjusting C: N: P stoichiometry, enhancing LMP and WUE • Showed a more beneficial effect on growth, nutrient condition, and water status of plant grown in S2	(Zhao et al. 2015)
Nursery and post-opencast coal mine field study	*Rhizophagus clarus*, *Gigaspora decipiens* and *Scutellospora* sp. (AMF strains)	*Albezia saman* and *Paraserianthes falcataria*	• 3–80% of AM colonization, increased shoot growth, P content and shoot dry weight under nursery conditions • The inoculated seedlings in field conditions showed better results	(Wulandari et al. 2016)
Greenhouse pot experiment followed by field trial	NFB (*Rhizobia* sp., *Azospirillum* sp.) and VAM fungi spores	*Acacia mangium*, *A. cracicarpa*, *Cassia siamia*, *Dendrocalamus strictus*, *Dalbergia sissoo*, *G. sepium*, *Pterocarpus santalinum*, *Sesbania grandiflora*, *Stylo hammata*, and *S. scabra*	• Changes in plant growth under both experiment conditions reflect the rise in phosphate solubilization and mobilization by VAM fungi and atmospheric N fixation by *Rhizobia* • *S. grandiflora* and *G. sepium* showed the highest increase in growth and nodulation, spore population, and percentage root colonization • T6 that included the VAM fungi inoculation was considered the best combination	(Arshi 2017)

(*Continued*)

TABLE 11.2 (*Continued*)
Microbes Used for Ecological Restoration of Coal Mine Spoil and Coal Mine-Affected Areas

Type of Experiment	Microbial Strains	Plant Species	Beneficial Effects	References
Pot experiment	*Acaulospora colombiana, Acaulospora morrowiae, Acaulospora scrobiculata, Dentiscutata heterogama, Gigaspora margarita,* and *Rhizophagus clarus* (AMF strains)	Vetiver (*Chrysopogon zizanioides*)	• *G. margarita, R. clarus*, and *A. morrowiae* showed better results in terms of vetiver growth • Other benefits include high biomass production, enhanced P uptake, and increased vetiver capacity in absorbing and tolerating trace elements	(Meyer et al. 2017)
Two-compartment microcosms experiment	*Glomus mosseae* (AMF strain) and *Pantoesstewarti* (PSB strain)	*Medicago sativa* L.	• AMF-PCB combined inoculation significantly improved AP, AGB, and BGB in root and hyphae compartments • Enhancement in mineralization of organic P and availability of AMF by PSB	(Bi et al. 2019b)

Abbreviations: LMP, leaf moisture percentage; WUE, water use efficiency; VAM, vesicular-arbuscular mycorrhiza; AMF, arbuscular mycorrhizal fungi; PCB, phosphate solubilizing bacteria; AP, available phosphorus.

and promoting nutrient mobilization to facilitate plant growth and development (Novo et al. 2018). PGPR includes a broad range of bacterial genera that are categorized according to their function and occurrence. Based on their roles, such as biofertilizers (increasing plant nutrient bioavailability), Phyto-stimulators (increasing plant growth and development through the release of phytohormones), rhizoremediation (breaking down organic contaminants), and biopesticides (control of diseases through antibiotic and antifungal synthesis) (Novo et al. 2018). Secondly, based on their occurrence, as suggested by Gray and Smith (2005), ePGPR (extracellular PGPR) and iPGPR (intracellular PGPR).

Azotobacter, Rhizobium (nitrogen-fixing bacteria), Pseudomonas, Bradyrhizobium, Azospirillum are some examples of PGPR (Novo et al. 2018). The PGPR helps through direct and indirect mechanisms, to improve the quality of plant growth. The direct mechanism involves enhancing non-pathogenic soilborne microbes for plant growth through different mechanisms such as biofertilization, rhizoremediation, phytostimulation, and stress management (Turgay and Bilen 2012). Moreover, indirect mechanisms occur when PGPR acts as biocontrol agents against phytopathogenic microbes and indirectly improves plants' growth (Turgay and Bilen 2012) through various mechanisms such as antibiotic development, signal interference, predation parasitism, etc. (Turgay and Bilen 2012).

Jambhulkar and Kumar (2019) had performed eco-restoration of coal mine spoil through an integrated biotechnological approach (IBA) developed at CSIR-NEERI. The IBA approach involved inoculation of *Azotobacter, Bradyrhizobium,* VAM spores of *Gigaspora* and *Glomus* species, and the amendment of ETP sludge (50 T/Ha) with suitable species of indigenous plants. The results showed improved physicochemical, nutritional (N, K, and P), and microbiological properties within three years, which led to the enrichment of the rhizosphere and ultimately facilitated abundant root development (Jambhulkar and Kumar 2019). Besides this, the decline in the heavy metal concentration such as Zn (49%), Fe (50%), Cu (47%), Cr (43%), Pb (40%), Mn (50%), Cd (35%), and Ni (51%) was also recorded. Other studies are also discussed in Table 11.2.

11.4.3.3 Role of Endophytes

The endophytes are microorganisms belonging to bacteria or fungi that stay within the inner tissues of plants without causing harm to the host (Deb et al. 2020). The endophytes even help enhance plant growth and development both in an abiotic or biotic stressed environment and provide many more beneficial effects than rhizobacteria and are able to tolerate high metal concentrations (Deb et al. 2020). Kumar et al. (2019) reported the remediation of degraded soils due to endophytic non-mycorhizal fungi. They involve the production of organic acids, siderophores, phytohormones, biosurfactants, enzymes, and growth hormones which help in nutrient and water uptake, osmolyte accumulation, osmotic adjustment, stomatal regulation, and associated nitrogen fixation as an additional advantage for host plants (Ma et al. 2011, 2016). Wężowicz et al. (2017) found that inoculation of both AMF with endophytic fungus significantly improved the *Verbascum lychnitis* growth in Pb-Zn mine waste substrate, whereas single inoculum with endophytes does not show positive results. Yamaji et al. (2016) reported that endophytic fungi had enhanced

C. barbinervis seedlings' growth at the Hitachi mine, increased K uptake in shoots, and reduced Cu, Ni, Zn, Cd, and Pb concentration in roots. Many works reported that DSE (Dark Septate Endophytic Fungi) are more stable in highly metal polluted regions compared with AM (*Arbuscular mycorrhiza)* (Kumar et al. 2019). Nutrients redistribution from senescent roots into the active roots by DSE helps in biogeochemical recycling, as reported by Kumar et al. 2019. Furthermore, melanin in the cell wall of DSE hyphae may decrease metal(loid)s toxicity (Cu, Zn, Pb, Cd) increasing the activity of antioxidant enzymes (SOD, CAT), which in turn decreases oxidative stress (Zhang et al. 2008; Ban et al. 2012).

11.5 CONCLUSION AND FUTURE PROSPECTS

Phytoremediation is a promising method for coal mine spoil restoration, with strong social support and many advantages over other physical and chemical approaches. Due to the elevated pollutants concentration and adverse pH, direct revegetation is not feasible in sulfidic coal mine tailing. Hence, appropriate soil amendments should be applied to improve mine tailings' spoil conditions to enhance plant establishment. It is expected to be an essential path in the research field of sulfidic mine reclamation. The analysis of geochemical processes in the rhizosphere will help guide amendment selection during phytoremediation; thus, potential phytoremediation applications could be more useful and environmentally friendly in terms of metal (loid) immobilization and ecological restoration. In practice, a single solution is neither feasible nor sufficient to reclaim highly degraded coal mine tailings. Henceforth, a combination of microbe-assisted phytoremediation will be necessary for highly integrated, efficient, cost-effective, and eco-sustainable technology.

However,, the AMF and DSE fungi have been discovered to coexist in plant roots (Kumar et al. 2019), suggesting further research is required to extract benefits for reclamation mine-affected areas. According to scientific reports, many spoil reclamations experiments have been conducted in the lab rather than on the field. This necessitates its expansion at the field level in order to determine its efficacy on coal mine spoil.

REFERENCES

Adriano, D.C., Page, A.L., Elseewi, A.A., Chang, A.C., Straughan, I. 1980. Utilization and disposal of fly ash and other coal residues in terrestrial ecosystems: A review. *Journal of Environmental Quality* 9(3): 333–344.

Ahirwal, J., Maiti, S.K. 2018. Assessment of soil carbon pool, carbon sequestration and soil CO_2 flux in unreclaimed and reclaimed coal mine spoils. *Environmental Earth Sciences* 77(1): 9.

Ahirwal, J., Maiti, S.K., Singh, A.K. 2016. Ecological restoration of coal mine-degraded lands in dry tropical climate: What has been done and what needs to be done? *Environmental Quality Management* 26(1): 25–36.

Arif, M., Ilyas, M., Riaz, M., Ali, K., Shah, K., Haq, I.U., Fahad, S. 2017. Biochar improves phosphorus use efficiency of organic-inorganic fertilizers, maize-wheat productivity and soil quality in a low fertility alkaline soil. *Field Crops Research* 214: 25–37.

Arshi, A. 2017. Reclamation of coalmine overburden dump through environmental friendly method. *Saudi Journal of Biological Sciences* 24(2): 371–378.

Asensio, V., Vega, F.A., Andrade, M.L., Covelo, E.F. 2013. Technosols made of wastes to improve physico-chemical characteristics of a copper mine soil. *Pedosphere* 23(1): 1–9.

Ban, Y., Tang, M., Chen, H., Xu, Z., Zhang, H., Yang, Y. 2012. The response of dark septate endophytes (DSE) to heavy metals in pure culture. *PLoS One* 7(10): e47968.

Bes, C.M., Pardo, T., Bernal, M.P., Clemente, R. 2014. Assessment of the environmental risks associated with two mine tailing soils from the La Unión-Cartagena (Spain) mining district. *Journal of Geochemical Exploration* 147: 98–106.

Bi, Y., Xiao, L., Liu, R. 2019a. Response of arbuscular mycorrhizal fungi and phosphorus solubilizing bacteria to remediation abandoned solid waste of coal mine. *International Journal of Coal Science & Technology* 6(4): 603–610.

Bi, Y., Xiao, L., Sun, J. 2019b. An arbuscular mycorrhizal fungus ameliorates plant growth and hormones after moderate root damage due to simulated coal mining subsidence: A microcosm study. *Environmental Science and Pollution Research* 26(11): 11053–11061.

Bi, Y., Zhang, Y., Zou, H. 2018. Plant growth and their root development after inoculation of arbuscular mycorrhizal fungi in coal mine subsided areas. *International Journal of Coal Science & Technology* 5(1): 47–53.

Chirakkara, R.A., Reddy, K.R. 2015. Phytoremediation of mixed contaminated soils: Enhancement with biochar and compost amendments. *IFCEE 2015*, 2687–2696.

Choudhury, A., Lahkar, J., Saikia, B.K., Singh, A.K.A., Chikkaputtaiah, C., Boruah, H.P.D. 2020. Strategies to address coal mine-created environmental issues and their feasibility study on northeastern coalfields of Assam, India: A Review. *Environment, Development and Sustainability* 1–43.

Conesa, H.M., Faz, Á., Arnaldos, R. 2006. Heavy metal accumulation and tolerance in plants from mine tailings of the semiarid Cartagena–La Unión mining district (SE Spain). *Science of the Total Environment* 366(1): 1–11.

Courtney, R.G., Jordan, S.N., Harrington, T.J.L.D. 2009. Physico-chemical changes in bauxite residue following application of spent mushroom compost and gypsum. *Land Degradation & Development* 20(5): 572–581.

Deb, V.K., Rabbani, A., Upadhyay, S., Bharti, P., Sharma, H., Rawat, D.S., Saxena, G. 2020. Microbe-assisted phytoremediation in reinstating heavy metal-contaminated sites: Concepts, mechanisms, challenges, and future perspectives. *Microbial Technology for Health and Environment*. Singapore: Springer, pp. 161–189.

Ezeokoli, O.T., Mashigo, S.K., Maboeta, M.S., Bezuidenhout, C.C., Khasa, D.P., Adeleke, R. A. 2020. Arbuscular mycorrhizal fungal community differentiation along a post-coal mining reclamation chronosequence in South Africa: A potential indicator of ecosystem recovery. *Applied Soil Ecology* 147: 103429.

Fasani, E., Manara, A., Martini, F., Furini, A., DalCorso, G. 2018. The potential of genetic engineering of plants for the remediation of soils contaminated with heavy metals. *Plant, Cell & Environment* 41(5): 1201–1232.

Fernández, S., Poschenrieder, C., Marcenò, C., Gallego, J.R., Jiménez-Gámez, D., Bueno, A., Afif, E. 2017. Phytoremediation capability of native plant species living on Pb-Zn and Hg-As mining wastes in the Cantabrian range, north of Spain. *Journal of Geochemical Exploration* 174: 10–20.

Gajić, G., Djurdjević, L., Kostić, O., Jarić, S., Mitrović, M., Stevanović, B., Pavlović, P. 2016. Assessment of the phytoremediation potential and an adaptive response of *Festuca rubra* L. sown on fly ash deposits: Native grass has a pivotal role in ecorestoration management. *Ecological Engineering* 93: 250–261.

Gajić, G., Pavlović, P. 2018. The role of vascular plants in the phytoremediation of fly ash deposits. *Phytoremediation: Methods, Management and Assessment* 151–236.

Gandhi, V., Priya, A., Priya, S., Daiya, V., Kesari, J., Prakash, K., ... & Kumar, N. 2015. Isolation and molecular characterization of bacteria to heavy metals isolated from soil samples in Bokaro Coal Mines, India. *Pollution* 1(3): 287–295.

Ghosh, D., Maiti, S.K. 2020. Can biochar reclaim coal mine spoil? *Journal of Environmental Management* 272: 111097.

Gomez-Ros, J.M., Garcia, G., Peñas, J.M. 2013. Assessment of restoration success of former metal mining areas after 30 years in a highly polluted Mediterranean mining area: Cartagena-La Unión. *Ecological Engineering* 57: 393–402.

Gray, E.J., Smith, D.L. 2005. Intracellular and extracellular PGPR: Commonalities and distinctions in the plant–bacterium signaling processes. *Soil Biology and Biochemistry* 37(3): 395–412.

Guo, W., Zhao, R., Fu, R., Bi, N., Wang, L., Zhao, W., Zhang, J. 2014. Contribution of arbuscular mycorrhizal fungi to the development of maize (*Zea mays* L.) grown in three types of coal mine spoils. *Environmental Science and Pollution Research* 21(5): 3592–3603.

Gupta, D.K., Huang, H.G., Corpas, F.J. 2013. Lead tolerance in plants: Strategies for phytoremediation. *Environmental Science and Pollution Research* 20(4): 2150–2161.

Haynes, R.J. 2009. Reclamation and revegetation of fly ash disposal sites: Challenges and research needs. *Journal of Environmental Management* 90(1): 43–53.

Jain, S., Baruah, B.P., Khare, P. 2014. Kinetic leaching of high sulphur mine rejects amended with biochar: Buffering implication. *Ecological Engineering* 71: 703–709.

Jain, S., Mishra, D., Khare, P., Yadav, V., Deshmukh, Y., Meena, A. 2016. Impact of biochar amendment on enzymatic resilience properties of mine spoils. *Science of the Total Environment*, 544: 410–421.

Jambhulkar, H.P. and Kumar, M.S. 2019. Eco-restoration approach for mine spoil overburden dump through biotechnological route. *Environmental Monitoring and Assessment* 191(12): 772.

Janoš, P., Vávrová, J., Herzogová, L., Pilařová, V. 2010. Effects of inorganic and organic amendments on the mobility (leachability) of heavy metals in contaminated soil: A sequential extraction study. *Geoderma* 159(3–4): 335–341.

Jha, A.K., Singh, J.S. 1991. Spoil characteristics and vegetation development of an age series of mine spoils in a dry tropical environment. *Vegetation* 97(1): 63–76.

Juwarkar, A.A., Jambhulkar, H.P. 2008. Phytoremediation of coal mine spoil dump through integrated biotechnological approach. *Bioresource Technology* 99(11): 4732–4741.

Khan, A., Sharif, M., Ali, A., Shah, S.N.M., Mian, I.A., Wahid, F., Ali, N. 2014. Potential of AM fungi in phytoremediation of heavy metals and effect on yield of wheat crop. *American Journal of Plant Sciences 2014*. doi: 10.4236/ajps.2014.511171

Kumar, V., Soni, R., Jain, L., Dash, B., Goel, R. 2019. Endophytic fungi: Recent advances in identification and explorations. *Advances in Endophytic Fungal Research* 267–281.

Kumari, S., Maiti, S.K. 2019. Reclamation of coalmine spoils with topsoil, grass, and legume: A case study from India. *Environmental Earth Sciences* 78(14), 429.

Li, X., Park, J.H., Edraki, M., Baumgartl, T. 2014. Understanding the salinity issue of coal mine spoils in the context of salt cycle. *Environmental Geochemistry and Health* 36(3): 453–465.

Liu, X., Bai, Z., Zhou, W., Cao, Y., Zhang, G. 2017. Changes in soil properties in the soil profile after mining and reclamation in an opencast coal mine on the Loess Plateau, China. *Ecological Engineering* 98: 228–239.

Ma, Y., Oliveira, R.S., Freitas, H., Zhang, C. 2016. Biochemical and molecular mechanisms of plant-microbe-metal interactions: Relevance for phytoremediation. *Frontiers in Plant Science* 7: 918.

Ma, Y., Prasad, M.N.V., Rajkumar, M., Freitas, H. 2011. Plant growth promoting rhizobacteria and endophytes accelerate phytoremediation of metalliferous soils. *Biotechnology Advances* 29(2): 248–258.

Maiti, S.K., Ahirwal, J. 2019. Ecological restoration of coal mine degraded lands: Topsoil management, pedogenesis, carbon sequestration, and mine pit limnology. *Phytomanagement of Polluted Sites*. Elsevier, pp. 83–111.

Manciulea, I., Dumitrescu, L., Bogatu, C., Draghici, C., Lucaci, D. 2017, October. Compost based on biomass wastes used as biofertilizers or as sorbents. *Conference on Sustainable Energy*. Cham: Springer, pp. 566–585.

Meyer, E., Londoño, D.M.M., de Armas, R.D., Giachini, A.J., Rossi, M.J., Stoffel, S.C.G., Soares, C.R.F.S. 2017. Arbuscular mycorrhizal fungi in the growth and extraction of trace elements by *Chrysopogon zizanioides* (vetiver) in a substrate containing coal mine wastes. *International Journal of Phytoremediation* 19(2): 113–120.

Ministry of Coal (MOC) 2019. Annual report 2019–20. Ministry of Coal, editor. Government of India. Available at: https://coal.nic.in/content/annual-report-2019–20. Accessed on 10 November 2020.

Miransari, M. 2010. Contribution of arbuscular mycorrhizal symbiosis to plant growth under different types of soil stress. *Plant Biology* 12(4): 563–569.

Mishra, V.K., and Shukla, R. 2018. Metal uptake potential of four methylotrophic bacterial strains from coal mine spoil, exploring a new possible agent for bioremediation. *Environmental Technology & Innovation* 11: 174–186.

Mitrović, M., Pavlović, P., Lakušić, D., Djurdjević, L., Stevanović, B., Kostić, O., Gajić, G. 2008. The potential of *Festuca rubra* and *Calamagrostis epigejos* for the revegetation of fly ash deposits. *Science of the Total Environment* 407(1): 338–347.

Mukhopadhyay, S., Maiti, S.K. 2011. Trace metal accumulation and natural mycorrhizal colonisation in an afforested coalmine overburden dump: A case study from India. *International Journal of Mining, Reclamation and Environment* 25(2): 187–207.

Mukhopadhyay, S., Maiti, S.K. 2011. Trace metal accumulation and natural mycorrhizal colonisation in an afforested coalmine overburden dump: A case study from India. *International Journal of Mining, Reclamation and Environment* 25(2): 187–207.

Mukhopadhyay, S., Masto, R.E. 2016. Carbon storage in coal mine spoil by Dalbergia sissoo Roxb. *Geoderma* 284: 204–213.

Mukhopadhyay, S., Masto, R.E., Tripathi, R.C., Srivastava, N.K. 2019. Application of soil quality indicators for the phytorestoration of mine spoil dumps. In *Phytomanagement of Polluted Sites*. Elsevier, pp. 361–388.

Nganje, T.N., Adamu, C.I., Ugbaja, A.N., Ebieme, E., Sikakwe, G.U. 2011. Environmental contamination of trace elements in the vicinity of Okpara coal mine, Enugu, Southeastern Nigeria. *Arabian Journal of Geosciences* 4(1–2): 199–205.

Ning, C.C., Gao, P.D., Wang, B.Q., LIN, W.P., Jiang, N.H., Cai, K.Z. 2017. Impacts of chemical fertilizer reduction and organic amendments supplementation on soil nutrient, enzyme activity and heavy metal content. *Journal of Integrative Agriculture* 16(8): 1819–1831.

Novo, L.A., Castro, P.M., Alvarenga, P., da Silva, E.F. 2018. Plant growth–promoting rhizobacteria-assisted phytoremediation of mine soils. *Bio-Geotechnologies for Mine Site Rehabilitation*. Elsevier, pp. 281–295.

Olivares, A.R., Carrillo-González, R., González-Chávez, M.D.C.A., Hernández, R.M.S. 2013. Potential of castor bean (*Ricinus communis* L.) for phytoremediation of mine tailings and oil production. *Journal of Environmental Management* 114: 316–323.

Palansooriya, K.N., Shaheen, S.M., Chen, S.S., Tsang, D.C., Hashimoto, Y., Hou, D., Ok, Y.S. 2020. Soil amendments for immobilization of potentially toxic elements in contaminated soils: A critical review. *Environment International* 134: 105046.

Pandey, V.C. 2013. Suitability of Ricinus communis L. cultivation for phytoremediation of fly ash disposal sites. *Ecological Engineering* 57: 336–341.

Pandey, V.C., Singh, N. 2014. Fast green capping on coal fly ash basins through ecological engineering. *Ecological Engineering* 73: 671–675.

Pankaj, U., Verma, R.S., Yadav, A., Verma, R.K. 2019. Effect of arbuscular mycorrhizae species on essential oil yield and chemical composition of commercially grown palmarosa (*Cymbopogon martinii*) varieties in salinity stress soil. *Journal of Essential Oil Research* 31(2): 145–153.

Parraga-Aguado, I., Querejeta, J.I., González-Alcaraz, M.N., Jiménez-Cárceles, F.J., Conesa, H.M. 2014. Usefulness of pioneer vegetation for the phytomanagement of metal (loid) s enriched tailings: Grasses vs. shrubs vs. trees. *Journal of Environmental Management* 133: 51–58.

Pavlović, P., Mitrović, M., Djurdjević, L. 2004. An ecophysiological study of plants growing on the fly ash deposits from the "Nikola Tesla–A" thermal power station in Serbia. *Environmental Management* 33(5): 654–663.

Pinto, A. P., de Varennes, A., Castanheiro, J. E., & Balsinhas, A. M. 2018. Fly ash and lime-stabilized biosolid mixtures in mine spoil reclamation. In *Bio-geotechnologies for mine site rehabilitation*. Elsevier, pp. 159–180.

Pourrut, B., Lopareva-Pohu, A., Pruvot, C., Garçon, G., Verdin, A., Waterlot, C., Douay, F. 2011. Assessment of fly ash-aided phytostabilisation of highly contaminated soils after an 8-year field trial: Part 2. Influence on plants. *Science of the Total Environment* 409(21): 4504–4510.

Ranđelović, D., Gajić, G., Mutić, J., Pavlović, P., Mihailović, N., Jovanović, S. 2016. Ecological potential of *Epilobium dodonaei* Village for restoration of metalliferous mine wastes. *Ecological Engineering* 95: 800–810.

Ribeiro, J., Da Silva, E.F., Li, Z., Ward, C., Flores, D. 2010. Petrographic, mineralogical and geochemical characterization of the Serrinha coal waste pile (Douro Coalfield, Portugal) and the potential environmental impacts on soil, sediments and surface waters. *International Journal of Coal Geology* 83(4): 456–466.

Rungwa, S., Arpa, G., Sakulas, H., Harakuwe, A., Timi, D. 2013. Phytoremediation–an eco-friendly and sustainable method of heavy metal removal from closed mine environments in Papua New Guinea. *Procedia Earth and Planetary Science* 6: 269–277.

Santibañez, C., de la Fuente, L.M., Bustamante, E., Silva, S., Leon-Lobos, P., Ginocchio, R. 2012. Potential use of organic-and hard-rock mine wastes on aided phytostabilization of large-scale mine tailings under semiarid Mediterranean climatic conditions: short-term field study. *Applied and Environmental Soil Science*. doi: https://doi.org/10.1155/2012/895817.

Santibáñez, C., Verdugo, C., Ginocchio, R. 2008. Phytostabilization of copper mine tailings with biosolids: Implications for metal uptake and productivity of *Lolium perenne*. *Science of the Total Environment* 395(1): 1–10.

Santos, E.S., Abreu, M.M., Magalhães, M.C.F. 2016. *Cistus ladanifer* phytostabilizing soils contaminated with non-essential chemical elements. *Ecological Engineering* 94: 107–116.

Singh, G., Pankaj, U., Chand, C., Verma, R.K. 2019. Arbuscular Mycorrhizal Fungi-Assisted Phytoextraction of Toxic Metals by *Zea mays* L. From Tannery Sludge. *Soil and Sediment Contamination: An International Journal* 28(8):729–746.

Singh, K. N., & Narzary, D. 2021. Heavy metal tolerance of bacterial isolates associated with overburden strata of an opencast coal mine of Assam (India). *Environmental Science and Pollution Research*, 28(44): 63111–63126.

Sopper, W.E. (Ed.). 1993. *Municipal Sludge Use in Land Reclamation*. CRC Press.

Spargo, A., Doley, D. 2016. Selective coal mine overburden treatment with topsoil and compost to optimise pasture or native vegetation establishment. *Journal of Environmental Management* 182: 342–350.

Teixeira, L.A.J., Berton, R.S., Coscione, A. R., Saes, L.A. 2011. Biosolids application on banana production: Soil chemical properties and plant nutrition. *Applied and Environmental Soil Science*. Article id 238185, 8. https://doi.org/10.1155/2011/238185

Tripathi, N., Singh, R.S., Hills, C.D. 2016. Soil carbon development in rejuvenated Indian coal mine spoil. *Ecological Engineering* 90: 482–490.

Tripathi, N., Singh, R.S., Nathanail, C.P. 2014. Mine spoil acts as a sink of carbon dioxide in Indian dry tropical environment. *Science of the Total Environment* 468: 1162–1171.

Turgay, O.C., Bilen, S. 2012. The role of plant growth-promoting rhizosphere bacteria in toxic metal extraction by *Brassica* spp. In *The Plant Family Brassicaceae*, Turgay, O.C. and Bilen, S. (Eds.). Dordrecht: Springer, pp. 213–237.

Vuppaladadiyam, S.S.V., Baig, Z.T., Soomro, A.F., Vuppaladadiyam, A.K. 2019. Characterisation of overburden waste and industrial waste products for coal mine rehabilitation. *International Journal of Mining, Reclamation and Environment* 33(8): 517–526.

Wężowicz, K., Rozpądek, P., Turnau, K. 2017. Interactions of arbuscular mycorrhizal and endophytic fungi improve seedling survival and growth in post-mining waste. *Mycorrhiza* 27(5): 499–511.

Wong, J.W., Ho, G.E. 1994. Effectiveness of acidic industrial wastes for reclaiming fine bauxite refining residue (red mud). *Soil Science* 158(2): 115–123.

Wong, J.W.C., Ho, G.E. 1993. Use of waste gypsum in the revegetation on red mud deposits: A greenhouse study. *Waste Management & Research* 11(3): 249–256.

World Coal Association (WCA) 2019. Available at: https://www.worldcoal.org/coal. Accessed on 10 November 2020.

World Energy Outlook (WEO) 2019. Available at: https://www.iea.org/reports/world-energy-outlook-2019. Accessed on 10 November 2020.

Wulandari, D., Cheng, W., Tawaraya, K. 2016. Arbuscular mycorrhizal fungal inoculation improves *Albizia saman* and *Paraserianthes falcataria* growth in post-opencast coal mine field in East Kalimantan, Indonesia. *Forest Ecology and Management* 376: 67–73.

Yamaji, K., Watanabe, Y., Masuya, H., Shigeto, A., Yui, H., Haruma, T. 2016. Root fungal endophytes enhance heavy-metal stress tolerance of *Clethra barbinervis* growing naturally at mining sites via growth enhancement, promotion of nutrient uptake and decrease of heavy-metal concentration. *PloS One* 11(12): e0169089.

Zhang, Y., Zhang, Y., Liu, M., Shi, X., Zhao, Z. 2008. Dark septate endophyte (DSE) fungi isolated from metal polluted soils: Their taxonomic position, tolerance, and accumulation of heavy metals in vitro. *Journal of Microbiology* 46(6): 624–632.

Zhao, R., Guo, W., Bi, N., Guo, J., Wang, L., Zhao, J., Zhang, J. 2015. Arbuscular mycorrhizal fungi affect the growth, nutrient uptake and water status of maize (*Zea mays* L.) grown in two types of coal mine spoils under drought stress. *Applied Soil Ecology* 88: 41–49.

12 Microbial Remediation of Hazardous Chemical Pesticides toward Sustainable Agriculture

Usha, M. Soniya Devi, and Vijay Kumar Mishra
Rani Lakshmi Bai Central Agricultural University, Jhansi, India
Radhey Shyam
Bihar Agriculture University, Sabour, India
Umesh Pankaj
Rani Lakshmi Bai Central Agricultural University, Jhansi, India

CONTENTS

DOI: 10.1201/9781003147091-12

12.1 INTRODUCTION

Pesticides have been used in agriculture to protect crops from pests, weeds, and diseases, but as much as 80 to 90% of applied pesticides hit non-target vegetation. After application, many harmful chemical pesticides may remain as residue in the environment which is potentially a grave risk to the agricultural ecosystem. Currently, the remediation techniques involve physical, chemical, and biological remediation as well as combined ways for the removal of contaminants. Bioremediation is the process by which living beings such as plants, algae and microorganisms are used to remediate, reduce or remove contamination from the environment. In other word the term bioremediation is defined as a "biological response to environmental abuse". Under this process primarily microorganisms is used to degrade the environmental contaminants into less toxic forms. Bioremediation is a combination of two words—bio for biological, and remediation which means to remediate or eliminate. Bioremediation may be of a different type, depending upon type of organism being used, i.e., phytoremediation (the use of plants to clean up the environment and mycoremediation (the use of fungi for the clean-up of contaminated environment).

Bioremediation is often characterized as environmentally friendly and a cost-effective approach for the removal of contaminants from the environment. For example, hydrocarbons are broken down by microorganisms and plants into greener products or are converted by other microbial species into water, CO_2, and other inorganic compounds. Hydrocarbon bioremediation, which involves heterogeneous and complex processes, is highly efficient but still takes a few weeks or even months to complete. Since effective bioremediation requires the action of microbial enzymes to transform contaminants into non-hazardous compounds, operating conditions must be optimized to allow microbial cells to quickly biodegrade them. The use of microorganisms is influenced by some factors, including the type of pollutant, the environmental conditions, as well as the presence of nitrogen and phosphorus sources. Temperature has a strong impact on oil hydrocarbon degradation, due to its influence on petroleum physical state and chemical composition, microbial metabolism, and composition of microbial consortium. An optimal temperature of hydrocarbon biodegradation is influenced by climate, site location and microbial species. Even though hydrocarbon degradation occurs over a wide temperature range, usually the higher the temperature, the faster the bioprocess. Other important factors that directly influence bioremediation are the medium pH, redox potential, moisture content, availability of oxygen and other compounds, soil/water composition, and pollutant solubility. Bioremediation of organic compound-polluted soil is influences by several factors such as:

1. Pesticide bioavailability
2. Rate of microbial uptake
3. Rate of enzymatic degradation
4. Rate of cell growth using the pesticide as carbon and energy source

Bioremediation approaches are done either *in situ* or *ex situ* remediation. *In situ* bioremediation can be described as the process whereby pesticides are biologically degraded under natural conditions to either carbon dioxide and water or an attenuated transformation product. It is a low cost, low maintenance, environmentally friendly, and sustainable approach for the clean-up of contaminated soils (Megharaj et al. 2011).

For *ex situ* bioremediation, contaminated soils have to be excavated and moved to another site for treatment, which may result in a high cost. Therefore, considering the large scale of agricultural land, *in situ* bioremediation is preferred to *ex situ* remediation for restoration of contaminated agricultural soils. Bioremediation processes include three different types (Figure 12.1).

1. Bioattenuation (or natural attenuation), which depends on the natural process of degradation.
2. Biostimulation, where intentional stimulation of pesticide degradation is achieved by addition of water, nutrients, and electron donors or acceptors.
3. Bioaugmentation, where microbes with proven capabilities of degrading or transforming pesticides are added (Madsen 1991).

The use of a particular bioremediation technology depends on several factors, such as site conditions, microbial activity (indigenous or exogenous strains), and the type, quantity, and toxicity of the pollutant chemical species present.

12.1.1 Bioattenuation

In bioattenuation, the biodegradation occurs naturally without human intervention and the process is determined by the metabolic potential of microorganisms to detoxify or transform the pesticide molecule, which is dependent on both accessibility

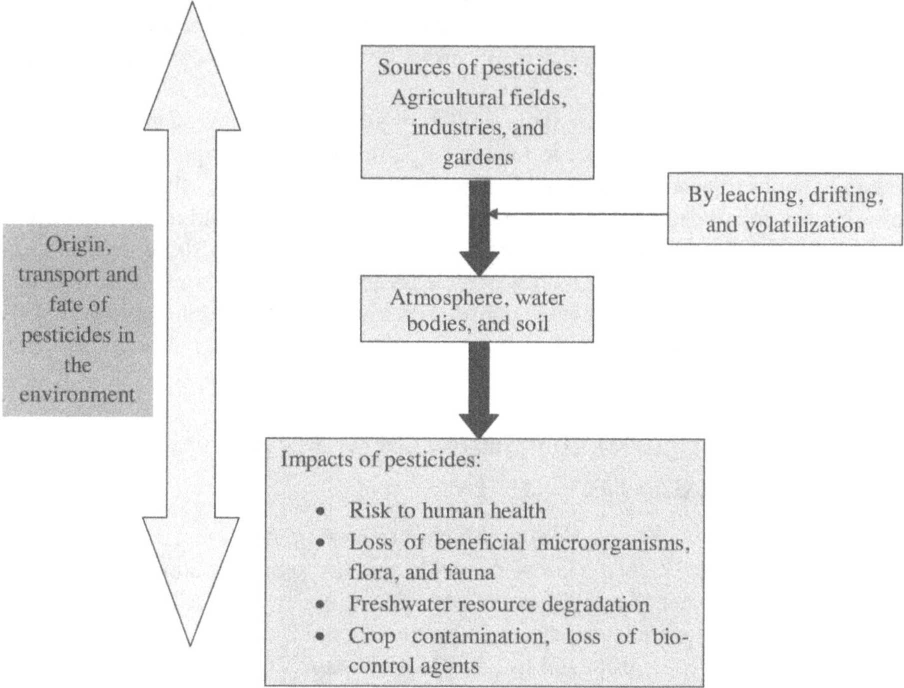

FIGURE 12.1 Microbial mediated bioattenuation, biostimulation and bioaugmentation processes for pesticides remediation.

and bioavailability. This process is often considered as the "do-nothing" solution, but it requires constant monitoring of the contaminant in soil. The time required for natural attenuation depends on the site conditions and the specific contaminant. Natural attenuation of a pesticide in some soils may take a very long time because of its persistent nature and lack of appropriate degrading microorganisms. The processes involved are largely due to biodegradation by soil microorganisms and to some extent by interaction with soil matrices.

12.1.2 Biostimulation

In the biostimulation process, the biodegradation can be stimulated by creating the correct environmental conditions for microorganisms in a soil, and the success of bioremediation depends on a number of soil physiochemical factors such as moisture, redox conditions, temperature, pH, organic matter, and nutrients that affect microbial activity and chemical diffusion in the soil (Hussain et al. 2009). The appropriate nutrient ratio of C: N: P is very important in the biostimulation process (Wolicka et al. 2009).

12.1.3 Bioaugmentation

In recent years, bioremediation technologies using bioaugmentation approaches have received attention within the scientific community and remediation industry. In a typical bioaugmentation approach, altered microorganisms are added to a contaminated site to accelerate detoxification and degradation. Such amended microorganisms can be a single bacterial strain or a syntropic bacterial consortium, which are isolated from natural environments or genetically engineered in the laboratory (Wackett et al. 2002). However, bioremediation via bioaugmentation approaches are more often met with failure due to lower competitiveness and adaptability relative to the indigenous microorganisms in the contaminated soils. On the other hand, immobilizing microbial degraders or enzymes on various carriers could make them more stable and resistant to environmental changes (Owsianiak et al. 2010). Bioremediation is therefore a promising technology for clean-up of pesticide-contaminated sites, but it is still in the developmental phase. Not all pesticides in contaminated soils are substrates for microbial metabolism; hence, bioremediation is limited to those chemicals that are biodegradable.

12.2 CURRENT SCENARIO OF CHEMICAL PESTICIDE USE IN THE INDIAN AGRO-ECOSYSTEM

At present, the Indian agro-system is using more pesticides and treating crops more frequently than ever before. Global pesticide use (in tonnes of active ingredients) increased by 46% between 1996 and 2016, according to the FAOSTAT database WHO (2019). The pesticide usage in India has surged hundreds of times over the previous seven decades, from 154 MT in 1953–1954 to 57,000 MT in 2016–2017. In 1994–1995, pesticide usage in India was around 80,000 MT as per the report produced by FAO (2018). The use of pesticides has been increasing immensely in India, and the waste generated from the pesticide industry has become a severe environmental problem due to the present insufficient and ineffective waste treatment technology involving physicochemical and biological treatments. As a result, recalcitrant pesticide residues remain in surface soil, leading to the toxicity of the soil-water environment. Bearing in mind the deadly effects of pesticides, it is necessary to eliminate these chemo-pollutants from the environment. Bioremediation, including phytoremediation and rhizoremediation, is estimated to be a useful cleaning method for soil contaminated with persistent organic pesticides. Conventional methods of pesticide detoxification have relied on landfills and incineration, which generate secondary contamination problems due to leaching of pesticides into the surrounding soil and groundwater supplies and produce potentially toxic by-product emissions. From this point of view, biodegradation is a much safer and economical process for detoxifying pesticides. Many techniques of dispersal, collection, removal, landfill disposal, and incineration simply dilute or sequester the contaminants or transfer them to another environment. By contrast, bioremediation may be viewed as a more effective and environmentally safe clean-up technology, since it results in the partial or complete bioconversion of the organic pollutants, such as pesticides, to microbial biomass and stable non-toxic endproducts (Baker and Herson 1994). The application of fungal technology (mycoremediation) for the clean-up of contaminated soil has held promise since 1985, when the white rot fungus *Phanerochaete chrysosporium* was found to be able to degrade a number of important environmental pollutants (Bumpus et al. 1985). Since then several strains of white-rot fungi have been demonstrated to attack many organopollutants. There are many advantages of bioremediation technology: (a) It harnesses natural biogeochemical processes. (b) It is a cost-effective alternative. (c) Toxic chemicals are destroyed or removed from the environment and not merely separated. (d) There is a low capital expenditure. (e) Less energy is required when compared with other technologies. (f) Less manual supervision is required.

12.3 EFFICIENT AVAILABLE TECHNOLOGIES FOR BIOREMEDIATION

Organic pollutants can be bioremediated using either *in situ* or *ex situ* systems. On the basis of this, bioremediation technologies have been divided into two categories.

12.3.1 *In Situ* Bioremediation Technique

In situ remediation involved the enhancement of biodegradation rate of organic contaminants within the affected area like: Soil, sediment, and surface water or groundwater environments. It is the clean-up approach that directly involves contact between microorganisms and the dissolved and sorbed contaminants for biotransformation (Alcalde et al. 2006). These types of technologies are less expensive, create less dust, and cause less release of volatile contaminants. Potential advantages of *in situ* bioremediation methods include minimal site disruption, simultaneous treatment of contaminated soil and groundwater, minimal exposure of public and site personnel, and low cost. But these techniques also have some disadvantages such as: The method is time consuming, as compared to other remedial methods, the seasonal variation of microbial activity under the influence of changing environmental conditions, and the problematic application of treatment additives (nutrients, surfactants, and oxygen). When native microorganisms lack biodegradable capacity, genetically engineered microorganisms may be added to the surface during *in situ* bioremediation of by-products of raw materials and wastes. Bacteria are the most commonly used agents in these biodegradation processes.

- **Biostimulation:** This involves the introduction of nutrients or substrates such as fertilizers, to stimulate the growth and metabolism of the indigenous species performing biodegradation of pollutants
- **Biosparging:** This involves the inoculation of air under pressure below the water level to increase the groundwater oxygen concentrations and enhance the rate of biological degradation of contaminants by indigenous microorganisms. Biosparging increases the mixing in the saturated zone and thereby increases the contact between the soil and groundwater

12.3.2 *Ex Situ* Bioremediation Techniques

This technique is usually aerobic and engages in the treatment of contaminated soil or sediments using solid or slurry phase systems. It actually involves the removal of waste material from polluted areas and their collection at a place distant from the contaminated site to facilitate microbial degradation. Based on the physical nature of the contaminant treatment, this is classified into two types:

a. Slurry phase bioremediation systems, involving treatment of solid liquid suspensions in bioreactors
b. Solid phase bioremediation system, including soil treatment units and soil piles, engineered bio piles and compost heaps

12.3.2.1 Slurry Phase Bioremediation

Slurry phase bioremediation is a controlled treatment technique that excavates soil or sediments that are mixed with water and then treated in bioreactor vessels. In this process the soil involves separation of stones and rubble from the contaminated soil to provide a low viscosity. After that, the soil is mixed with a predetermined amount of water to form slurry. Thus, slurry phase treatment is a triphasic system involving three major components: Water, air, and suspended particulate matter, including the desired microorganism. The concentration of water added depends on the concentration of pollutants, the rate of biodegradation, and the physical nature of the soil (USEPA 2006). In addition to the provision of adequate aeration and mixing, nutrients are universally added, together with surfactants or dispersants as required. Optimum pH and temperature conditions are also provided in bioreactor vessels. Effective bioremediation has been obtained with slurry phase systems for soil and sediments contaminated with a wide range of organic compounds including pesticides, petroleum hydrocarbons, pentachlorophenol, polychlorinated biphenyls, (PCBs), etc. A slurry bioreactor vessel apparatus is used to create a three-phase (solid, liquid, and gas) mixing condition to increase the rate of bioremediation of soil bound and water-soluble pollutants as a water slurry of the contaminated soil and indigenous microorganisms capable of degrading target contaminants. In general, the rate and extent of biodegradation are greater in a bioreactor system than *in situ* or in solid-phase systems because the contained environment is more convenient and hence more controllable and predictable. But side-by-side, there are disadvantages of bioreactor systems, such as the contaminated soil requires pre-treatment. Before placing it in a bioreactor, the contaminant is stripped from the soil via soil washing or vacuum extraction.

12.3.2.2 Solid Phase Bioremediation

The solid wastes include organic wastes, manures wastes, sewage sludge, and municipal solid wastes. The traditional cleaning practices involve the informal processing of organic materials and production of compost, which may be used as soil amendments. But in the solid phase bioremediation contaminated soil is excavated and placed into piles. Bacterial growth is stimulated through a network of pipes that are distributed throughout the piles. By pulling air through pipes, the necessary ventilation is provided for microbial respiration and the moisture introduced by spraying of water. Solid-phase systems require a large space and clean-ups require more time to complete than the slurry-phase processes. Some solid-phase treatment processes include land farming chemical groups on mineral surfaces, reactive organic compounds, and inorganic metals.

Microbial bioremediation can take place under aerobic as well as anaerobic conditions. In aerobic conditions, microorganisms use the available atmospheric oxygen for their metabolic functions in order to produce carbon dioxide and water through pesticide degradation. However, under anaerobic conditions, due to the absence of oxygen, microorganisms use these chemical compounds in the soil as substrate, breaking them down to obtain the energy they need. Field et al. (1995) reviewed the intrinsic chemical considerations that limit the biodegradability of aromatic pollutants in aerobic and anaerobic environments.

12.3.3 Phytoremediation

It is an innovative technology that is gaining recognition as a cost-effective and aesthetically-pleasing method of remediating contaminated sites. Due to the fact that herbicides are designed to kill plants, the use of phytoremediation to remediate them can be a difficult and complicated task. Many studies have been done to determine the effectiveness of remediating persistent pollutants with various plant species, and more results are frequently being reported. A significant amount of work has been conducted to examine the ability of plants to remediate heavy metal contaminated soils. Plants are often capable of the uptake and storage of significant concentrations of some heavy metals and other compounds in their roots, shoots, and leaves, referred to as phytoextraction. The plants are then harvested and disposed of in an approved manner, such as in a hazardous waste landfill. This technique results in up to a 95% reduction in waste volume over the equivalent concentration of contaminated soil. The plants that are capable of this type of remediation are referred to as hyperaccumulators. Types of plants that appear promising for this form of remediation include the mustard plant, alpine pennycress, broccoli, and cabbage. Phytotransformation occurs when plants transform organic contaminants into less toxic, less mobile, or more stable forms. This process includes phytodegradation, which is the metabolism of the organic contaminant by the plant enzymes and phytovolatilization, which is the volatilization of organic contaminants as they pass through the plant leaves. The releases of these pollutants into the air results in the exchange of one form of pollution for another. Phytostabilization immobilizes the contaminants and reduces their migration through the soil by absorbing and binding leachable constituents to the plant structure. This process effectively reduces the bioavailability of the harmful contaminants. Almost any vegetation present at contaminated sites will contribute to phytostabilization (Arthur and Coats 1998).

At the soil-root interface, known as the rhizosphere, there is a very large and very active microbial population. Often the plant and microbial populations provide needed organic and inorganic compounds for one another. The rhizosphere environment is high in microbial abundance and rich in microbial metabolic activity, which has the potential to enhance the rate of biodegradation of contaminants by the microorganisms. Generally, the plant is not directly involved in the biodegradation process. It serves as a catalyst for increasing microbial growth and activity, which subsequently increases the biodegradation potential. However, the rhizosphere can be limited in its remediation potential because it does not extend far from the root. This process is often referred to as phytostimulation, or plant-assisted bioremediation. Currently, a significant amount of research is being conducted on the interaction between microorganisms and plants in the rhizosphere and the potential to use this for the remediation of pesticide-contaminated media. According to preliminary studies, enhanced degradation of atrazine, metolachlor, and trifluralin have been observed in contaminated soils where plants of the *Kochia* sp. have been planted. The increased degradation occurs in the rhizosphere of this herbicide-tolerant plant, suggesting that rhizosphere interactions between the plant and microorganisms have led to the increased degradation of the pesticides present (Coats and Anderson 1997). Additional studies using the *Kochia* sp. have been conducted by these researchers and also show promise for the phytoremediation of pesticide-contaminated soils and groundwater (Arthur and Coats 1998). In laboratory studies, the quick-growing and deep-rooted poplar tree has been shown to be successful in the remediation of groundwater (Poplar trees may remove contaminants 1998). Its rapid growth requires high volumes of water that are pulled from the saturated zone. Contaminated groundwater is absorbed by the plant and the pollutants are subsequently transformed into organic molecules for plant growth. This technique has already proven successful for the remediation of atrazine-contaminated soil and groundwater (Poplar trees may remove contaminants 1998). Hybrid poplars are being used in the remediation of a farm chemical site with high levels of nitrate, atrazine and arochlor in the groundwater. Approximately 1.5 acres are being treated; results to date show reduction of both the nitrates and herbicides in the groundwater (Phytoremediation of Organics Action Team 2000). Phytoremediation is still relatively unproven and its capabilities are still being discovered.

12.4 BIOREMEDIATION OF DIFFERENT GROUPS OF PESTICIDES

12.4.1 Bioremediation of Organophosphorous Pesticides (OPs)

Pesticides have become an inevitable part of the modern environment as they are widely used in agriculture, household, and the public health sector and, hence, are extensively distributed throughout most ecosystems. Currently, OP pesticides are the most commercially favored group of pesticides, with large application all over the world. Depending on their fate (Figure 12.2), these organophosphorus compounds may become bioavailable for microbial degradation. Environmental microbes, such as *Aspergillus*, *Pseudomonas*, *Chlorella*, and *Arthrobacter*, are capable of coupling a variety of physical and biochemical mechanisms for the degradation of organophosphate pesticides, including adsorption, hydrolysis of P–O alkyl and aryl bonds, photodegradation, and enzymatic mineralization. Enzymes, such as esterase, diisopropyl

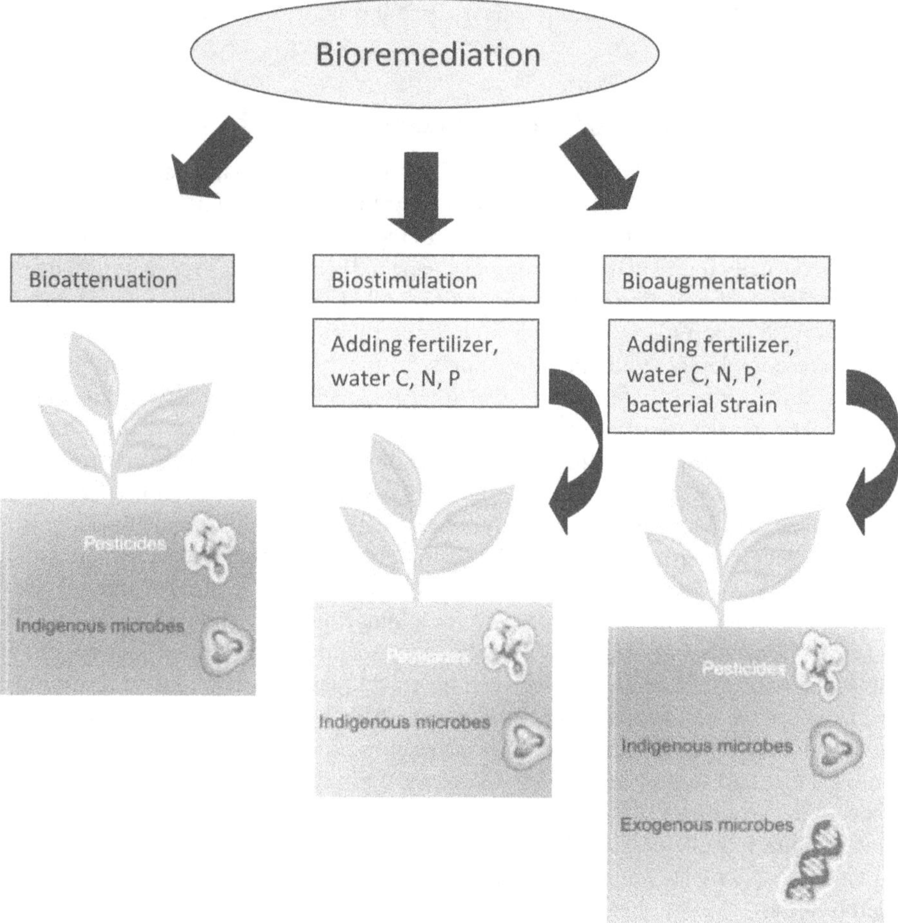

FIGURE 12.2 Impact of pesticide application on soil, water and agricultural produce.

fluorophosphatase, phosphotriesterase, somanase, parathion hydrolase, and paraoxonase have been isolated from microbes to study and understand the catabolic pathways involved in the biotransformation of these xenobiotic compounds. This review highlights various aspects of biodegradation of organophosphate pesticides, along with biological and molecular characterization of some organophosphate pesticide-degrading bacteria.

Scientists have been exploring the bacterial diversity, particularly of contaminated areas, in the search for indigenous bacteria that can utilize and degrade a wide range of pollutants (Stroud et al. 2007). Biotransformation of organic contaminants in the natural environment has been extensively studied to understand microbial physiology, ecology, and evolution due to their bioremediation potential. Among the organophosphate pesticides, chlorpyrifos, dichlorvos, diazinon, fenamiphos, fenitrothion, isofenophos, parathion, phorate, malathion, parathion-methyl, monocrotophos, profenophos, and others have been used extensively and their environmental fate and bacterial degradation approaches have been studied extensively. Some of the organophosphate pesticides with bacterial/fungal isolates capable of degrading them are detailed in (Table 12.1).

12.4.2 Bioremediation of Organochlorine

Most organochlorine compounds, such as Dieldrin and endrin are persistent organic pollutants that cause serious environmental problems. Although these compounds have been prohibited over the past decades in most countries around the world, they are still routinely found in the environment, especially in the soil in agricultural fields. Bioremediation, including phytoremediation and rhizoremediation, is expected to be a useful cleanup method for this soil contamination. Organochloride pesticides (OCs) are completely synthetic compounds. In the 1970s, these were widely used, mainly in the United States. Because of their wide structural variety and divergent chemical properties they have a broad range of applications. Because of the effects of these compounds on the environment, the use of many derivatives are controversial. Although their use has been banned in many countries, they are still used in developing countries. Many xenobiotic compounds, especially the organochloride pesticides, are recalcitrant and resistant to biodegradation (Chaudhry and Chapalamadugu 1991; Diaz 2004; Dua et al. 2002). The rate of breakdown of most of them is slow and they can remain in the environment for a long time after application and also in organisms long after exposure. They act by hyperexcitation in the nervous system of the animals, causing acute toxic effects; death frequently occurs due to respiratory failure after the disruption of nervous system function. Organochloride pesticides are accumulated in the organisms and causes chronic health effects, like cancer and neurological and teratogenic effects (Vaccari et al. 2006). In general, these highly toxic and carcinogenic compounds persist in the environment for many years. Some of the most widely used organochlorine pesticides include, dieldrin, aldrin, alpha-BHC, beta-BHC, deltaBHC, gamma-BHC (Lindane), heptachlor, endosulfan, methoxychlor, aroclor, and dichloro diphenyl trichloro ethane (DDT) (Menone et al. 2001; Patnaik 2003). In 1939, the use of organochlorine pesticides started when Paul Hermann Müller realized that the DDT was an efficient insecticide (Matolcsy et al. 1988).

TABLE 12.1
Biodegradation of Different Organophosphate Pesticides (OPs) by Microbes

Insecticides	Microbial Strain	References
Acephate	*Exiguobacterium* sp., *Rhodococcus* sp.	Phugare et al. 2012
Cadusafos	*Pseudomonas putida*	Abo-Amer 2012
	Sphingomonas sp., *Flavobacterium* sp.	Karpouzas et al. 2005
Chlorfenvinphos	*Arthrobacter* sp., *Mycobacterium* sp.	Seo et al. 2007
Chlorpyrifos	*Bacillus aryabhattai*	Pailan et al. 2015
	Stenotrophomonas sp.	Deng et al. 2015
	Lactobacillus brevis, Lactobacillus plantarum	Zhang et al. 2014
	Cupriavidus sp.	Lu et al. 2013
	Serratia marcescens	Cycon et al. 2013
	Sphingobacterium sp.	Abraham and Silambarasan 2013
	P. putida, Klebsiella sp., *P. stutzeri, P. aeruginosa*	Sasikala et al. 2012
	S. chattanoogensis, S. olivochromogenes	Briceno et al. 2012
	Cladosporium cladosporioides	Gao et al. 2012
	P. aeruginosa, P. nitroreducens, P. putida	Latifi et al. 2012
	Bacillus sp., *Pseudomonas* sp.	Madhuri and Rangaswamy 2009
Chlorpyrifosmethyl	*Burkholderia cepacia*	Kim and Ahn 2009
Coumaphos	*Leuconostoc mesenteroides, L. brevis, L. plantarum, L. sakei*	Cho et al. 2009
Dichlorvos	*Proteus vulgaris, Vibrio* sp., *Serratia* sp., *Acinetobacter* sp.	Agarry et al. 2013
	Bacillus sp., *Pseudomonas* sp.	Madhuri and Rangaswamy 2009
	Bacillus sp.	Pawar and Mali 2014
Diazinon	*Stenotrophomonas* sp.	Deng et al. 2015
	Lactobacillus brevis	Zhang et al. 2014
	Serratia marcescens	Abo-Amer 2011
Dimethoate	*P. pseudoalcaligenes, Exiguobacterium aurantiacum, Bacillus* sp., *Micrococcus luteus, Aspergillus niger*	L´opez et al. 2005
Fenitrothion	*Lactobacillus brevis*	Zhang et al. 2014
	Serratia marcescens	Cycon et al. 2013
Parathion	*Stenotrophomonas* sp.	Deng et al. 2015
	Leuconostoc mesenteroides, L. brevis, L. plantarum, L. sakei	Cho et al. 2009
	Serratia marcescens	Cycon et al. 2013
Phorate	*Ralstonia eutropha, P. aeruginosa, E. cloacae*	Rani and Juwarkar 2012
	Bacillus sp., *Pseudomonas* sp.	Madhuri and Rangaswamy 2009
	Rhizobium sp., *Pseudomonas* sp., *Proteus* sp.	Bano and Musarrat 2003
Triazophos	*Stenotrophomonas* sp.	Deng et al. 2015
Malathion	*Bacillus amyloliquefaciens, Pseudomonas* sp.	Baishya and Sharma 2014

(Continued)

TABLE 12.1 (*Continued*)
Biodegradation of Different Organophosphate Pesticides (OPs) by Microbes

Insecticides	Microbial Strain	References
	Lactobacillus brevis	Zhang et al. 2014
	Pseudomonas sp.	Jilani 2013
	Pseudomonas sp., *P. putida, Micrococcus lylae, Pseudomonas aureofaciens, Acetobacter liquefaciens*	Goda et al. 2010
	Enterobacter aerogenes, Bacillus thuringiensis	Mohamed et al. 2010
Isofenphos	*Arthrobacter* sp.	Ohshiro et al. 1997
Monocrotophos	*Arthrobacter atrocyaneus, Bacillus megaterium*	Bhadbhade et al. 2002
	Aspergillus fumigatusa, A. nigera, Paracoccus sp.	Pandey et al. 2014
Parathion-methyl	*Stenotrophomonas* sp.	Deng et al. 2015
	Agrobacterium sp.	Wang et al. 2012
	Bacillus sp., *Pseudomonas* sp.	Madhuri and Rangaswamy 2009
	Flavobacterium sp.	Ortiz-Hernaández et al. 2010
	Pseudomonas sp., *Azospirillum* sp.	Foster et al. 2004
Ethion	*Stenotrophomonas* sp.	Deng et al. 2015
Profenofos	*P. putida, Burkholderia gladioli*	Malghani et al. 2009
Quinalphos	*Staphylococcus* sp., *Bacillus licheniformis*	Baishya and Sharma 2014

During the World War II, for prevention of epidemics of typhus transmitted by lice, DDT powder was pulverized on the population's skin. In other countries, this insecticide was also used to control malaria-bearing mosquitoes (Konradsen et al. 2004). Although most organochlorines were banned from some countries, OCs pesticides are still widely studied due to their recalcitrant nature. Both biotic and abiotic factors determine the fate of pesticides in the environment. The rate that different pesticides undergo biodegradation varies widely. Some pesticides like DDT and dieldrin are recalcitrant and they accumulate in the environment for a long time and enter into food chains and accumulate for decades after their application to the soil (Kannan et al. 1994). The in situ degradation of organochlorine pesticides by pure cultures has been proven. Among microorganisms, bacteria comprise the major group involved in organochlorine degradation, especially soil habitants belonging to genera *Bacillus*, *Pseudomonas*, *Arthrobacter* and *Micrococcus* (Langlois et al. 1970) (Table 12.2). Matsumura et al. (1968) reported evidence of the breakdown of dieldrin in the soil by a *Pseudomonas* sp. Studies with fungi have also evidenced the biodegradation of organochlorine pesticides. Ortega et al. (2011) evaluated marine fungi collected off the coast of São Sebastião, North of São Paulo State, Brazil for the biodegradation of organochlorine. The fungi strains were obtained from marine sponges. The fungi like *Penicillium miczynskii*, *Aspergillus sydowii*, *Trichoderma* sp., *Penicillium raistrickii*, and *Bionectria* sp. were previously tested in solid culture medium for their degradation capability.

TABLE 12.2
Biodegradation of Different Organochloride (OC) Pesticides by Microbes

Pesticides	Microbial Species	References
DDT	*Aerobacter aerogenes*	Ortega et al. 2011
	Trichoderma viridae, Pseudomonas sp., *Micrococcus* sp., *Arthrobacter* sp., *Bacillus* sp.	Wedemeyer 1967
	Pseudomonas sp.	Patil et al. 1970
Dieldrin	*Pseudomonas* sp.	Matsumura et al. 1968
Endosulfan	*Pseudomonas aeruginosa, Burkholderia cepaeia, Arthrobacter* sp. *KW, Aspergillus niger*	Bhalerao and Puranik 2007
Lindane	*Pleurotus ostreatus, Streptomyces* sp.	Benimeli et al. 2008
DDE	*Phanerochaete chrysosporium*	Bumpus et al. 1993
DDD	*Phlebia* sp.	Xiao et al. 2010
Heptachlor O	*Pseudomonas* sp., *Micrococcus* sp., *Bacillus* sp.	Patil et al. 1970
Toxaphene O	*Pseudomonas* sp., *Micrococcus* sp. *Arthrobacter* sp., *Bacillus* sp.	
Aldrin O	*Pseudomonas* sp.	
Endrin O		
Dieldrin O		Matsumura et al. 1968

Abbreviations: DDT, dichloro-diphenyl-trichloroethane; DDE, dichlorodiphenyldichloroethylene; DDD, dichlorodiphenyldichloroethane.

12.4.3 Bioremediation of Carbamates

Numerous factors are involved in pesticide biodegradation such as the physicochemical mechanism, photochemical mechanism, microbial degradation and bioremediation. The biodegradation of carbamates by diverse microbial consortia that metabolize the pesticides has been studied extensively (Onunga et al. 2015; Satish et al. 2017). Mostly, the studies did not undermine the involvement of abiotic processes in the microbial degradation (Rangasamy et al. 2017). The use of pesticides over the past few decades has caused numerous microbes to develop mechanisms to degrade toxic compounds, including pesticides, by means of different mechanisms, approaches, and enzymatic pathways. Pesticides are mainly used to protect damage to crops and fruits while these chemicals are applied directly to soil. When these pesticides are used in soil, they undergo different processes such as volatilization, degradation, sorption, or surface transport to other places (Li et al. 2016). Studies showed that soils with a history of pesticide application mostly have shorter half-lives, compared with soil that has no history of pesticide application (Cycon et al. 2017). Moreover, the chemicals considered as non-degradable become biodegradable after several years in soil. Thus, the soil is a major source of microorganisms that can degrade pesticides. Other microbial sources include: Microbes from pesticide industry wastewater and sediment, activated sludge, sewage slurry, surface waters and their sediments. Many groups of microorganisms are characterized by the growth and degradation ability of pesticides (Ishag et al. 2016). Isolation and characterization of microbes for pesticides degradation bring about new tools to restore environments polluted with pesticides.

Several microbial species are capable of degrading pesticide such as *Pseudomonas*, *Flavobacterium*, *Achromobacterium* sp., *Sphingomonas* sp., *Arthrobacter* and *Bacillus* species have been isolated and characterized in an effort to know their mechanism to remove pesticides (Table 12.3). Degradation of carbofuran was first

TABLE 12.3
Biodegradation of Different Carbamate Pesticide by Microbes

Pesticides	Microbial Species	Environmental Media	Mechanism Involved	References
Carbofuran	*Consortia*	Soil	Degradation	Tien et al. 2017
	Pseudomonas sp.	Soil	Degradation	Devi et al. 2017
	Pseudomonas and *Alcaligenes*	Soil	Degradation	Omolo et al. 2012
	Enterobacter cloacae strain	Soil	Degradation	Fareed et al. 2017
	TA7	Soil	Degradation	Seo et al. 2007
	Novosphingobium sp.	Soil	Degradation	Nguyen et al. 2014
	Aspergillus sp.	Soil	Degradation	Devi et al. 2017
Carbaryl	*Bacillus, Morganella, Pseudomonas, Aeromonas, Corynebacterium* sp.	Soil	Degradation	Hamada et al. 2015
	Micrococcus sp.	Soil	Degradation	Doddamani and Ninnekar 2001
	Enterobacter cloacae sp.	Soil	Degradation	Fareed et al. 2017
	Paecilomyces	Soil	Degradation	Chan-Cupul et al. 2016
Propoxur	*Corynebacterium kutscheri, Staphylococcus aureus, Bacillus pasteurii* and *Aeromonas* sp.	MSW	Degradation	Anusha et al. 2009
	Consortia	Sediment	Degradation	Dewi et al. 2015
	Pseudomonas sp.	Soil	Degradation	Kamanavalli and Ninnekar 2000
Methomyl	*Escherichia coli*	Soil	Degradation	Kulkarni et al. 2018
	Pseudomonas sp.	Soil	Degradation	Kulkarni et al. 2014
	Stenotrophomonas sp.	Soil	Degradation	Mohamed 2009
Oxamyl	*Pseudomonas* sp.	Soil	Transformation	Rousidou et al. 2016
	Micrococcus luteus	Soil	Degradation	Mohamed 2017
	Aminobacter sp.	Soil	Degradation	Osborn et al. 2010
	Trichoderma sp.	Soil	Degradation	Helal et al. 2015
	Aspergillus niger	Soil	Degradation	Aemmr et al. 2013
Aldicarb	*Enterobacter cloacae*	Soil	Degradation	Fareed et al. 2017
	Stenotrophomonas maltophilia	Soil	Degradation	Aemmr et al. 2013
	Consortia	Sediment	Transformation	Kazumi and Capone 1995
Carbendizim	*Sphingomonas* sp.	Soil	–	Xiao et al. 2013
	Rhodococcus sp.	Soil	Degradation	Xiao et al. 2013
	Trichoderma sp.	Soil	Degradation	Tian and Chen 2009
	Nocardioides soli	Soil	–	Sun et al. 2014

carried out through degradation to carbofuran phenol and subsequently degraded to 2-hydroxy-3- (3-methylpropan-2-ol) phenol by *Sphingomonas* sp. and *Arthrobacter* sp. strains (Chaudhry and Ali 1988; Kim et al. 2004). Carbaryl is another carbamate that is widely used and poisonous to living organism (Anguiano et al. 2017). Bacteria identified as *Aeromonas*, *Pseudomonas*, *Bacillus* and *Morganella* genera isolated from Gaza Strip utilized carbaryl as a source of nutrient. *Bacillus* and *Morganella* degraded carbaryl up to 94.6% and 87.3% degradation rates (Hamada et al. 2015). In another study, *Pseudomonas* sp. C5pp isolated from soil was described to have mineralize carbaryl by means of 1-napthol, salicylate and gentisate (Trivedi et al. 2016). The genomic investigation of C5pp strain shows that carbaryl pesticides catabolic genes are arranged into three operons; middle, lower and upper. The study mentioned the role of horizontal gene transfer events in the evolution of carbaryl degradative pathway in C5pp isolate. Table 12.3 depicted the summary of microbial species used for biodegradation of carbamate pesticides, in which different bacterial and fungal strains from different sources were studied by different scholars.

Microorganisms have the capability to degrade multiplicity of environmental pollutants, including carbamate pesticides. It is highly significant to understand the biochemical bases for the development of new degradative capacities of microorganisms involved in pesticide degradation (Sogorb and Vilanova 2002). The enzyme-based degradation method is potentially capable of qualitatively faster action than traditional microbial remediation methods (Nguyen et al. 2014). Several enzymes that hydrolyzed carbamates compounds are either esterases or amidases. The hydrolysis of carbamate compound is influence by the chemical structure of the side chains and substrate (Lei et al. 2017). The parent compound is often detoxified by the hydrolytic process. Carbamate insecticide enzymes mediated hydrolysis typically yields an alcohol and methylamine, as well as carbondioxide (Lei et al. 2017). Several carbamate hydrolyzed enzymes responsible for degradation have been isolated; as well, carboxylesterases from various bacteria were previously reported (Ufarte et al. 2017; Wang et al. 2017).

12.4.4 Bioremediation of Neonicotinoids

Neonicotinoids are a relatively new and highly popular class of insecticides, with registrations in more than 120 countries since their introduction in the late 1980s (Jeschke et al. 2011; Simon-Delso et al. 2015). They share a common mode of action: Mimicking the action of neural transmitters affecting the central nervous system, resulting in paralysis and death (Godiyal 2013; Shivanandappa and Rajashekar 2014). These pesticides accounted for about 24% of the total world insecticides market in 2007 (Jeschke et al. 2011) and have been developed from the compound, nithiazine, a compound that is similar to the historical insecticide nicotine (Table 12.4). Neonicotinoid pesticides can be divided into three main classes: Chloropyridinyl compounds (imidacloprid, nitenpyram, acetamiprid, thiacloprid), chlorothiazolyl compounds (thiamethoxam, clothianidine), and tetrahydrofuryl compounds (dinotefuran). Imidacloprid was the first neonicotinoid insecticide introduced to the market, by Bayer in 1992, and is the most widely used insecticide worldwide.

Neonicotinoids are relatively water soluble and are absorbed through all parts of plants; thereby acting as systemic insecticides to protect the different phases of plant

TABLE 12.4
Bacterial Strains Capable of Degrading Neonicotinoid Pesticides

Insecticides	Microbial Strain	References
Imidacloprid	*Bacillus aerophilus*	Akoijam and Singh 2015
	Bacillus alkalinitrilicus	Sharma, Singh and Gupta 2014
	Bacillus sp.	Shaikh et al. 2014
	Brevundimonas sp.	Shetti and Kaliwal 2012
	Bacillus weihenstephanensis	Shetti, Kaliwal and Kaliwal 2014
	Burkholderia cepacia	Gopal et al. 2011
	Klebsiella pneumoniae BCH-1	Phugare et al. 2013
	Leifsonia sp. PC-21	Anhalt et al. 2007
	Mycobacterium sp. strain MK6	Kandil et al. 2015
	Ochrobactrum sp.	Hu et al. 2013
	Pseudoxanthomonas indica	Ma et al. 2014
	Pseudomonas sp. 1G	Pandey et al. 2009
	Rhizobium sp.	Sabourmoghaddam, et al. 2014
Acetamiprid	*Ensifer meliloti* CGMCC7333	Zhou et al. 2014b
	Ochrobactrum sp. D-12	Wang et al. 2013a
	Pigmentiphaga sp. AAP-1	Wang et al. 2013b
	Pigmentiphaga sp. D-2	Yang et al. 2013
	Pseudomonas sp. FH2	Yao and Min 2006
	Pseudoxanthomonas sp. AAP-7	Wang et al. 2013a
	Rhodococcus sp. BCH-2	Phugare and Jadhav 2014
	Stenotrophomonas sp. THZ-XP	Tang et al. 2012
	Stentrophomonas maltophila CGMCC1.178	Chen et al. 2008; Dai et al. 2010
Thiacloprid	*Ensifer meliloti* CGMCC7333	Ge et al. 2014
	Stenotrophomonas maltophilia CGMCC1.178	Zhao et al. 2009
	Variovorax boronicumulans J1	Zhang et al. 2012
Thiomethoxam	*Ensifer adhaerens* TMX-23	Zhou et al. 2013
	Pseudomonas sp.1G	Pandey et al. 2009

life from germination to maturity. The use of neonicotinoids is continually increasing because of their unique mode of action and the low levels of neonicotinoid resistance among insect pests. These pesticides are highly effective against sap feeding and chewing insect pest species such as whitefly, aphid, leaf hoppers, plant hoppers, and the beetles (Jeschke et al. 2011). Versatility is a feature of the neonicotinoids, whose physicochemical properties make them suitable for many application formulations, including foliar, seed treatment, soil drench, and stem injection (Elbert et al. 2008; Rouchaud et al. 1994). Seed treatment with these pesticides, in particular, has proven to be a very effective plant protection strategy resulting not only in increased efficiency of protection but also in reduced labor and costs for crop producers. About 60% of neonicotinoid applications are seed treatments, which are especially useful for transgenic plants expressing *Bacillus thuringiensis* (Bt) toxin genes, as the treatment protects the plant seedlings prior to the production of sufficient Bt toxin to provide effective pest resistance (Jeschke et al. 2011).

There is little published research on bacterial biodegradation of the neonicotinoid group of pesticides, with the exception of imidacloprid, for which there are 12 published studies. A few research papers have discussed the biodegradation of acetamiprid, thiacloprid, and thiamethoxam (Table 12.4). Bacterial biodegradation can be broadly separated into two categories: Biodegradation by pure bacterial cultures and that by microbial consortia. Bacterial biodegradation can be either catabolic, with the neonicotinoid providing a sole source of either carbon or nitrogen for growth, or cometabolic, whereby biodegradation relies on supplementation with additional carbon and/or nitrogen sources. Neonicotinoid metabolites can vary widely, depending on the chemical structure of the pesticide and the catabolic activity of a degrading microorganism under a particular set of environmental conditions (Hussain et al. 2009). Several neonicotinoids metabolites are more toxic and persistent than the original pesticides.

12.4.5 Bioremediation of Synthetic Pyrethroids

Pyrethroid insecticides have been used to control pests in agriculture, forestry, horticulture, public health, as well as indoor homes for more than 20 years. Because pyrethroids were considered to be a safer alternative to organophosphate pesticides (OPs), their applications significantly increased when the use of OPs was banned or limited. Although, pyrethroids have agricultural benefits, their widespread and continuous use is a major problem as they pollute terrestrial and aquatic environments and affect non-target organisms. Since pyrethroids are not degraded immediately after application and because their residues are detected in soils, there is an urgent need to remediate pyrethroid-polluted environments. Various remediation technologies have been developed for this purpose; however, bioremediation, which involves bioaugmentation and/or biostimulation and is a cost-effective and eco-friendly approach, has emerged as the most advantageous method for cleaning-up pesticide-contaminated soils. This review presents an overview of the microorganisms that have been isolated from pyrethroid-polluted sites, characterized and applied for the degradation of pyrethroids in liquid and soil media. The paper is focused on the microbial degradation of the pyrethroids that have been most commonly used for many years such as allethrin, bifenthrin, cyfluthrin, cyhalothrin, cypermethrin, deltamethrin, fenpropathrin, fenvalerate, and permethrin. Special attention is given to the bacterial strains from the genera *Achromobacter, Acidomonas, Bacillus, Brevibacterium, Catellibacterium, Clostridium, Lysinibacillus, Micrococcus, Ochrobactrum, Pseudomonas, Serratia, Sphingobium, Streptomyces*, and the fungal strains from the genera *Aspergillus, Candida, Cladosporium, and Trichoderma*, which are characterized by their ability to degrade various pyrethroids (Table 12.5). Moreover, current knowledge on the degradation pathways of pyrethroids, the enzymes that are involved in the cleavage of pesticide molecules, the factors/conditions that influence the survival of strains that are introduced into soil and the rate of the removal of pyrethroids are also discussed. This knowledge may be useful to optimize the environmental conditions of bioremediation and may be crucial for the effective removal of pyrethroids from polluted soils (Cycon et al. 2013).

TABLE 12.5
Biodegradation of Different Synthetic Pyrethroid Pesticides by Microbes

Insecticides	Microbial Strain	References
Permethrin	*Achromobacter* sp. (SM-2)	Maloney et al.1988
	Bacillus cereus (SM-3)	
	Pseudomonas fluorescens (SM-1)	
Allethrin	*Acidomonas* sp.	Paingankar et al. 2005
Cypermethrin	*Acinetobacter calcoaceticus* MCm5	Akbar et al. 2015a
Bifenthrin	*Ochrobactrum anthropi* JCm 1	
Cyhalothrin	*Bacillus cereus* Y1	
Deltamethrin	*Sphingomonas* sp. RCm6	Zhang et al. 2016
Cypermethrin	*Azoarcus indigens* HZ5	Ma et al. 2013
	Bacillus sp. AKD1	Tiwary and Dubey 2016
	Bacillus sp. ISTDS2	Sundaram et al. 2013
	Bacillus sp. SG2	Bhatt et al. 2016
	Bacillus amyloliquefaciens AP01	Lee et al. 2016
	Bacillus subtillis BSF01	Xiao et al. 2015
	Bacillus thuringiensis ZS-19	Chen et al. 2015
	Micrococcus sp. CPN1	Tallur et al. 2008
	Pseudomonas fluorescens (SM-1)	Grant et al. 2002; Grant and Betts 2004
	Rhodococcus sp. JCm5	Akbar et al. 2015b
	Serratia sp., *S. plymuthica*	Grant et al. 2002; Grant and Betts 2004
Fenpropathrin	*Bacillus* sp. DG02	Chen et al. 2012a, 2014
	Clostridium sp. ZP3	Zhang et al. 2011
Cyfluthrin	*Brevibacterium aureum* DG-12	Chen et al. 2013
	Lysinibacillus sphaericus FLQ-11–1	Hu et al. 2014
Fenvalerate	*Catellibacterium* sp. CC-5	Zhao et al. 2013
	Stenotrophomonas sp. ZS-S-01	Chen et al. 2011

12.4.6 Bioremediation of Miscellaneous Pesticides

The chemicals released from the plants may enhance xenobiotic degradation and it may be beneficial to use plants in the remediation of contaminated soils. There are three general mechanisms:

1. The rhizosphere may allow selective enrichment of degrader organisms that have densities too low to significantly degrade xenobiotics in root-free soil
2. The rhizosphere may enhance growth-linked metabolism or stimulate microbial growth by providing a natural substrate when the concentration of xenobiotics is low or unavailable
3. The rhizosphere is rich in natural compounds that may induce cometabolism of xenobiotics in certain microorganisms that carry degradative genes or plasmids. This may permit initial degradation of xenobiotics that would otherwise be unavailable as carbon sources. Bioremediation of inorganic pollutants is

supported by adsorption on solid surfaces such as glass, entrapment in polymeric gels, encapsulation, or intermolecular cross-linking. Although preparing supports can be time consuming and expensive, the support can generally be reused. Enzymatic treatment holds great promise in the bioremediation of contaminated soil and water. Spatial and temporal heterogeneity in O_2 distribution in the rhizosphere environment usually (but not always) provides microbes with localized environments that are anaerobic and have low redox potential, thereby favoring RDE reactions (Barkovskii 2001). Moreover, most of the terminal electron acceptors, such as nitrate, ferric iron, sulfate, and quinones are abundant in the rhizosphere. The bioavailability of hydrophobic contaminants determines the rate of xenobiotic transformation and mineralization.

12.5 CONCLUSION AND FUTURE PROSPECTS

Bioremediation is becoming a more popular method of remediating inorganic pollutant contamination as it has many advantage that make it a proper and successful technology. In this chapter, various strategies of bioremediation techniques and mechanisms have been discussed for advancing new ideas to remove pollutants from the contaminated sites. Bioremediation is a highly interdisciplinary area of research that requires background knowledge of agriculture and allied sectors like soil/water chemistry, plant biology, ecology and soil/water microbiology, and environmental engineering. Additional studies should be conducted to better understand plant physiology, biochemistry, and uptake of these contaminants and proper evaluation of possible synergistic effects of multiple contaminations.

Currently, biotechnology is a powerful tool used in bioremediation to improve the pollutant removal efficiencies, but it is limited to the lab conditions or at a very small scale. Moreover, the application of transgenic technology and plant-microbe interactions are feasible strategies for the improvement of plants for pollutant tolerance. Hence, it is better to create or find an appropriate plant/microbe system for environmental cleanup. More field-scale studies are needed to prove the effectiveness of bioremediation processes for pesticide-contaminated soils.

REFERENCES

Abo-Amer, A.E. 2011. Biodegradation of diazinon by *Serratia marcescens* DI101 and its use in bioremediation of contaminated environment. *Journal of Microbiology and Biotechnology*, 21: 71–80.

Abo-Amer, A.E. 2012. Characterization of a strain of *Pseudomonas putida* isolated from agricultural soil that degrades cadusafos (an organophosphorus pesticide). *World Journal of Microbiology and Biotechnology*, 28: 805–814.

Abraham, J., Silambarasan, S. 2013. Biodegradation of chlorpyrifos and its hydrolyzing metabolite 3,5,6-trichloro-2-pyridinol by *Sphingobacterium* sp. JAS3. *Process Biochemistry*, 48: 1559–1564.

Aemmr, A., Abo-El-Seoud, M.A., Ibrahim, G.M., Kassem, B.W. 2013. Stimulating of biodegradation of oxamyl pesticide by low dose gamma irradiated fungi. *Plant Pathology and Microbiology*, 4(9): 1–5. doi:10.4172/2157-7471.1000201.

Agarry, S.E., Olu-Arotiowa, O.A., Aremu, M.O., Jimoda, L.A. 2013. Biodegradation of dichlorovos (organophosphate pesticide) in soil by bacterial isolates. *Journal of Natural Sciences Research*, 3: 12–16.

Akbar, S., Sultan, S., Kertesz, M. 2015a. Determination of cypermethrin degradation potential of soil bacteria along with plant growth-promoting characteristics. *Current Microbiolgy*, 70:75–84. doi: 10.1007/s00284-014-0684-7

Akbar, S., Sultan, S., Kertesz, M. 2015b. Bacterial community analysis of cypermethrin enrichment cultures and bioremediation of cypermethrin contaminated soils. *Journal of Basic Microbiology*, 55: 819–829. doi: 10.1002/jobm.201400805.

Akoijam. R., Singh, B. 2015. Biodegradation of Imidacloprid in sandy loam soil by *Bacillus aerophilus. International Journal of Environmental Analytical Chemistry* 95: 730–743.

Alcalde, M., Ferrer, M., Plou, F.J., Ballesteros, A. 2006. *Trends in Biotechnology* 24: 281–287.

Anguiano, O.L., Vacca, M., Araujo, M.E.R., Montagna, M., Venturino, A., Ferrari, A. 2017. Acute toxicity and esterase response to carbaryl exposure in two different populations of amphipods *Hyalella curvispina. Aquatic Toxicology* 188:72–79. doi:10.1016/j.aquatox.2017.04.013.

Anhalt, J.C., Moorman, T.B., Koskinen, W.C. 2007. Biodegradation of imidacloprid by an isolated soil microorganism. *Journal of Environmental Science and Health, Part-B*, 42: 509–514.

Anusha, J., Kavitha, P., Louella, C., Chetan, D., Rao, C. 2009. A study on biodegradation of propoxur by bacteria isolated from municipal solid waste. *International Journal of Biotechnology Applications*, 1(2): 26–31.

Arthur, E.L., Coats, J.R. 1998. Phytoremediation. In P.C. Kearney, T. Roberts (eds.), *Pesticide Remediation in Soils and Water*, John Wiley & Sons, New York, pp. 251–283.

Baishya, K., Sharma, H. P. 2014. Isolation and characterization of organophosphorus pesticide degrading bacterial isolates. *Archives of Applied Science Research*, 6:144–149.

Baker, K.H., Herson, D.S. 1994. *Bioremediation. McGraw-Hill Inc*, New York.

Bano, N., Musarrat, J. 2003. Isolation and characterization of phorate degrading soil bacteria of environmental and agronomic significance. *Letters in Applied Microbiology*, 36: 349–353.

Benimeli, C.S., Fuentes, M.S., Abate, C.M., Amoroso, M.J. 2008. Bioremediation of lindane-contaminated soil by *Streptomyces* sp. M7 and its effects on *Zea mays* growth. *International Biodeterioration & Biodegradation*, 61:233–239.

Bhadbhade, B.J., Sarnaik, S.S., Kanekar, P.P. 2002. Biomineralization of an organophosphorus pesticide, monocrotophos, by soil bacteria. *Journal of Applied Microbiology*, 93: 224–234.

Bhalerao, T.S., Puranik, P.R. 2007. Biodegradation of organochlorine pesticide, endosulfan, by a fungal soil isolate, *Aspergillus niger. International Biodeterioration & Biodegradation*, 59(4):315–321.

Bhatt, P., Sharma, A., Gangola, S., Khati, P., Kumar, G., Srivastava, A. 2016. Novel pathway of cypermethrin biodegradation in a *Bacillus* sp. strain SG2 isolated from cypermethrin-contaminated agriculture field. *BioTech*, 6: 45. doi: 10.1007/s13205-016-0372-3.

Briceno, G., Fuentes, M.S., Palma, G., Jorquera, M.A., Amoroso, M.J., Diez, M.C. 2012. Chlorpyrifos biodegradation and 3,5,6-trichloro-2-pyridinol production by actinobacteria isolated from soil. *International Biodeterioration & Biodegradation*, 73: 1–7.

Bumpus, J. A., Tien, M., Wright, D., Aust, S.D. 1985. *Science*, 228(4706): 1434–1436.

Bumpus, J.A., Powers, R.H., Sun, T. 1993. Biodegradation of DDE (1,1-dichloro-2,2-bis(4-hlorophenyl)ethane) by *Phanerochaete chrysosporium. Mycological Research*, 97: 95–98.

Chan-Cupul, W., Heredia-Abarca, G., Rodriguez-Vazquez, R. 2016. Atrazine degradation by fungal coculture enzyme extracts under different soil conditions. *Journal of Environmental Science and Health, Part B*, 51(5): 298–308. doi:10.1080/03601234.2015.1128742.

Chaudhry, G.R., Ali, A.N. 1988. Bacterial metabolism of carbofuran. *Applied and Environmental Microbiology*, 54(6): 1414–1419.

Chaudhry, G.R., Chapalamadugu, S. 1991. Biodegradation of halorgenated organic compounds microbiological reviews. *Microbiology and Molecular Biology*, 55(1):59–79.

Chen, S., Chang, C., Deng, Y., An, S., Dong, Y.H., Zhou, J. 2014. Fenpropathrin biodegradation pathway in *Bacillus* sp. DG-02 and its potential for bioremediation of pyrethroid-contaminated soils. *Journal of Agricultural and Food Chemistry*, 62, 2147–2157. doi: 10.1021/jf404908j.

Chen, S., Deng, Y., Chang, C., Lee, J., Cheng, Y., Cui, Z. 2015. Pathway and kinetics of cyhalothrin biodegradation by *Bacillus thuringiensis* strain ZS-19. *Scientific Reports,* 5: 8784. doi: 10.1038/srep08784.

Chen, S., Dong, Y. H., Chang, C., Deng, Y., Zhang, X. F., Zhong, G. 2013. Characterization of a novel cyfluthrin-degrading bacterial strain *Brevibacterium aureum* and its biochemical degradation pathway. *Bioresource Technology*, 132: 16–23. doi: 10.1016/j.biortech.2013.01.002.

Chen, S., Hu, W., Xiao, Y., Deng, Y., Jia, J., Hu, M. 2012a. Degradation of 3-phenoxybenzoic acid by a *Bacillus* sp. *PLoS ONE*, 7: e50456. doi: 10.1371/journal.pone.0050456.

Chen, S., Yang, L., Hu, M., Liu, J. 2011 Biodegradation of fenvalerate and 3-phenoxybenzoic acid by a novel *Stenotrophomonas* sp. strain ZS-S-01 and its use in bioremediation of contaminated soils. *Applied Microbiology Biotechnology*, 90: 755–767. doi: 10.1007/s00253-010-3035-z.

Chen, T., Dai, Y.J., Ding, J.F. 2008. N-demethylation of neonicotinoid insecticide acetamiprid by bacterium *Stenotrophomonas maltophilia* CGMCC 1.1788. *Biodegradation*, 19: 651–658.

Cho, K.M., Math, R.K., Islam, S.M.A., Lim, W.J., Hong, S.Y., Kim, J.M., Yun, M.G., Cho, J.J., Yun. H.D. 2009. Biodegradation of chlorpyrifos by lactic acid bacteria during kimchi fermentation. *Journal of Agricultural and Food Chemistry*, 57: 1882–1889.

Coats, J.R., Anderson, T.A. 1997. *The Use of Vegetation to Enhance Bioremediation of Surface Soils Contaminated with Pesticide Wastes.* US EPA. Office of Research and Development. Washington, DC.

Cycon, M., Mrozik, A., Piotrowska-Seget, Z. 2017. Bioaugmentation as a strategy for the remediation of pesticide-polluted soil: A Review. *Chemosphere*, 172: 52–71. doi: 10.1016/j.chemosphere.2016.12.129.

Cycon, M., Zmijowska, A., W´ojcik, M., Piotrowska-Seget, Z. 2013. Biodegradation and bioremediation potential of diazinondegrading *Serratia marcescens* to remove other organophosphorus pesticides from soils. *Journal of Environmental Management*, 117: 7–16.

Dai, Y.J., Ji, W.W., Chen, T. 2010. Metabolism of the neonicotinoid insecticides acetamiprid and thiacloprid by the yeast *Rhodotorula mucilaginosa* strain IM-2. *Journal of Agricultural and Food Chemistry*, 58: 2419–2425.

Deng, S.Y., Chen, Y., Wang, D.S., Shi, T.Z., Wu, X.W., Ma, X., Li, X.Q., Hua, R.M., Tang, X. Y., Li, Q.X. 2015. Rapid biodegradation of organophosphorus pesticides by *Stenotrophomonas* sp. G1. *Journal of Hazardous Materials*, 297: 17–24.

Devi, K.Y., Iyer, P. 2017. Isolation & characterization of carbofuran pesticide degrading microorganisms from various field areas in Tamil Nadu. *Research Journal of Life Sciences, Bioinformatics, Pharmaceutical and Chemical Sciences*, 2(5): 270–281.

Dewi, T.K., Imamuddin, H., Antonius, S. 2015. Study of propoxur degrading bacteria isolated from various sampling site of rice field from Ngawi. *KnE Life Sciences*, 2(1): 658–661.

Diaz, E. 2004. Bacterial degradation of aromatic pollutants: A paradigm of metabolic versatility. *International Microbiology,* 7(3): 173–180.

Doddamani, H.P., Ninnekar, H.Z. 2001. Biodegradation of carbaryl by a *Micrococcus* species. *Current Microbiology*, 43(1): 69–73. doi:10.1007/s002840010262.

Dua, M., Singh, A., Sethunathan, N., Johri, A.K. 2002. Biotechnology and Bioremediation: Successes and Limitations. *Applied Microbiology and Biotechnology*, 59(2–3):143–152.

Elbert, A., Haas, M., Springer, B., Thielert, W., Nauen, R. 2008. Applied aspects of neonicotinoid uses in crop protection. *Pest Management Science*, 64:1099–1105.

FAO. 2018. Pesticide Use Data-FAOSTAT. Retrieved from http://www.fao.org/faostat/en/#data/RP.

Fareed, A., Zaffar, H., Rashid, A., Maroof Shah, M., Naqvi, T.A. 2017. Biodegradation of N-methylated carbamates by free and immobilized cells of newly isolated strain *Enterobacter cloacae* strain TA7. *Bioremediation Journal*, 21(3–4): 119–127.

Field, J., Stams, A.M., Kato, M., Schraa, G. 1995. Enhanced Biodegradation of aromatic pollutants in cocultures of anaerobic and aerobic bacterial consortia. *Antonie Leeuwenhoek*, 67 (1): 47–77.

Foster, L.J.R., Kwan, B.H., Vancov, T. 2004. Microbial degradation of the organophosphate pesticide, Ethion. *FEMS Microbiology Letters*, 240: 49–53.

Gao, Y., Chen, S.H., Hu, M.Y., Hu, Q.B., Luo, J.J., Li, Y.N. 2012. Purification and characterization of a novel chlorpyrifos hydrolase from *Cladosporium cladosporioides* Hu-01. *PLoS One*, 7: e38137.

Ge, F., Zhou, L.Y., Wang, Y. 2014. Hydrolysis of the neonicotinoid insecticide thiacloprid by the N2-fixing bacterium *Ensifer meliloti* CGMCC 7333. *International Biodeterioration & Biodegradation*, 93:10–7.

Goda, S.K., Elsayed, I.E., Khodair, T.A., El-Sayed, W., Mohamed, M.E. 2010. Screening for and isolation and identification of malathion-degrading bacteria: Cloning and sequencing a gene that potentially encodes the malathion-degrading enzyme, carboxylestrase in soil bacteria. *Biodegradation*, 21: 903–913.

Godiyal, S. 2013. Neonicotinoids Pesticide Continue to Cause the Decline of Bee Population in U.S. http://www.bibliotecapleyades. net/ciencia/ciencia bees42.htm.

Gopal, M., Dutta, D., Jha, S.K. 2011. Biodegradation of imidacloprid and metribuzin by *Burkholderia cepacia* strain CH9. *Pesticide Research Journal*, 23:36–40.

Grant, R.J., Betts, W.B. 2004. Mineral and carbon usage of two synthetic pyrethroid degrading bacterial isolates. *Journal of Applied Microbiology*, 97: 656–662. doi: 10.1111/j.1365-2672.2004.02358.x.

Grant, R.J., Daniell, T.J., Betts, W.B. 2002. Isolation and identification of synthetic pyrethroid-degrading bacteria. *Journal of Applied Microbiology*, 92:534–540. doi: 10.1046/j.1365-2672.2002.01558.x.

Hamada, M., Matar, A., Bashir, A. 2015. Carbaryl degradation by bacterial isolates from a soil ecosystem of the Gaza Strip. *Brazilian Journal of Microbiology*, 46(4): 1087–1091. doi:10.1590/S1517- 838246420150177.

Helal, I.M., Abo-El-Seoud, M.A. 2015. Fungal biodegradation of pesticide vydate in soil and aquatic system. *In 4th International Conference on Radiation Sciences and Applications Taba, Egypt.* (pp. 87–94).

Hu, G., Zhao, Y., Liu, B. 2013. Isolation of an indigenous imidacloprid degrading bacterium and imidacloprid bioremediation under simulated *in situ* and *ex situ* conditions. *Journal of Microbiology and Biotechnology*, 23: 1617–1626.

Hu, G.P., Zhao, Y., Song, F.Q., Liu, B., Vasseur, L., Douglas, C. 2014. Isolation, identification and cyfluthrin-degrading potential of a novel *Lysinibacillus sphaericus* strain, FLQ-11–1. *Research in Microbiology*, 165:110–118. doi: 10.1016/j.resmic.2013.11.003.

Hussain, S., Siddique, T., Arshad, M., Saleem, M. 2009. Bioremediation and phytoremediation of pesticides: Recent advances. *Critical Reviews in Environmental Science and Technology*, 39: 843–907.

Ishag, A.E., Abdelbagi, A.O., Hammad, A.M., Elsheikh, E.A., Elsaid, O.E., Hur, J.H., Laing, M.D. 2016. Biodegradation of chlorpyrifos, malathion, and dimethoate by three strains of bacteria isolated from pesticide-polluted soils in Sudan. *Journal of Agricultural and Food Chemistry*, 64(45): 8491–8498. doi:10.1021/acs.jafc.6b03334.

Jeschke, P., Nauen, R., Schindler, M. 2011. Overview of the status and global strategy for neonicotinoids. *Journal of Agricultural and Food Chemistry*, 59:2897–908.

Jilani, S. 2013. Comparative assessment of growth and biodegradation potential of soil isolate in the presence of pesticides. *Saudi Journal of Biological Sciences*, 20: 257–264.

Kamanavalli, C., Ninnekar, H. 2000. Biodegradation of propoxur by *Pseudomonas* species. *World Journal of Microbiology and Biotechnology*, 16(4):329–331.

Kandil, M. M., Trigo, C., Koskinen, W. C., Sadowssky, Michael J. 2015. Isolation and characterization of a novel imidacloprid-degrading *Mycobacterium* sp. strain MK6 from an Egyptian soil. *Journal of Agricultural and Food Chemistry*, 63(19):4721–4727.DOI: 10.1021/acs.jafc.5b00754.

Kannan, K., Tanabe, S., Willians, R. J., Tatsukawa, R. 1994. Persistent organochlorine residues in foodstuffs from Australia, Papua New Guinea and the Solomon Islands: Contamination levels and dietary exposure. *Science of the Total Environment*, 153: 29–49.

Karpouzas, D.G., Fotopoulou, A., Menkissoglu-Spiroudi, U., Singh, B.K. 2005. Non-specific biodegradation of the organophosphorus pesticides, cadusafos and ethoprophos, by two bacterial isolates. *FEMS Microbiology Ecology*, 53: 369–378.

Kazumi, J., Capone, D.G. 1995. Microbial aldicarb transformation in aquifer, lake, and salt marsh sediments. *Applied and Environmental Microbiology*, 61(8): 2820–2829.

Kim, I.S., Ryu, J.Y., Hur, H.G., Gu, M.B., Kim, S.D., Shim, J.H. 2004. *Sphingomonas* sp. strain SB5 degrades carbofuran to a new metabolite by hydrolysis at the furanyl ring. *Journal of Agricultural and Food Chemistry*, 52(8): 2309–2314. doi:10.1021/jf035502l.

Kim, J.R., Ahn, Y.J. 2009. Identification and characterization of chlorpyrifos-methyl and 3,5,6-trichloro-2-pyridinol degrading *Burkholderia* sp. strain KR100. *Biodegradation*. 20: 487–497.

Konradsen, F., Vander Hoek, W., Amerasinghe, F.P., Mutero, C., Boelee, E. 2004. Engineering and malaria control: Learning from the past 100 years. *Acta Tropica*. 89(2): 99–108.

Kulkarni, A.G., Kaliwal, B. 2014. Bioremediation of methomyl by soil isolate—*Pseudomonas aeruginosa*. *Journal of Environmental Science, Toxicology and Food Technology*, 8(12): 1–10.

Kulkarni, A.G., Kaliwal, B.B. 2018. Bioremediation of Methomyl by *Escherichia coli*. In: Bidoia, E.D., Montagnolli, R.N. (Eds.), *Toxicity and Biodegradation Testing*, New York, NY: Humana Press, 75–86.

L´opez, L., Pozo, C., Rodelas, B., Calvo, C., Juárez, B., Mart´ınez-Toledo, M.V., González-L´opez, J. 2005. Identification of bacteria isolated from an oligotrophic lake with pesticide removal capacities. *Ecotoxicology*, 14: 299–312.

Langlois, B.E., Collins, J.A., Sides, K.G. 1970. Some factors affecting degradation of organochlorine pesticide by bacteria. *Journal of Dairy Science*, 53(12): 1671–1675.

Latifi, A.M., Khodi, S., Mirzaei, M., Miresmaeili, M., Babavalian, H. 2012. Isolation and characterization of five chlorpyrifos degrading bacteria. *African Journal of Biotechnology*, 11: 3140–3146.

Lee, Y.S., Lee, J.H., Hwang, E.J., Lee, H.J., Kim, J.H., Heo, J.B. 2016. Characterization of biological degradation cypermethrin by *Bacillus amyloliquefaciens* AP01. *Journal of Applied Biological Chemistry*, 59: 9–12. doi: 10.3839/jabc.2016.003.

Lei, J., Wei, S., Ren, L., Hu, S., Chen, P. 2017. Hydrolysis mechanism of carbendazim hydrolase from the strain *Microbacterium* sp. djl-6F. *Journal of Environmental Sciences*, 54: 171–177. doi:10.1016/j. jes.2016.05.027.

Li, J., Zhang, S., Wu, C., Li, C., Wang, H., Wang, W., Ye, Q. 2016. Stereoselective degradation and transformation products of a novel chiral insecticide, paichongding, in flooded paddy soil. *Journal of Agricultural and Food Chemistry*, 64(40): 7423–7430. doi:10.1021/acs.jafc.6b02787.

Lu, P., Li, Q.F., Liu, H.M., Feng, Z.Z., Yan, X., Hong, Q., Li, S.P. 2013. Biodegradation of chlorpyrifos and 3,5,6-trichloro-2-pyridinol by *Cupriavidus* sp. DT-1. *Bioresource Technology*, 127: 337–342.

Ma, Y., Chen, L., Qiu, J. 2013. Biodegradation of beta-cypermethrin by a novel *Azoarcus indigens* strain HZ5. *Journal of Environmental Science and Health, Part B*, 48: 851–859. doi: 10.1080/03601234.2013.795843.

Ma, Y., Zhai, S., Mao, S.Y. 2014. Co-metabolic transformation of the neonicotinoid insecticide imidacloprid by the new soil isolate *Pseudoxanthomonas indica* CGMCC 6648. *J. Environ. Sci. Heal. B.*, 49: 661–670.

Madhuri, R.J., Rangaswamy, V. 2009. Biodegradation of selected insecticides by *Bacillus* and *Pseudomonas* sp. in groundnut fields. *Toxicology International*, 16: 127–132.

Madsen, E. L. 1991. Determining *in Situ* biodegradation—facts and challenges. *Environmental Science & Technology*, 25: 1663–1673.

Malghani, S., Chatterjee, N., Yu, H.X., Luo, Z.J. 2009. Isolation and identification of profenofos degrading bacteria. *Brazilian Journal of Microbiology*, 40: 893–900.

Maloney, S. E., Maule, A., Smith, A. R. 1988. Microbial transformation of the pyrethroid insecticides: Permethrin, deltamethrin, fastac, fenvalerate, and fluvalinate. *Applied and Environmental Microbiology*, 54: 2874–2876.

Matolcsy, G., Nadasy, M., Andriska, V. 1988. Chlorinated hydrocarbons. Pesticide chemistry. ISBN 978-044-4989-03-1, New York, USA. *Studies in Environmental Science*, 32: 83.

Matsumura, F., Boush, G.M., Tai, A. 1968. Breakdown of dieldrin in the soil by a microorganism. *Nature*, 219(5157): 965–967.

Megharaj, M., Ramakrishnan, B., Venkateswarlu, K., Sethunathan, N., Naidu, R. 2011 Bioremediation approaches for organic pollutants: A critical perspective. *Environment International*, 37: 1362–1375.

Menone, M.L., Bortolus, A., Aizpún de Moreno, J.E., Moreno, V.J., Lanfranchi, A.L., Metcalfe, T.L., Metcalfe, C.D. 2001. Organochlorine pesticides and PCBs in a southern Atlantic coastal lagoon watershed, Argentina. *Archives of Environmental Contamination and Toxicology*, 40(3): 355–362.

Mohamed, E.A. 2017. Oxamyl utilization by micrococcus luteus OX, isolated from Egyptian soil. *International Journal of Applied Environmental Sciences*, 12(5): 999–1008.

Mohamed, M.S. 2009. Degradation of methomyl by the novel bacterial strain *Stenotrophomonas maltophilia* M1. *Electronic Journal of Biotechnology*, 12(4): 6–7.

Mohamed, Z.K., Ahmed, M.A., Fetyan, N.A., Elnagdy, S.M. 2010. Isolation and molecular characterisation of malathion degrading bacterial strains from waste water in Egypt. *Journal of Advanced Research*, 1: 145–149.

Nguyen, T.P.O., Helbling, D.E., Bers, K., Fida, T.T., Wattiez, R., Kohler, H.P.E., De Mot, R. 2014. Genetic and metabolic analysis of the carbofuran catabolic pathway in *Novosphingobium* sp. KN65. 2. *Applied Microbiology and Biotechnology*, 98(19): 8235–8252. doi:10.1007/s00253-014-5858-5.

Ohshiro, K., Ono, T., Hoshino, T., Uchiyama, T. 1997. Characterization of isofenphos hydrolases from *Arthrobacter* sp. Strain B-5. *Journal of Fermentation and Bioengineering*, 83: 238–245.

Omolo, K.M., Magoma, G., Ngamau, K., Muniru, T. 2012. Characterization of methomyl and carbofuran degrading-bacteria from soils of horticultural farms in Rift Valley and Central Kenya. *African Journal of Environmental Science and Technology*, 6(2): 104–114.

Onunga, D.O., Kowino, I.O., Ngigi, A.N., Osogo, A., Orata, F., Getenga, Z.M., Were, H. 2015. Biodegradation of carbofuran in soils within Nzoia River Basin, Kenya. *Journal of Environmental Science and Health, Part B*, 50(6): 387–397. doi:10.1080/03601234.2015.1011965.

Ortega, N.O., Nitschke, M., Mouad, A.M., Landgraf, M.D., Rezende, M.O.O., Seleghim, M.H.R., Sette, L.D., Porto, A.L.M. 2011. Isolation of Brazilian marine fungi capable of growing on DDD pesticide. *Biodegradation*, 22: 43–50.

Ortiz-Hernández, M.L., Sánchez-Salinas, E. 2010. Biodegradation of the organophosphate pesticide tetrachlorvinphos by bacteria isolated from agricultural soils in Mexico. *Revista Internacional de Contaminacion Ambiental*, 26: 27–38.

Osborn, R., Haydock, P., Edwards, S. 2010. Isolation and identification of oxamyl-degrading bacteria from UK agricultural soils. *Soil Biology and Biochemistry*, 42(6): 998–1000.

Owsianiak, M., Dechesne, A., Binning, P.J., Chambon, J.C., Sorensen, S.R., Smets, B.F. 2010. Evaluation of bioaugmentation with entrapped degrading cells as a soil remediation technology. *Environmental Science & Technology*, 44: 7622–7627.

Pailan, S., Gupta, D., Apte, S., Krishnamurthi, S., Saha, P. 2015. Degradation of organophosphate insecticide by a novel *Bacillus aryabhattai* strain SanPS1, isolated from soil of agricultural field in Burdwan, West Bengal, India. *International Biodeterioration & Biodegradation*, 103: 191–195.

Paingankar, M., Jain, M., Deobagkar, D. 2005. Biodegradation of allethrin, a pyrethroid insecticide, by an *Acidomonas* sp. *Biotechnology Letters*, 27: 1909–1913. doi: 10.1007/s10529-005-3902-3.

Pandey, B., Baghel, P.S., Shrivastava, S. 2014. To study the bioremediation of monocrotophos and to analyze the kinetics effect of Tween 80 on fungal growth. *Indo American Journal of Pharmaceutical Research*, 4: 925–930.

Pandey, G., Dorrian, S.J., Russell, R.J. 2009. Biotransformation of the neonicotinoid insecticides imidacloprid and thiamethoxam by *Pseudomonas* sp. 1G. *Biochemical and Biophysical Research Communications*, 380: 710–714.

Patil, K.C., Matsumura, F., Boush, GM. 1970. Degradation of Endrin, Aldrin and DDT by soil microorganisms. *Journal of Applied Microbiology*, 19(5): 879–881.

Patnaik, P. 2003. *A Comprehensive Guide to the Hazardous Properties of Chemical Substances*, John Wiley & Sons, Inc., New Jersey, USA.

Pawar, K.R., Mali, G.V. 2014. Biodegradation of the organophosphorus insecticide dichlorvas by *Bacillus* species isolated from grape wine yard soils from Sangli District, M.S., India. *International Research Journal of Environmental Sciences*, 3: 8–12.

Phugare, S.S., Gaikwad, Y.B., Jadhav, J.P. 2012. Biodegradation of acephate using a developed bacterial consortium and toxicological analysis using earthworms (*Lumbricus terrestris*) as a model animal. *International Biodeterioration & Biodegradation*, 69: 1–9.

Phugare, S.S., Jadhav, J.P. 2014. Biodegradation of acetamiprid by isolated bacterial strain *Rhodococcus* sp. BCH2 and toxicological analysis of its metabolites in silkworm (*Bombax mori*). *CLEAN–Soil Air Water*, 43(2): 296–304.DOI: 10.1002/clen.201200563.

Phugare, S.S., Kalyani, D.C., Gaikwad, Y.B. 2013. Microbial degradation of imidacloprid and toxicological analysis of its biodegradation metabolites in silkworm (*Bombyx mori*). *Chemical Engineering Journal*, 230: 27–35.

Phytoremediation of Organics Action Team. 2000. *Phytoremediation Site Profiles: Cantrall.* US EPA. Washington, DC.

Poplar trees may remove contaminants. 1998. *Environmental News Network*. www.enn.com/enn-newsarchive/1998/09/093098/poplar30.asp June 29, 2000.

Rangasamy, K., Athiappan, M., Devarajan, N., Parray, J.A. 2017. Emergence of multi drug resistance among soil bacteria exposing to insecticides. *Microbial Pathogenesis*, 105:153–165. doi:10.1016/j. micpath.2017.02.011.

Rani, R., Juwarkar, A. 2012. Biodegradation of phorate in soil and rhizosphere of *Brassica juncea* (L.) (Indian Mustard) by a microbial consortium. *International Biodeterioration & Biodegradation*, 71: 36–42.

Rouchaud, J., Gustin, F., Wauters, A. 1994. Soil biodegradation and leaf transfer of insecticide imidacloprid applied in seed dressing in sugar beet crops. *Bulletin of Environmental Contamination and Toxicology*, 53:344–350.

Rousidou, K., Chanika, E., Georgiadou, D., Soueref, E., Katsarou, D., Kolovos, P., Karpouzas, D. G. 2016. Isolation of oxamyl-degrading bacteria and identification of cehA as a novel oxamyl hydrolase gene. *Frontiers in Microbiology*, 7: 1–12. doi:10.3389/fmicb.2016.00616.

Sabourmoghaddam, N., Zakaria, MP., Oma, R.D. 2014. Evidence for the microbial degradation of imidacloprid in soils of Cameron Highlands. *Journal of the Saudi Society of Agricultural Sciences*,14: 182–188. DOI: 10.1016/j.jssas.03.002.

Sasikala, C., Jiwal, S., Rout, P., Ramya, M. 2012. Biodegradation of chlorpyrifos by bacterial consortium isolated from agriculture soil. *World Journal of Microbiology and Biotechnology*, 28: 1301–1308.

Satish, G.P., Ashokrao, D.M., Arun, S.K. 2017. Microbial degradation of pesticide: A Review. *African Journal of Microbiology Research*, 11(24): 992–1012. doi:10.5897/ajmr2016.8402.

Seo, J.S., Keum, Y.S., Harada, R. M., Li, Q.X. 2007. Isolation and characterization of bacteria capable of degrading polycyclic aromatic hydrocarbons (PAHs) and organophosphorus pesticides from PAH-contaminated soil in Hilo, Hawaii. *Journal of Agricultural and Food Chemistry* 55: 5383–5389.

Shaikh, N.S., Kulkarni, S.V., Mulani, M.S. 2014. Biodegradation of imidacloprid, the new generation neurotoxic insecticide. *International Journal of Innovative Research in Science Engineering and Technology* 3: 16301–16307.

Sharma, S., Singh, B., Gupta, V.K. 2014. Biodegradation of imidacloprid by consortium of two soil isolated *Bacillus* sp. *Bulletin of Environmental Contamination and Toxicology*, 93: 637–642.

Shetti, A.A., Kaliwal, B.B. 2012. Biodegradation of imidacloprid by soil isolate *Brevunimonas* sp. MJ15. *International Journal of Current Research*, 4: 100–106.

Shetti, A.A., Kaliwal, R.B., Kaliwal, B.B. 2014. Imidacloprid induced intoxication and its biodegradation by soil isolate *Bacillus weihenstephanensis*. *British Biotechnology Journal*, 4: 957–969.

Shivanandappa, T., Rajashekar, Y. 2014. Mode of action of plant derived natural insecticides. In: Singh, D. (Ed.), *Advances in Plant Biopesticides*. New Delhi, India: Springer, 317–322.

Simon-Delso, N., Amaral-Rogers, V. et al. 2015.. Systemic insecticides (neonicotinoids and fipronil): Trends, uses, mode of action and metabolites. *Environmental Science and Pollution Research*. 22: 5–34.(this issue). doi:10.1007/s11356-014-3470-y.

Sogorb, M.A., Vilanova, E. 2002. Enzymes involved in the detoxification of organophosphorus, carbamate and pyrethroid insecticides through hydrolysis. *Toxicology Letters*, 128(1–3): 215–228.

Stroud, J.L, Paton, G.I., Semple, K.T. 2007. Microbe-aliphatic hydrocarbon interactions in soil: Implications for biodegradation and bioremediation. *Journal of Applied Microbiology*, 102: 1239–1253.

Sun, L.N., Zhang, J., Gong, F.F., Wang, X., Hu, G., Li, S.P., Hong, Q. 2014. Nocardioides soli sp. nov., a carbendazim-degrading bacterium isolated from soil under the long-term application of carbendazim. *International Journal of Systematic and Evolutionary Microbiology*, 64(6): 2047–2052. doi:10.1099/ijs.0.057935.

Sundaram, S., Das, M.T., Thakur, I.S. 2013. Biodegradation of cypermethrin by *Bacillus* sp. in soil microcosm and *in-vitro* toxicity evaluation on human cell line. *International Biodeterioration & Biodegradation*, 77: 39–44. doi: 10.1016/j.ibiod.2012.11.008.

Tallur, P.N., Megadi, V.B., Ninnekar, H.Z. 2008. Biodegradation of cypermethrin by *Micrococcus* sp. strain CPN 1. *Biodegradation*, 19: 77–82. doi: 10.1007/s10532-007-9116-8.

Tang, H.Z., Li, J., Hu, H.Y. 2012. A newly isolated strain of *Stenotrophomonas* sp. hydrolyzes acetamiprid, a synthetic insecticide. *Process Biochemistry*, 47: 1820–1825.

Tian, L., Chen, F. 2009. Characterization of a carbendazim-degrading Trichoderma sp. T2–2 and its application in bioremediation. *Wei Sheng Wu Xue Bao*, 49(7): 925–930.

Tien, C.J., Huang, H.J., Chen, C.S. 2017. Accessing the carbofuran degradation ability of cultures from natural river biofilms in different environments. *CLEAN–Soil, Air, Water*, 45(5): 1–10.

Tiwary, M., Dubey, A.K. 2016. Cypermethrin bioremediation in presence of heavy metals by a novel heavy metal tolerant strain, *Bacillus* sp. AKD1. *International Biodeterioration & Biodegradation*, 108: 42–47. doi: 10.1016/j.ibiod.2015.11.025.

Trivedi, V.D., Jangir, P.K., Sharma, R., Phale, P.S. 2016. Insights into functional and evolutionary analysis of carbaryl metabolic pathway from *Pseudomonas* sp. strain C5pp. *Scientific Reports.*, 6: 1–12.

Ufarté, L., Laville, E., Duquesne, S., Morgavi, D., Robe, P., Klopp, C., Potocki-Veronese, G. 2017. Discovery of carbamate degrading enzymes by functional metagenomics. *PLoS One*, 12(12): 1–21. doi: 10.1371/journal.pone.0189201.

United States Environmental Protection Agency (USEPA). 2006. *In situ* and *ex situ* biodegradation technologies for remediation of contaminated sites. EPA/625/R-06/015.

Vaccari, D.A., Strom, P.F., Alleman, J.E. 2006. *Environmental Biology for Engineers and Scientists*. John Wiley & Sons, Inc.

Wackett, L.P., Sadowsky, M.J., Martinez, B., Shapir, N. 2002. Biodegradation of atrazine and related s-triazine compounds: From enzymes to field studies. *Applied Microbiology and Biotechnology*, 58: 39–45.

Wang, C., Qiu, J., Yang, Y., Zheng, J., He, J., Li, S. 2017. Identification and characterization of a novel carboxylesterase (FpbH) that hydrolyzes aryloxyphenoxypropionate herbicides. *Biotechnology Letters*, 39(4): 553–560. doi:10.1007/s10529-016-2276-z.

Wang, G., Chen, X., Yue, W. 2013a. Microbial degradation of acetamiprid by *Ochrobactrum* sp. D-12 isolated from contaminated soil. *PLoS One*, 8: e82603.

Wang, G.L., Zhao, Y.J., Gao, H. 2013b. Co-metabolic biodegradation of acetamiprid by *Pseudoxanthomonas* sp. AAP-7 isolated from a long-term acetamiprid-polluted soil. *Bioresource Technology,* 150: 259–265.

Wang, S. H., Zhang, C., Yan, Y.C. 2012. Biodegradation of methyl parathion and *p*-nitrophenol by a newly isolated *Agrobacterium* sp. strain Yw12. *Biodegradation*, 23: 107–116.

Wedemeyer, G. 1967. Dechlorination of 1,1,1-Trichloro-2,2-bis(pchlorophenyl) ethane by *Aerobacter aerogene. Journal of Applied Microbiology*, 15(3): 569–574.

WHO. 2019. Global situation of pesticide management in agriculture and public health: Report of a 2018 WHO-FAO survey. World Health Organization.

Wolicka, D., Suszek, A., Borkowski, A., Bielecka, A. 2009. Application of aerobic microorganisms in bioremediation *in situ* of soil contaminated by petroleum products. *Bioresources Technology,* 100: 3221–3227.

Xiao, P., Mori, T., Kamei, I., Kondo, R. 2010. Metabolism of organochlorine pesticide heptachlor and its metabolite heptachlor epoxide by white-rot fungi, belonging to genus *Phlebia. FEMS Microbiology Letters*, 314(2): 140–146.

Xiao, W., Wang, H., Li, T., Zhu, Z., Zhang, J., He, Z., Yang, X. 2013. Bioremediation of Cd and carbendazim co-contaminated soil by Cd-hyperaccumulator *Sedum alfredii* associated with carbendazim-degrading bacterial strains. *Environmental Science and Pollution Research*, 20(1): 380–389. doi:10.1007/s11356- 012-0902-4.

Xiao, Y., Chen, S., Gao, Y., Hu, W., Hu, M., Zhong, G. 2015. Isolation of a novel beta-cypermethrin degrading strain *Bacillus subtilis* BSF01 and its biodegradation pathway. *Applied Microbiology and Biotechnology*, 99: 2849–2859. doi: 10.1007/s00253-014-6164-y.

Yang, H., Wang, X., Zheng, J. 2013. Biodegradation of acetamiprid by *Pigmentiphaga* sp. D-2 and the degradation pathway. *International Biodeterioration & Biodegradation*, 85: 95–102.

Yao, X.H., Min, H. 2006. Isolation, characterization and phylogenetic analysis of a bacterial strain capable of degrading acetamiprid. *Journal of Environmental Sciences-China.*,18: 141–146.

Zhang, C., Wang, S., Yan, Y. 2011. Isomerization and biodegradation of beta-cypermethrin by *Pseudomonas aeruginosa* CH7 with biosurfactant production. *Bioresource Technology*, 102: 7139–7146. doi: 10.1016/j.biortech.2011.03.086.

Zhang, H., Zhang, Y., Hou, Z., Wang, X., Wang, J., Lu, Z. 2016. Biodegradation potential of deltamethrin by the *Bacillus cereus* strain Y1 in both culture and contaminated soil. *International Biodeterioration & Biodegradation*, 106: 53–59. doi: 10.1016/j.ibiod.2015.10.005.

Zhang, H.J., Zhou, Q.W., Zhou, G.C. 2012. Biotransformation of the neonicotinoid insecticide thiacloprid by the bacterium Variovorax boronicumulans strain J1 and mediation of the major metabolic pathway by nitrile hydratase. *Journal of Agricultural Food Chemistry*, 60: 153–159.

Zhang, Y.H., Xu, D., Liu, J.Q., Zhao, X.H. 2014. Enhanced degradation of five organophosphorus pesticides in skimmed milk by lactic acid bacteria and its potential relationship with phosphatase production. *Food Chemistry*, 164: 173–178.

Zhao, H., Geng, Y., Chen, L., Tao, K., Hou, T. 2013. Biodegradation of cypermethrin by a novel *Catellibacterium* sp. strain CC-5 isolated from contaminated soil. *Canadian Journal of Microbiology*, 59: 311–317. doi: 10.1139/cjm-2012-0580.

Zhao, Y.J., Dai, Y.J., Yu, C.G. 2009. Hydroxylation of thiacloprid by bacterium *Stenotrophomonas maltophilia* CGMCC1.1788. *Biodegradation*, 20: 761–768.

Zhou, G.C., Wang, Y., Zhai, S. 2013. Biodegradation of the neonicotinoid insecticide thiamethoxam by the nitrogenfixing and plant growth-promoting rhizobacterium *Ensifer adhaerens* strain TMX-23. *Applied Microbiology and Biotechnology*, 97: 4065–4074.

Zhou, G.C., Zhang, Y., Sun, S.L. 2014b. Degradation of the neonicotinoid insecticide acetamiprid via the N-carbamoylimine derivative (IM-1–2) mediated by the nitrile hydratase of the nitrogen-fixing bacteria *Ensifer meliloti* CGMCC733. *Journal of Agricultural and Food Chemistry*, 62: 9957–9964.

13 Plant-Soil-Microbe Interaction for Organic Production of Oilseeds

Kulasekaran Ramesh, Sankari Meena K., A. Aziz Qureshi, and P. Ratna Kumar
ICAR–Indian Institute of Oilseeds Research
Hyderabad, India

CONTENTS

13.1 INTRODUCTION

Organic crop production concerns the use of natural materials for plant nutrition in association with limited eco-friendly pesticides, along with ecological engineering concepts. Use of inorganic nutrient sources and/or pesticides, even though they have contributed to food production, established significant adverse effects on soil fertility, as well as macro- and microfauna biodiversity. The issue has assumed center stage since it has been established that the use of inorganic chemical nutrients has decelerated the soil microbial growth vigor (Wang et al. 2018), led to a reduction in composition (Fierer et al. 2012), impacted the diversity of soil microbial population, particularly the bacterial communities (Li et al. 2014; Huang et al. 2019), and could weaken the plant-microbe linkages since carbon from plant biomass is sparsely available to microbes (Huang et al. 2019). Relative abundance of Protecbacteria and Actinobacteria was positively correlated with the chemical N application rate, while a reverse in Chloroflexi in sesame (Wang et al. 2018) and Acidobacteria (Ramirez et al. 2012) has been noticed. Hence, the typical role of microbes is considered to be an acceptable solution to plant nutrition, and microbial biocontrol approaches have been

DOI: 10.1201/9781003147091-13

recognized for pest and disease management globally. Soil microbiota, which primarily involves beneficial bacteria and fungi, has diverse modes of action. Culturable microbes isolated and cultured from this source are reported to contain little or no toxic compounds to non-target species as well as human beings; albeit the successful use depends on sound awareness of the physiological reaction of plants to the bioagents. The plant-microbe generally establish an intimate, dynamic relationship that allows them to coexist (Nihorimbere et al. 2011) through synergistic effect (Mendes et al. 2011) to boost plant health. To further understand the processes, an understanding of the Tritrophic relationship between the soil, plant and the microbes is required.

In essence, soil microbes have a central role to play in controlling soil health and preserving plant quality. The techniques that manipulate soil microbiota are required to make the conducive soil into a suppressive one. This can also be accomplished by supplementing exogenous microbiota provided by selective biocontrol agents, organic amendments, for example, compost. Restoration of soil fertility and promotion of soil suppressiveness have been reported as a result of organic manure applications. Furthermore, integration of organic amendments and biocontrol agents effectively manage several soilborne pathogens. Research conducted elsewhere in the world has indicated that rhizosphere inhabiting microorganisms are a potential and effective management tactic to manage pests and diseases (Kerry 2000), besides serving as a supply of plant nutrients. This chapter explores the microbial biodiversity of oilseed crops, their mechanism of nutrient availability and abiotic and biotic stress mitigation in oilseed crops, and the tritrophic interactions between soil, plant and microbes.

13.2 MICROBIAL DIVERSITY IN OILSEED CROP SOIL ECOLOGY

The soil ecosystem around the plant root system is regarded as one of the most active interfaces on earth (Berendsen et al. 2012), where nutrient related interactions takes place between plants, soil and microbes, particularly the bacterial community in the rhizosphere (Jiménez et al. 2020). Depending on the terrestrial microhabitat of the organisms, their behavior and interaction with other organisms varies. Among the different microbes, rhizobacteria and mycorrhizal fungi are the most influential class in the rhizosphere, reported to take part in soil aggregation at various levels (Rillig and Mummey 2006). Diversity of microbiomes has been well documented in oilseed based production systems. Groundnut and soybean, two important leguminous oilseed crops inherently associated with bacterial strains through nodule-bacteria symbiosis to fix atmospheric nitrogen. *Chloroflexi, Acidobacteria, Actinobacteria*, and *Proteobacteria* (Levy et al. 2018) are the common phylae associated with oilseed crops. Among them, *Acidobacteria* is the most abundant phylae found in the terrestrial ecosystem (Barns et al. 1999). A wide variety of bacterial species associated with soybean rhizosphere is enriched with bacterial species through the nodule-bacteria association (Ziegler et al. 2013), which produces siderphores for aiding phosphate solubilization and nitrogen fixation. Other than Bradyrhhizobium, rhizosphere fungal communities with highest abundance of Ascomycota and Basidiomycota were noticed (Sugiyama, 2019). A few specific microbes found in groundnut rhizosphere are *Alcaligenes faecalis, Agrobacterium tumefeciens, Azotobacter chrocooccum, Aerobacter aerogenes, Bacillus subtilis, B. cereus, Citrobacter freundii, Clostridium perfringens, Enterobacter aerogenes, Flavobacterium lutescens, Klebsiella pneumoniae, Micrococcus varians, M. luteus, Proteus vulgaris, Pseudomonas fluorescens, P. aeruginosa, Streptococcus faecalis, S. pyogenes, Rhizobium japonicum and Serratia marcescens* in a sandy loam soil (Babadoko and Pacy 2011).

Irrespective of the nutrient source, the population of Proteobacteria was multiplied linearly with plant-derived carbon substrates (Philippot et al. 2013), needing further exploration (Wang et al. 2018) in sesame soils. A little over 189 plant growth promoting endophytic bacteria have been reported in Sunflower (leaves, stems, flowers and roots) in its short span of a life cycle (Bashir et al. 2020) of one-fourth of a year. Bacterial strains belonging to *Klebsiella, Enterobacter, Novosphingobium, Asticcacaulis, Grimontella, Microbacterium, Pantoea, Acinetobacter, Chryseobacterium, Herbaspirillum, Variovorax, Moraxella, Mitsuaria, Shinella, Sphingobium, Serratia* and *Xanthomonas* (Ambrosini et al. 2012) as well as *Enterobacter* and *Burkholderia* were noticed in sunflower. Recently, phosphorus mobilizer *Alcaligenes faecalis* has also been isolated by Shahid et al. (2012) from the roots of sunflower. It is of particular interest to note that microbiomes associated with soybean (Xu et al. 2009) and canola (Farina et al. 2012) remained dynamic at various growth stages of the plant, possibly due to nutritional balances (Chaparro et al. 2013). A high abundance of Proteobacteria, Actinobacteria and Bacteroidetes were noticed in Canola (Cordero et al. 2020). *Bacillusbrevis, B. amyloliquefaciens, B. megaterium, Erwinia amylovora, P. fluorescens, P. chlororaphis, P. putida, P. marginalis, P. syringae, Micrococcus luteus, Comamonas acidovorans* (Graner et al. 2003), *Agrobacterium, Burkholderia, Enterobacter* (Dunfield and Germida, 2003), *Serratia plymuthica and S. proteamaculans* (Abuamsha et al. 2011) have been reported in oilseed rape. For example, canola harbored 20% more microbes at the rosette stage over the normal population and decreased as the plant approach senescence stage. Analysis of canola microbiome revealed the abundance of *Enterobacter, Burkholderia, Agrobacterium* and *Pseudomonas* genera (Farina et al. 2012), with the majority belonging to the *Proteobacteria* phylum.

Soybean, accommodated more bacterial communities during the seedling stage than maturity stage with abundance of the phyla *Actinobacteria, Acidobacteria, Proteobacteria* (Wang et al. 2020) *Bacteroidetes, Nitrospirae, Firmicutes* and *Verrucomicrobia* (Xu et al. 2009). *Acidobacteria* plays a pivotal role in carbon cycling through degradation of plant parts, including cellulose and lignin. Notwithstanding this fact, their relationship in the rhizosphere is yet to be explored (Ward et al. 2009). *Proteobacteria* and *Actinobacteria* genera have been demonstrated to exhibit traits associated with the plant disease suppression (Mendes et al. 2011), while *Bacteroidetes* have a significant role in nitrogen cycling by denitrification (Chaparro et al. 2012).

13.3 PLANT-SOIL-MICROBE: TRITROPHIC INTERACTION

13.3.1 Soil-Microbe Interactions

Soil is a warehouse of microbes to the tune of 1×10^4 to 5×10^5, can be detected per gram of soil, the diversity of which varies with soil texture and nutrient composition (Schloss and Handelsman 2006). The majority of the complex rhizosphere microbiome food webs are reliant on the plant released nutrients (Mendes et al. 2013). Soil pH, nutrient content and temperature do play a key role in the soil microbial composition and population. It has also been reported that soil deficit in micronutrients render the plants more susceptible to the pathogens infection. For example, lowering the concentration of secondary (Ca and Mg) and micro nutrients (Na and Fe) and increasing the concentration of K, Cu and Zn in the soil was found to minimize the root rot incidence (*Macrophomina phaseolina*) in sesame (Narayanaswamy and Gokulakumar 2010). Manganese is one of the key elements that make the plants immune to the pathogens through inhibition of the induction of aminopeptidase, an enzyme essential for fungal growth (Dordas 2008). Soil deficit in manganese (Mn) decrease the metabolism of nitrogen and affects photosynthetic rate of the plants. This can be corrected by the addition of *B. subtilis* to the soil that reduce Mn^{4+} to Mn^{2+}, which can easily be taken up by the plants. When entered into the plant system, the reduced form of Mn strengthens the plant defense mechanism (Hoagland and Arnon 1938).

The beneficial effects of arbuscular mycorrhizal fungi (Gianinazzi et al. 2010) may influence CO_2 fixation by plants, through an increase in the "sink effect" and supply of photo-assimilates to the roots (Begum et al. 2019). Soilborne fungi deposits metabolites into the soil and affects the beneficial microbes' density. Addition of micronutrients to the soil eliminates this issue by suppressing the effect of pathogenic species in the soil. The investigation of Duffy and Defago (1997) confirmed that whenever the external supplementation limited the soil micronutrient zinc, it resulted in a drastic reduction in the composition of fusaric acid produced by *Fusarium oxysporum* f.sp. *radices-lycopersici* and enhanced the antibiotic production of *P. fluorescens* CHAo.

Soil salinity is another factor limiting the oilseed productivity that can be managed through the application of salt tolerant PGPR (Rodriguez and Redman 2008). Under salinity, due to the elevated level of Na^+, growth of the plant and uptake of K will be affected (Islam et al. 2015). Application of salt tolerant PGPR protects the plants from harmful effects of excessive concentration of Na^+ by binding Na^+ in the root zone through the production of exopolysaccharides and biofilms (Qurashi and Sabri 2012). Application of PGPR under salt stress conditions is reported to improve the availability of N, P, K, Zn, Fe, Cu and Mn, macro- and micronutrients. The key appliance of disease suppression by the microbes, antibiosis and competition could not be easily transferred from one soil unit to another, whereas the unique clamp-down effect triggered by parasitism and predation is easily transferable between soils, as it involves only a few microbial taxonomic classes. Thus, soil properties play a significant role in the suppression of soilborne disease causing microbes, either directly or indirectly, due to the close association of several factors. Knowledge of soil factors is very important for the adaption of biocontrol treatments in various soil environments. It will also provide a basis for the inclusion of biocontrol agents with the best agronomic practices to exploit and manipulate the soil microbiota for improvised disease control.

13.3.2 Mechanisms

While contact between microbes and plants occur at every phase of plant growth and development by signaling molecules, when soil is also taken into account, biochemical, physiological and molecular processes also come into the picture. In principle, plants, typically transfer about 20 to 40% of phytosynthetically fixed carbon to their roots in order to sustain root growth. Biochemically, plants produce an array of compounds with different molecular weights, as released as root exudates, which are used as cues for attraction of soil microbes (Haichar et al. 2014). Thus, the root system plays a crucial role in interacting with the microbe-soil complex, functions as a chemical factory and mediates this interaction. In return, these microbes interact with the plant in various ways, as the plant released volatile compounds are made up of CO_2, alcohol and aldehydes (Dudareva et al. 2004). To access nitrogen from the soil, plants are reliant on soil microbes, which possess the metabolic routes to depolymerize and convert the organic forms of N (Bonkowski 2004; Richardson et al. 2009) under organically managed conditions.

The chemical composition of root secretions and its type can alter the microbial diversity of the soil and may prevent the population spurt of harmful microbes; hence, the importance of root and microbe secretion is increasingly recognized (De-la-Peña and Loyola-Vargas 2014). The plant roots may also secrete strigolactones, a terpenoid and flavonoids (Bais et al. 2006; Venturi and Fuqua 2013; Massalha et al. 2017), besides carbohydrates, organic acids, vitamins and are reported to attract and support the beneficial bacteria development such as *Enterobacter cloacae* in the spermosphere and rhizosphere for their carbon requirement (Madhaiyan et al. 2010). As a complementary interaction, the microbes release volatile phytohormones promoting plant growth and bestow immunity to the plants. Rhizodeposits produced account from 5 to 30% of the total plant fixed C (Bekku et al. 1997; Hutsch et al. 2002; Dennis et al. 2010) into their direct surroundings; spermosphere, phyllosphere, rhizosphere and mycorrhizosphere (Frey-Klett et al. 2007; Raaijmakers et al. 2009; Vorholt 2012) are found to be the main sources of organic carbon that attract microbes toward them (Whipps 1990). Plants identify those compounds released by the microbes and change their growth and defense responses accordingly (Bais et al. 2006).

The compounds released by the plants may either be low molecular weight compounds (sugars and simple polysaccharides such as fructose, glucose, maltose, mannose, etc., amino acids such as arginine, aspartic, cysteine and glutamine; organic acids viz., acetic acid, ascorbic, benzoic and malic acids and phenolic compounds) or high molecular weight compounds (mucilage and proteins), which serve as signaling molecules to attract microbes since microbes use these compounds as carbon source for their nutrition (Schulz and Dickschat 2007). ATP-binding cassette (ABC) transporters have a significant role to play in the development of exudate compounds (Badri et al. 2008), which are primarily released through diffusion, vesicle transportation and ion exchange (Neuman and Romheld 2007) from the plant roots. It is an interesting phenomenon that when a pathogen infests a plant, it damages root cells and allows the mucilage to escape, which raises the amount of carbon in the rhizosphere, and attracts the beneficial microbes (Campbell and Greaves 1990) toward the roots. Any change in the composition of root exudates can, therefore, alter the diversity of soil microbial consortia. Chemotaxis is yet another process through which the roots release carbohydrates and amino acids to attract the beneficial microbes (Somers et al. 2004).

Plants physiologically change their rhizosphere under unfavorable conditions to allow the microbes to colonize the root zone. Under waterlogged conditions, soil could not hold oxygen within its particles. Physiological alteration of the plants during such adverse conditions contributes oxygen to the rhizosphere, which are otherwise utilized by the rhizosphere microbes. Under these oxygen rich environments, iron loving bacteria uses the excess free oxygen to oxidize the soluble mineral FeS, and Fe plaque gets precipitated on the root surfaces as oxidized coatings (Uren 2007). Due to the accumulation of iron on the surface of rhizosphere, plants can easily get Fe out of the soil. The other physiological adaption includes acidification of the rhizosphere, which is possible through the reduction of pH a variety of ways, for example, reduction of insoluble manganese oxide by the roots and the production of organic acids (Figure 13.1). These metabolites serve as signaling molecules that help the colonization of microbes by the rhizosphere (Berg and Smalla 2009). Microbes also use quorum-sensing mechanisms to change their metabolism. Plants secrete a number of chemicals that mimic the quorum signals provided by the bacteria that lead to the alteration of bacterial activity in the rhizosphere (Bauer and Mathesius 2004). Since rhizosphere microbes *viz.*, *Pseudomonas*, *Bacillus*, *Paenibacillus*, actinomycetes and arbuscular mycorrhizal fungi and (AMF) enhance nutrient absorption (Cummings, 2009), as well as resistance to biotic (De Vleesschauwer and Hofte 2009) and abiotic stresses (Zhang et al. 2008;Berendsen et al. 2012; Selvakumar et al. 2012) termed rhizosphere soil microbiomes and the associated microbial community as the second genome of the plant.

Production of the iron chelating compound for e.g. siderophores, commonly produced by fungi as well as bacteria, is pronounced in fluorescent pseudomonads known for the suppression of soilborne pathogenic microbes. Uptake of ferric ion

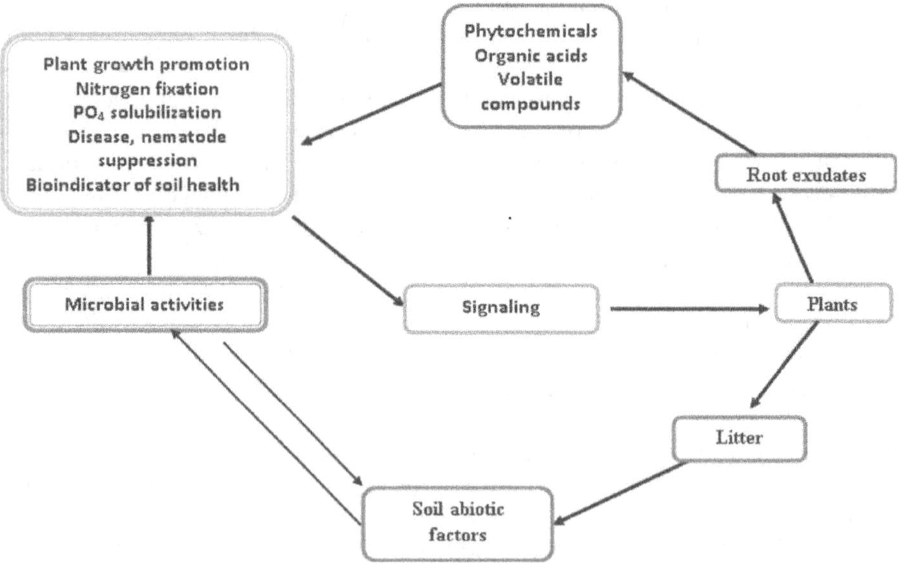

FIGURE 13.1 Role of soil microorganism in plant growth and molecular signaling.

(Fe^{3+}) from the soil into their cells renders the iron inaccessible to the pathogens, which creates a competition (Kloepper et al. 1999) between *Pseudomonas* and pathogens for nutrients. The 3-hydroxy-2-butanone, an elicitor compound production by the *Bacillus* spp. provided a persuasive effect on the infestation of the pathogens (Rudrappa et al. 2008). Another significant plant-soil-microbe interaction is through fixation of atmospheric nitrogen (von der Weid et al. 2002) and solubilization of phosphorus (Deepa et al. 2010).

The essential components of soil nitrogen cycle are mineralization and immobilization of nitrogen by soil microbes. Microbial enzymes breakdown organic nitrogenous molecules to produce ammonia (NH_3), which increases the quantity of plant-available nitrogen in soils. A proportion of the nitrogen in the transformed organic material is utilized by microbial tissue, while the rest is liberated (mineralized) as inorganic nitrogen. Some of the genus of nitrogen fixing bacteria includes *Alcaligenes, Beijerinckia, Campylobacter, Derxia,* (Rennie 1980), *Azospirillium,* (Li and Castellano 1987), *Klebsiella, Pantoae, Bacillus, Azotobacter* (Kennedy et al. 1997), *Pseudomonas,* and *Burkholderia.* In general, summer recorded the largest population of ammonifying bacteria in soil under diverse crops, while winter reported the lowest. Insoluble IP, such as $Ca_3(PO_4)_2$, $AlPO_4$, and $FePO_4$, as well as insoluble/soluble organic phosphorus (OP), such as phytate, which accounts for 80% of soil OP, require phosphorus-solubilizing microorganisms (PSM) to convert into orthophosphate, which may be taken by plants and bacteria. The majority of P in soils exists in inorganic insoluble form ($Ca_3(PO_4)_2$) and organic insoluble/soluble form (phytate and nucleic acid) and, hence, only a small amount can be directly digested by plants (Neal et al., 2017). To solubilize the insoluble inorganic phosphates as well as insoluble/soluble organic phosphates, phosphorus-solubilizing microorganisms (PSM) are essential, which convert insoluble form into orthophosphate, which may be taken by the plants. Fertilizer use degrades soil quality and decreases phosphorus availability (Liu et al. 2018). Application of phosphorus solubilizing microorganisms with multiple P source using abilities gives a new strategy to improving soil quality in order to reach the goal of sustainable agriculture.

PGPR plays a crucial mechanism of reducing the quantum of ethylene in the plants during stress condition since its production (Glick et al. 2007) accelerates due to the accumulation of 1-aminocyclopropane-1-carboxylic acid (ACC), a precursor in the biosynthetic pathway of ethylene (Zapata et al. 2017). This interrupts normal functioning of the plant through impairment in the root growth (Figure 13.1). When PGPR are integrated, they release ACC deaminase and disintegrate and α-ketobutyrate (Singh et al. 2015), thus reducing ethylene concentration in the root zone and root growth is not compromised (Glick et al. 2007). Furthermore, in the plant roots, synthesis of indole acetic acid, activates the enzyme ACC synthase to transform S-adenosylmethionine to ACC (Gravel et al. 2007), which is subsequently absorbed by PGPR through root exudation and ultimately disintegrates through bacterial ACC deaminase, thus preventing the plant from suffering the negative effects of ethylene (Yaish et al. 2015). Yang et al. (2008) were of the optimistic view that PGPR usage might reduce the external application of fertilizers.

A wide range of microorganisms use the mechanism of competition toward the ecological niches and food through which they contribute to the reduction of pathogen infestation. Compounds released by root exudates in response to the chemical signals from soil microbes are a function of soil type (Rovira 1969), biotic, abiotic components (Tang et al. 1995) and plant age (De-la-Pena et al. 2010). As they are associated by identifying the compounds secreted by the roots against particular beneficial microbes, it may be possible to culture the beneficial microbes on their own in the roots that provide protection from deleterious species.

13.4 SOIL-PLANT-MICROBE: POSSIBLE INTERVENTIONS

13.4.1 Management Interventions

In developing a successful organic farming protocol, development of a complex microbiome network in a phylogenetic context is the key for a high degree of negative correlation between the biocontrol microbes and disease causing organisms. Utilization of PGPR, which could be a major component of integrated nutrient management program (Yang et al. 2008), has been articulated in the recent past. Endophytic asymbiotic diazotrophic bacteria reported to fix atmospheric nitrogen in canola (Puri et al. 2016) could well be integrated in the organic package for nitrogen supplementation. Disease suppression is a significant phenomenon of healthy soil induced by soil microbes through the combined antagonistic mechanisms of the microbes. In healthy soil, soil microbiota interact with soil organic matter to achieve synergistic results in disease suppression and strengthening the soil structure (Bronick and Lal 2005) through increased microbial activity (Bulluck et al. 2002). The presence of various microbial biomes in the soil provide basal protection against a wide range of biotic stresses and hence become an essential component in the determination of soil health (Kibblewhite et al. 2008). Supplementation of selected beneficial microbiota, along with organic substrates for their proliferation increase suppression against soilborne pathogens (Puglisi et al. 2009) with a rider that efficacy vary, depending on the soil pathogen density (Ruano-Rosa and Mercado-Blanco 2015). In contrast, organic amendments could improve disease severity (Bonanomi et al. 2007) under rare circumstances. Synergistic interaction of microbes is a natural phenomenon seen in organic soil through which many soilborne pathogen are suppressed (Wan et al. 2017).

Compost is typically rich in nutrients with bountiful microbes possibly triggering the process of resistance against various soilborne pathogens. Few notable references include resistance to *F. oxysporum* f.sp. *melonis* in vegetable by sewage sludge compost (Cotxarrera et al. 2002) and *V. dahliae* of cotton by olive mill compost (Aviles and Borrero 2017) through enhanced microbial activity (Berendsen et al. 2012). These soils are typically colonized by a wide group of microorganisms from which several strains have been documented to induce resistance against abiotic stresses (Melero-Vara et al. 2011). Efficiency of beneficial microbes in the soil is a function of physicochemical properties of organic matter such as pH, electrical conductivity, presence of macro and micronutrients content and the pathogen load in the soil (Hadar and Papadopoulou 2012). Use efficiency of cheap sources of P, namely rock phosphates, was improved along with organic amendments and P solubilizers viz., *Aspergillus awamorii, P. striata and B. polymyxa*in meeting P nutrition of soybean and mustard crops (Qureshi and Narayanasamay 1999 and Qureshi and Narayanasamy 2005). Despite conflicting information, integrated management strategies with the combined use of organic amendments, biocontrol agents and biofertilizers will certainly provide control of various soilborne pathogens (Cretoiu et al. 2013; Zhang et al. 2014). Pre-fumigation of soil combined with the incorporation of organic amendments and/or bio fertilizers induce shifts in the soil microbiota population, which play a crucial role in the control of soilborne pathogens (Deng et al. 2019).

Hence, pre fumigation with composts, biochar and undecomposed plant residues followed by the addition of plant residues, either alone or with organic amendments or biocontrol agents (Himmelstein et al. 2014; Torres et al. 2015), increases disease suppressiveness, thereby suggesting possible recovery in the fertility of the soil. Evidence shows that soil pre-fumigation with ammonium bicarbonate, along with compost, impaired *Ralstonia solanacearum*, through alteration of the beneficial bacterial and fungal population in the soil and spurt in new beneficial taxa (Antoniou et al. 2017). Addition of compost added macro- and micronutrients to the soil (Evanylo et al. 2008) also improved organic carbon content in the soil (Hemmat et al. 2010), its structure (Celik et al. 2004) and water-holding capacity (Caravaca et al. 2002), thereby raising the crop productivity (Zaccardelli et al. 2013) through containment of the soilborne pathogens (De Corato 2020). Compost may enhance soil microbiota (Zhen et al. 2014), decrease it (Martinez Blanco et al. 2013) or leave it unaffected (Nair and Ngouajio 2012). Addition of green compost, either agro or agro industries wastes, accounted for a greater diversity of microorganism functioning against *Verticillium*, *Rhizoctonia*, *Pythium* and *Phytophthora.* Composted animal manure has been reported to suppress *Fusarium* wilt, as it contains *Trichoderma*, *Aspergillus, Penicillium, Streptomyces, Pseudomonas* and *Bacillus* (Xiong et al. 2017). However, to be a successful biocontrol agent, its stability under field conditions is essential for evaluating its efficacy (Sommermann et al. 2018). However, it is postulated that long term supplementation of plant residues combined with biocontrol agents could manipulate soil microbiota (Yang et al. 2019)

Incorporation of organic amendments certainly modifies the soil microbiota community (De Corato 2020). Adding compost helps the growth of abundant soil microbiota populations. Amplicon sequence of compost derived microorganisms showed a wide abundance and diversity of bacterial and fungal species (De Corato et al. 2019) in the composted soil. Sustainable production of oilseed to satisfy rising demand is a herculean task. Biotic and abiotic stress exerted on the crop contributes to a greater rise in the crop yield reductions (Selvakumar et al. 2012). Inoculation of oilseed crops with beneficial microbes improves crop production. Among the numerous microorganisms, plant growth promoting rhizobacteria is a versatile community of complex bacterial taxa capable of promoting the supply of plant nutrients, growth and disease suppression (Morrison et al. 2017). Manipulation of soil microbiome in the oilseed ecosystem would support the plant by increasing plant production and decreasing the biotic and abiotic stresses imposed on them. Inoculation of *P. fluorescens* strain, LBUM677 in Canola (*Brassica napus*), Soybean (*Glycine max*) and Corn gromwell (*Buglossoides arvensis*) was reported to improve oil yield (Cumberford and Hebard 2015), in addition to plant growth promotion effects.

Generally, oilseed crops grown under stressful conditions suffer iron deficiency. Inoculation of PGPR allows the plants to extract more micronutrient element iron from the soil environment in order to reduce the harmful effects of stress. Recently, inoculation of microbes in the rhizosphere has been suggested to minimize heavy metals and salt stress by enhancing the antioxidant enzyme (Islam et al. 2015). *Paenibacillus polymyxa* P2b-2R, an endophytic diazotroph isolated from lodge pole pine tree tested on canola increased biomass by 27% 2 months after bacterial inoculation which could fix around 20% of atmospheric nitrogen (Puri et al. 2016). Inclusion of plant growth promoting microbes in the crop packages helps the plant through the systemic activation of defense mechanisms against pests and pathogens. Application of PGPRs are reported to activate the defense genes such as chitinase, proteinase inhibitors and lipoxygenase to protect the plant from infection (Saravanakumar et al. 2007) in peanut and dual inoculation of sulfur oxidizing bacteria, along with *Ochrobactrum intermedium* (Paulucci et al. 2015) was also beneficial. Further complementary interaction between S-oxidizing bacteria and rhizobia promoted the oil content of groundnut grown in sulfur deficit soils (Anandham et al. 2007). Antagonistic rhizobacteria must be predominant in rhizosphere soils of sunflower (Rangeshwaran and Prasad 2000) in order to contain Sclerotium rot of sunflower.

The sunflower crop is mainly prone to water and salinity stress besides the accumulation of heavy metals which limit crop productivity (Ali et al. 2013). Sunflower growth is also impaired by the deficiency of nitrogen (Dordas and Sioulas, 2008). Jalilian et al. (2012) observed improved productivity of sunflowers when microbes were inoculated in combination with N fertilizers, in addition to upgrading plant water status (Singh et al. 2015). Inoculation of PGP bacteria strains, *Pseudomonas fluorescens* biotype F and *P. fluorescens* CECT 378T were reported to alleviate salinity-induced stress in sunflowers (Kiani et al. 2016). Similarly, the PGPR strains, *Ralstonia eutropha* (B1) and *Chrysiobacterium humi* (B2) protected the crop from metal toxicity by preventing the crop from the uptake of heavy metals viz., Zn and Cd (Ana et al. 2013).

Inoculation of *B. licheniformis* MML2501, *Bacillus* sp. MML2551 and *P. aeruginosa* MML2212 reduced sunflower necrosis disease (SND), which otherwise reduced sunflower productivity by 90% (Bhat et al. 2002; Lavanya et al. 2005). Wherever salinity is a threat to soybean production impairing crop yield (Rahman et al. 2008), either inoculation of PGPR improved crop growth (Kang et al. 2014) through the regulation of stress hormones or through PGPR interaction with *Rhizobium* spp. (Drogue et al. 2013). The beneficial association of PGPR strains *viz*., *Streptomyces* (Tokala et al. 2002), *Azotobacter* (Wu et al. 2012), *Bacillus* (Atieno et al. 2012), *Serratia* (Zahir et al. 2016), *Azospirillum* (Aung et al. 2013) has been reported, in addition to *Bradyrhizobium diazoefficiens* USDA110 and *B. japonicum* THA6 (Prakamhang et al. 2015). An improvement in the number of root hairs and nodules and secretions of flavonoids was found when Rhizobia was co-inoculated with PGPR over rhizobia inoculation alone (Ramans et al. 2008). Hence, it has been proposed that inoculation of mixture of suitable strains would be more successful for soybean than the inoculation of single bacterial strains.

When performing artificial inoculation experiments, emphasis should be given to the impact of artificially inoculated microbes on the modification of native microbial populations. This knowledge is important firstly, to study the adverse effects of inoculated microbiota on natural soil microbial communities and secondly, to consider the indirect effect of inoculated PGPR strains on the plants. Experimental and technical methodology used to describe the effect of PGPR on soil microbiome is an important consideration to be taken into account when analyzing the findings of such studies. Supplementation of bioagents could mobilize soil nutrients and reduce the need of fertilizer application. Soil microbiota, through the production of secondary metabolite compounds, such as 2,4-DAPG (Raaijmakers et al. 1997), iturin and surfactin (Kinsella et al. 2009), tubercidin, phosphalactomycin, candicidin (Shekhar et al. 2006) and lytic enzymes such as β-1,3-glucanase, chitobiase, hydrolases and chitinase (De La Cruz et al. 1993) could allow a strong binding with the plants toward the suppression of pathogens. As an added benefit, growth hormones, auxin (Da Mota et al. 2008) and cytokinin (Timmusk et al. 1993) are also produced, which impart indirect resistance to the plants against abiotic and biotic stresses.

13.4.2 Determinants

Shaping of the rhizospheric microbiome is dictated by the plant species and the healthy soil (Berg and Smalla 2009; Bakker et al. 2012) underpins the significance of plant-soil-microbe interactions. Furthermore, edaphic features such as soil nutrients, pH and moisture content may have a major effect in the alteration of rhizosphere microbial population (Bakker et al. 2020). Management strategies that are being adopted to enhance nutrient release and to fight pests and diseases could also improve the composition of the soil microbiome population (Lumini et al. 2011). Agricultural practices that ratify and promote beneficial microbes can contribute to improved yield of the crops over the long term. Crop rotation and intercropping have played a major role in altering soil microbial populations. Long term monoculture of a single crop promotes the abundance of soilborne pathogens and decreases crop yield due to a decline in soil chemical properties and soil microbiota composition. Diverse distribution frequencies and transmission mechanisms of biocontrol agents had a direct effect on the rhizosphere. Conventional agricultural practices, which depend primarily on the use of chemical insecticides and fungicides (Krauss et al. 2011), minimize soil microbiome levels. On the other hand, the practice of organic farming would promote microbial diversity (Sugiyama et al. 2010).

13.5 CONCLUSION AND FUTURE PROSPECTS

Increased demand for the production of organic foods has called for more environmentally sustainable agricultural activities. The existence of a healthy microbiota can be the key to acquiring healthier plants. Recently, several studies have started to illustrate the significance of plant-microbiome interaction and its effect on plant health and productivity. Involvement of microbiota will increase natural soil suppression against soilborne pathogens whenever soil properties deteriorate, rendering a decline in crop yields. A significant group of microorganisms play a vital role in enhancing plant health in oilseed environment. A wide range of microbes found in the soil rhizosphere provides an opportunity for genetic variation and exploitation of microbes. Both bacterial and fungal microbiomes facilitate the growth and development of various oilseed crops, both directly and indirectly under normal and stressful conditions. Root exudates serve as signaling molecules and play a crucial role in the manipulation of microbes. Further research should be carried out to determine the composition of root exudate that will enable the culture of particular beneficial microbes to produce more productive plants. In addition, the ecological function of soil microbiomes should be further explained in order to obtain a deeper understanding of the microbial population network inhabiting the soil. Agronomic practices appeared to be the best methods for increasing soil suppression by microbiomes. As a result, best agronomic methods to increase soil suppression can be put in place that can transfer wild soil microbiota to beneficial microbial consortia. It can be inferred that by following effective agronomic methods such as the addition of compost, crop rotation, intercropping and soil pre fumigation with eco-friendly molecules, problems in the suppression of soilborne diseases can be resolved through the manipulation of soil microbiomes in healthy agroecosystems. Manipulation of microbes by organic farming has been shown to improve the productivity of oilseed crops and reduce their biotic stress. While plant-microbe interactions are very reliable, the underlying mechanisms still need to be studied to optimize and popularize the use of soil microbial cultures in an effective way. In general, a significant change in microbial community composition is imminent after the addition of an amendment, be it organic or inorganic; hence, suitable mechanisms need to be developed for complementary use of organic and/or microbial fertilization, in addition to chemical fertilization. Probing into the interaction of plants and microbes at a molecular level with the aid of stable isotopes and biochemical markers can enhance the understanding for the identification of chemicals involved in the interaction process. Detection and application of a variety of beneficial microbial strains could also play a central role as biofertilizers and biopesticides, which will reduce the use of fertilizers and agrochemicals in oilseed crops.

REFERENCES

Abuamsha, R., Salman, M., Ehlers, R.U. 2011. Differential resistance of oilseed rape cultivars (*Brassica napus* sp. *oleifera*) to *Verticillium longisporum* infection is affected by rhizosphere colonization with antagonistic bacteria, *Serratia plymuthica* and *Pseudomonas chlororaphis. Biocontrol* 156: 101–112.

Ali, M., Ali, M., Ramezani, A., Far, S.M., Sadat, K., Moradi-Ghahderijani, M., Jamian, S.S. 2013. Application of silicon ameliorates salinity stress in sunflower (*Helianthus annuus* L.) plants. *International Journal of Agriculture and Crop Science* 6: 1367–1372.

Ambrosini, A., Beneduzi, A., Stefanski, T., Pinheiro, F.G., Vargas, L.K., Passaglia, L.M.P. 2012. Screening of plant growth promoting rhizobacteria isolated from sunflower (*Helianthus annuus* L.). *Plant and Soil* 356: 245–264.

Ana, P.G.C.M., Moreira, H., Franco, A.R., Rangel, A.O.S.S., Castro, P.M.L. 2013. Inoculating *Helianthus annuus* (sunflower) grown in zinc and cadmium contaminated soils with plant growth promoting bacteria effects on phytoremediation strategies. *Chemosphere* 92(1): 74–83.

Anandham, R., Sridar, R., Nalayini, P., Sa T. 2007. Potential for plant growth promotion in groundnut (*Arachis hypogaea* L.) cv. ALR-2 by co-inoculation of sulfur-oxidizing bacteria and Rhizobium. *Microbiological Research* 162(2): 139–53.

Antoniou, A, Tsolakidou, M.D., Stringlis, I.A., Pantelides, I.S. 2017. Rhizosphere microbiome recruited from suppressive compost improves plant fitness and increases protection against vascular wilt pathogens of tomato. *Frontiers in Plant Science.* 8: 2022, 1–16. https://doi.org/10.3389/fpls.2017.02022.

Atieno, M., Herrmann, L., Okalebo, R. Lesueur, D. 2012. Efficacy of different formulation of *Bradyrhizobium japonicum* and effect of co-inoculation of *B. subtilis* with two different strains of *Bradyrhizobium japonicum. World Journal of Microbiol Biotechnology* 28(7): 2541–2550.

Aung, T.T., Tittabutr, P., Boonkerd, N., Teaumroong, N. 2013. Co-inoculation effects of *Bradyrhizobium japonicum* and *Azospirillum* sp. on competitive nodulation and rhizosphere eubacterial community structures of soybean under rhizobia-established soil conditions. *African Journal of Biotechnology* 12: 2850–2862.

Aviles, M., Borrero, C. 2017. Identifying characteristics of *V. dahliae* wilt suppressiveness in olive mill composts. *Plant Disease* 101(9): 1568–77.

Babadoko, A M., Pacy O.G. 2011. Rhizosphere bacterial flora of groundnut (*Arachis hypogeae*). *Advances in Environmental Biology* 5(10): 3196–3202.

Badri, D.V., Vargas, L.V.M., Broeckling, C.D., De la Pena, C., Jasinski, M., Santelia, D., Martinoia, E., Sumner, L.W., Banta, L.M., Stermitz, F., Vivanco, J.M. 2008. Altered profile of secondary metabolites in the root exudates of Arabidopsis ATP-binding cassette transporter mutants. *Plant Physiology* 146: 762–771.

Bais, H.P., Weir, T.L., Perry, L.G., Gilroy, S., Vivanco, J.M. 2006. The role of root exudates in rhizosphere interactions with plants and other organisms. *Annual Review of Plant Biology* 57: 233–266.

Bakker, A.H.M.P., Berendsen, R.L., Van Pelt, J.A. et al. 2020.The soilborne identity and microbiome-assisted agriculture: Looking back to the future. *Molecular Plant* 13(10): 1395–1402.

Bakker, M.G., Manter, D.K., Sheflin, A.M., Weir, T.L. Vivanco, J.M. 2012. Harnessing the rhizosphere microbiome through plant breeding and agricultural management. *Plant Soil* 360: 1–13.

Barns, S.M., Takala, S.L., Kuske, C.R. 1999. Wide distribution and diversity of members of the bacterial kingdom Acidobacterium in the environment. *Applied Environmental Microbiology* 65(4): 1731–1737.

Bashir, S., Iqbal, A., Hasnain, S. 2020.Comparative analysis of endophytic bacterial diversity between two varieties of sunflower *Helianthus annuus* with their PGP evaluation. *Saudi Journal of Biological Science.* 27(2): 720–726.

Bauer, W.D., Mathesius, U. 2004. Plant responses to bacterial quorum sensing signals. *Current Opinion in Plant Biology* 7(4): 429–433.

Begum, N., Qin, C., Ahanger, M.A., Raza, S., Khan, M.I., Ashraf, M., Ahmed, N., Zhang, L. 2019. Role of arbuscular mycorrhizal fungi in plant growth regulation: Implications in abiotic stress tolerance. *Frontiers in Plant Science* 19: 1068.

Bekku, Y., Kimura, M., Ikeda, H., Koizumi, H. 1997. Carbon input from plant to soil through root exudation in *Digitaria adscendens*, and *Ambrosia artemisiifolia. Ecol Res* 12: 305–312.

Berendsen, R.L., Pieterse, C.M.J., Bakker, P.A.H.M. 2012. The rhizosphere microbiome and plant health. *Trends in Plant Science* 17(8): 478–86.

Berg, G., Smalla, K. 2009. Plant species and soil type cooperatively shape the structure and function of microbial communities in the rhizosphere. *FEMS Microbiol Ecology* 68: 1–13.

Bhat, A.I., Jain, R.K., Chaudhary, V., Reddy, M.K. 2002. Sequence conservation in the coat protein gene of tobacco streak virus isolates causing necrosis disease in cotton, mung bean, sunflower and sunhemp in India. *Indian Journal of Biotechnology* 1(4): 350–356.

Bonanomi, G., Antignani, V., Pane, C., Scala, F. 2007. Suppression of soil borne fungal diseases with organic amendments. *Journal of Plant Pathology* 89: 311–324.

Bonkowski, M. 2004. Protozoa and plant growth: The microbial loop in soil revisited. *New Phytology* 162: 617–631.

Bronick, C., Lal, R. 2005. Soil structure and management: A Review. *Geoderma* 124: 3–22.

Bulluck, L.R., Brosius, M., Evanylo, G.K., Ristaino, J.B. 2002. Organic and synthetic fertility amendments influence soil microbial, physical and chemical properties on organic and conventional farms. *Applied Soil Ecology* 19: 147–160.

Campbell, R., Greaves, M. P. 1990. Anatomy and community structure of the rhizosphere. In The Rhizosphere. Ed. J M Lynch (Essex: John Wiley & Sons Ltd.), 11–34.

Caravaca, F., Hernandez, T., Garcia, C., Roldan, A. 2002. Improvement of rhizosphere aggregate stability of afforested semiarid plant species subjected to mycorrhizal inoculation and compost addition. *Geoderma* 108: 133–144.

Celik, I., Ortas, I., Kilic, S. 2004. Effects of compost, mycorrhiza, manure and fertilizer on some physical properties of a Chromoxerert soil. *Soil and Tillage Research* 78: 59–67.

Chaparro, J.M., Badri, D.V., Bakker, M.G., Sugiyama, A., Manter, D.K., Vivanco, J.M. 2013. Root exudation of phytochemicalsin Arabidopsis follows specific patterns that are developmentally programmed and correlate with soil microbial functions. *PLoS One* 8(2): e55731.

Chaparro, J.M., Sheflin, A.M., Manter, D.K., Vivanco, J.M. 2012. Manipulating the soil microbiome to increase soil health and plant fertility. *Biology and Fertility of Soils* 48: 489–499.

Cordero, J., de Freitas, J.R., Germida, J.J. 2020. Bacterial microbiome associated with the rhizosphere and root interior of crops in Saskatchewan. *Canadian Journal of Microbiology* 66, 71–85.

Cotxarrera, L., Trillas, M.I., Steinberg, C., Alabouvette, C. 2002. Use of sewage sludge compost and *Trichoderma asperellum* isolates to suppress *Fusarium oxysporum* f sp *melonis* wilt of tomato. *Soil Biology and Biochemistry* 34: 467–476.

Cretoiu, M.S., Korthals, G.W., Visser, J.H.M., Van Elsas, J.D. 2013. Chitin amendment increases soil suppressiveness toward plant pathogens and modulates the actinobacterial and oxalobacteraceal communities in an experimental agricultural field. *Applied and Environ Microbiology* 79: 5291–5301.

Cumberford, G., Hebard, A. 2015. Ahiflower oil: A novel non-GM plant-based omega-3+6 source. *Lipid Technology* 27: 207–210.

Cummings, S.P. 2009. The application of plant growth promoting rhizobacteria (PGPR) in low input and organic cultivation of graminaceous crops; potential and problems. *Environmental Biotechnology* 5: 43–50.

Da Mota, F.F., Gomes, E.A., Seldin, L. 2008. Auxin production and detection of the gene coding for the auxin efflux carrier (AEC) protein in *Paenibacillus polymyxa*. *Journal of Microbiology* 46(3): 257–264.

De Corato, U. 2020. Disease-suppressive compost enhances natural soil suppressiveness against soilborne plant pathogens: A critical review. *Rhizosphere* 13: 100192.

De Corato, U., Patruno, L., Avella, N., Lacolla, G., Cucci, G. 2019. Composts from green sources show an increased suppressiveness to soilborne plant pathogenic fungi: Relationships between physicochemical properties, disease suppression, and the microbiome. *Crop Protection* 124: 104870.

De la Cruz, J.M.R., Lora, J.M., Hidalgo-Galiego, A., Dominguez, F., Pintor-Toro, J.A., Llobell, A., Benitez, T. 1993. Carbon source control on β-gluc Masciarelli seschitobiase and chitinase from *Trichoderma harzianum*. *Annual Review of Microbiology* 159: 316–322.

De Vleesschauwer, D., Hofte, M. 2009. Rhizobacteria-induced systemic resistance. In: Advances in botanical research. Ed. L.C. Van Loon (Burlington: Elsevier), 51: 223–281.

Deepa, C., Dastager, S.G., Pandey, A. 2010. Plant growth-promoting activity in newly isolated *Bacillus thioparus* (NII-0902) from Western ghat forest, India. *World Journal of Microbiology and Biotechnology* 26(12): 2277–2283.

De-la-Peña, C., Badri, D.V., Lei, Z., Watson, B.S., Brandao, M.M., Silva-Filho, M.C., Sumner, L.W., Vivanco, J.M. 2010. Root secretion of defense related proteins is development dependent and correlated with flowering time. *Journal of Biological Chemistry* 285: 30654–30665.

De-la-Peña, C., Loyola-Vargas, V.M. 2014. Biotic interactions in the rhizosphere: A diverse cooperative enterprise for plant productivity. *Plant Physiology* 166: 701–719.

Deng, X., Zhang, N., Shen, Z., Zhu, C., Li, R., Falcao Salles, J., Shen, Q. 2019. Rhizosphere bacteria assembly derived from fumigation and organic amendment triggers the direct and indirect suppression of tomato bacterial wilt disease. *Applied Soil Ecology* 147: 103364.

Dennis, P.G., Miller, A.J., Hirsch, P.R. (2010) Are root exudates more important than other sources of rhizodeposits in structuring rhizosphere bacterial communities? *FEMS Microbiol Ecology* 72: 313–327.

Dordas, C. 2008. Role of nutrients in controlling plant diseases in sustainable agriculture. A review. *Agronomy for Sustainable Development* 28(1), 33–46.

Dordas, C.A., Sioulas, C. 2008. Safflower yield, chlorophyll content, photosynthesis, and water use efficiency response to nitrogen fertilisation under rainfed conditions. *Industrial Crops and Products* 27: 75–85.

Drogue, B., Combes-Meynet, E., Moenne-Loccoz, Y., Wisniewski-Dye, F., Combaret, P.C. 2013. Control of the cooperation between plant growth promoting rhizobacteria and crops by rhizosphere signals. In: Molecular Microbial Ecology of the Rhizosphere. Eds. F.J. de Bruijn (NJ, USA: John Wiley & Sons, Inc), 281–294.

Dudareva, N., Pichersky, E., Gershenzon, J. 2004. Biochemistry of plant volatiles. *Plant Physiology* 135(4):1893–1902.

Duffy, B.K. and Defago. 1997. Improves biocontrol of *Fusarium* crown and root rot of tomato by *pseudomonas* fluorescens and represses the production of pathogen metabolites inhibitory to bacterial antibiotic biosynthesis. *Phytopathology* 87(12): 1250–1257.

Dunfield, K.E., Germida, J.J. 2003. Seasonal changes in the rhizosphere microbial communities associated with field-grown genetically modified canola (*Brassica napus*). *Applied Environmental Microbiology* 69(12): 7310–7318.

Evanylo, G., Sherony, C., Spargo, J., Starner, D., Brosius, M., Haering, K. 2008. Soil and water environmental effects of fertilizer, manure and compost based fertility practices in an organic vegetable cropping system. *Agriculture, Ecosystem and Environment* 127: 50–58.

Farina, R.A., Beneduzi, A., Ambrosini, A., Campos, S.B., Lisboa, B.B., Wendisch, V., Vargas, L.K., Passaglia, L.M.P. 2012. Diversity of plant growth promoting rhizobacteria communities associated with the stages of canola growth. *Applied Soil Ecology* 55: 44–52.

Fierer, N., Lauber, C.L., Ramirez, K.S., Zaneveld, J., Bradford, M.A., Knight, R., 2012.Comparative metagenomic, phylogenetic and physiological analyses of soil microbial communities across nitrogen gradients. *ISME Journal* 6, 1007–1017.

Frey-Klett, P., Garbaye, J., Tarkka, M. 2007. The mycorrhiza helper bacteria revisited. *New Phytology* 176: 22–36

Gianinazzi, S., Gollotte, A., Binet, M.-N., van Tuinen, D., Redecker, D., Wipf, D. 2010. Agroecology: The key role of arbuscular mycorrhizas in ecosystem services. *Mycorrhiza* 20(8): 519–30.

Glick, B.R., Todorovic, B., Czarny, J., Cheng, Z., Duan, J., McConkey, B. 2007. Promotion of plant growth by bacterial ACC deaminase. *Critical Reviews in Plant Sciences* 26 (5–6): 227–242.

Graner, G., Persson, P., Meijer, J., Alstrom, S. 2003. A study on microbial diversity in different cultivars of *Brassica napusin* relation to its wilt pathogen, *Verticillium longisporum. FEMS Microbiology Letters* 224: 269–276.

Gravel V., Antoun H., Tweddell R.J. 2007. Growth stimulation and fruit yield improvement of greenhouse tomato plants by inoculation with *Pseudomonas putida* or *Trichoderma atroviride*: Possible role of indole acetic acid, (IAA). *Soil Biology and Biochemistry* 39: 1968–1977.

Hadar, Y., Papadopoulou, K.K. 2012. Suppressive composts: Microbial ecology links between abiotic environments and healthy plants. *Annual Review of Phytopathology* 50: 133–153.

Haichar, F.Z., Santaella, C., Heulin, T., Achouak, W. 2014. Root exudates mediated interactions belowground. *Soil Biology and Biochemistry* 77: 69–80.

Hemmat, A., Aghilinategh, N., Rezainejad, Y., Sadeghi, M. 2010. Long term impacts of municipal solid waste compost, sewage sludge and farmyard manure application on organic carbon, bulk density and consistency limits of a calcareous soil in central Iran. *Soil and Tillage Research* 108: 43–50.

Himmelstein, J.C., Maul, J.E., Everts, K.L. 2014. Impact of five cover crop green manures and actinovate on fusarium wilt of watermelon. *Plant Disease* 98: 965–972.

Hoagland, D.R., Arnon, D.I. 1938. The water culture method for growing plants without soil. *Circular* 347. Berkeley, California: University of California Agricultural Experiment Station.

Huang, R.,McGrath, S.P., Hirsch, P.R., Clark, I.M., Storkey, J., Wu, L., Zhou, J., Liang, Y. 2019. Plant–microbe networks in soil are weakened by century-long use of inorganic fertilizers. *Microbial Biotechnology* 12(6): 1464–1475.

Hutsch, B.W., Augustin, J., Merbach, W. 2002. Plant rhizodeposition-an important source for carbon turnover in soils. *Journal of Plant Nutrition and Soil Science* 165: 394–407.

Islam, F., Yasmeen, T., Ali, S., Ali, B., Farooq, M.A., Gill, R.A., 2015. Priming-induced antioxidative responses in two wheat cultivars under saline stress. *Acta Physiologia Plantarum* 37: 153.

Jalilian, J., Modarres-Sanavya, S.A., Saberali, S.F., Asilan, K.S. 2012. Effects of the combination of beneficial microbes and nitrogen on sunflower seed yields and seed quality traits under different irrigation regimes. *Field Crops Research* 127: 26–34.

Jiménez, J.A., Novinscak, A., Filion, M. 2020. Inoculation with the Plant Growth-Promoting Rhizobacterium *Pseudomonas fluorescens* LBUM677 Impacts the Rhizosphere Microbiome of Three Oilseed Crops. *Frontiers in Microbiology* 11: 569366.

Kang, Y., Shen, M., Yang, X., Cheng, D., Zhao, Q. 2014. A plant growth-promoting rhizobacteria (PGPR) mixture does not display synergistic effects, likely by biofilm but not growth inhibition *Microbiology* 83(5): 666–673.

Kerry, B.R. 2000. Rhizoshere interaction and the exploitation of microbial agents for the biological control of plant parasitic nematodes. *Annual Review of Phytopathology* 38: 423–441.

Kiani, M., Gheysari, M., Mostafazadeh-Fard, B., Majidi, M.M., Karchani, K., Hoogenboom, G. 2016. Effect of the interaction of water and nitrogen on sunflower under drip irrigation in an arid region. *Agricultural Water Management* 171:162–172.

Kibblewhite, M.G., Ritz, K., Swift, M.J. 2008. Soil health in agricultural systems. *Transactions of the Royal Society of Biological Sciences* 363: 685–701.

Kinsella, K., Schulthess, C.P., Morris, T.F., Stuart, J.D. 2009.Rapid quantification of Bacillus subtilis antibiotics in the rhizosphere. *Soil Biology and Biochemistry* 41: 374–379.

Kloepper, J.W., Rodriguez– Ubana, R., Zehnder, G.W., Murphy, J.F., Sikora, E., Fernandez, C. 1999. Plant root–bacterial interactions in biological control of soilborne diseases and potential extension to systemic and foliar diseases. *Australasian Plant Pathology* 28: 21–26.

Krauss, J., Gallenberger, I., Steffan-Dewenter, I. 2011. Decreased functional diversity and biological pest control in conventional compared to organic crop fields. *PLoS One* 6(5): e19502.

Lavanya, N., Ramiah, R.M., Sankaralingam, A., Renukadevi, P., Velazhahan, R. 2005. Identification of hosts for Ilarvirus associated with sunflower necrosis disease. *Acta Phytopathol Entomol. Hungarica* 40(1): 31–34.

Levy, A., Gonzalez, I.S., Mittelviefhaus, M. et al. 2018. Genomic features of bacterial adaptation to plants. *Nature Genetics* 50: 138–150.

Li, J.G., Ren, G.D., Jia, Z.J., Dong, Y.H., 2014. Composition and activity of rhizosphere microbial communities associated with healthy and diseased greenhouse tomatoes. *Plant Soil* 380: 337–347.

Liu H., Shi Z., Li J., Zhao P., Qin S., Nie Z. 2018. The impact of phosphorus supply on selenium uptake during hydroponics experiment of winter wheat (*Triticum aestivum*) in China. *Frontiers in Plant Science.* 9:373. doi: 10.3389/fpls.2018.00373.

Lumini, E., Vallino, M., Alguacil, M.M., Romani, M., Bianciotto, V. 2011. Different farming and water regimes in Italian rice fields affect arbuscular mycorrhizal fungal soil communities. *Ecological Applications* 21: 1696–1707.

Madhaiyan M., Poonguzhali, S, Lee, J.-S., Saravanan, V.S., Lee, K.-C., Santhanakrishnan, P. 2010. *Enterobacter arachidis* sp. a plant growth-promoting diazotrophic bacterium isolated from rhizosphere soil of groundnut. *International Journal of Systematic and Evolutionary Microbiology* 60(7): 1559–156.

Martinez-Blanco, J., Lazcano, C., Christensen, T.H., Munoz, P., Rieradevall, J., Moller, J., Anton, A., Boldrin, A. 2013. Compost benefits for agriculture evaluated by life cycle assessment. A review. *Agronomy for Sustainable Development* 33: 721–732.

Massalha, H., Korenblum, E., Tholl, D., Aharoni, A. 2017. Small molecules below-ground: The role of specialized metabolites in the rhizosphere. *Plant Journal* 90: 788–807.

Melero-Vara, J.M., Lopez-Herrera, C.J., Prados-Ligero, A.M., Vela-Delgado, M.D., Navas-Becerra, J.A., Basallote-Ureba, M.J. 2011. Effects of soil amendment with poultry manure on carnation *Fusarium* wilt in greenhouses in southwest Spain. *Crop Protection* 30: 970–976.

Mendes, R., Garbeva, P., Raaijmakers, J.M. 2013.The rhizosphere microbiome: Significance of plant beneficial, plant pathogenic, and human pathogenic microorganisms. *FEMS Microbiology Reviews* 37(5): 634–663.

Mendes, R., Kruijt, M., De Bruijn, I., Dekkers, E., Van der Voort, M., Schneider, J.H.M., Piceno, Y.M., DeSantis, T.Z., Andersen, G.L., Bakker, P.A.H.M., Raaijmakers, J.M. 2011. Deciphering the rhizosphere microbiome for disease-suppressive bacteria. *Science* 332: 1097–1100.

Morrison, C.K., Arseneault, T., Novinscak, A., Filion, M. 2017. Phenazine-1-carboxylic acid production by *Pseudomonas fluorescens* LBUM636 alters *Phytophthora infestans* growth and late blight development. *Phytopathology* 107: 273–279.

Nair, A., Ngouajio, M. 2012. Soil microbial biomass, functional microbial diversity, and nematode community structure as affected by cover crops and compost in an organic vegetable production system. *Applied Soil Ecology* 58: 45–55.

Narayanaswamy, R., Gokulakumar, B. 2010. ICP-AES analysis in sesame on the root rot disease incidence. *Archives of Phytopathology and Plant Protection* 43(10): 940–948.

Neal A.L., Rossmann M. et al. 2017. Land-use influences phosphatase gene microdiversity in soils. *Environmental Microbiology.* 19(7):2740–2753.

Neuman, G., Romheld, V. 2007. The release of root exudates as affected by the plant physiological status. In: The Rhizosphere Biochemistry and Organic Substances at the Soil-Plant Interface. Eds. R. Pinton, et al. (Boca Raton: CRC), 23–72.

Nihorimbere, V., Ongena, M., Smargiassi, M., Thonart, P. 2011. Beneficial effect of the rhizosphere microbial community for plant growth and health. *Biotechnology, Agronomy and Society and Environment* 5(2): 327–337.

Paulucci, N.S., Gallarato, L.A., Reguera, Y.B., Vicario, J.C., Cesari, A.B., Garcia de Lema, M.B., Dardanelli, M.S. 2015. *Arachis hypogaea* PGPR isolated from Argentine soil modifies its lipids components in response to temperature and salinity. *Microbiological Research* 173: 1–9.

Philippot, L., Raaijmakers, J.M., Lemanceau, P., van der Putten, W.H., 2013. Going back to the roots: The microbial ecology of the rhizosphere. *Nature Reviews Microbiology* 11: 789–799.

Prakamhang, J., Tittabutr, P., Boonkerd, N., Teaumroong, N. 2015. Proposed some interactions at molecular level of PGPR coinoculated with *Bradyrhizobium diazoefficiens* USDA110 and *B. japonicum* THA6 on soybean symbiosis and its potential of field application. *Applied Soil Ecology* 85: 38–49

Puglisi, E., Fragoulis, G., Ricciuti, P., Cappa, F., Spaccini, R., Piccolo, A., Trevisan, M., Crecchio. 2009. Effects of a humic acid and its size-fractions on the bacterial community of soil rhizosphere under maize (*Zea mays* L.). *Chemosphere* 77 (6): 829–837.

Puri, A., Padda, K.P., Chanway, C.P. 2016. Evidence of nitrogen fixation and growth promotion in canola (*Brassica napus* L.) by an endophytic diazotroph *Paenibacillus polymyxa* P2b-2R. *Biology and Fertility of Soils* 52(1): 119–125.

Qurashi, A.W., Sabri, A.N. 2012. Bacterial exopolysaccharide and biofilm formation stimulate chickpea growth and soil aggregation under salt stress. *Brazilian Journal of Microbiology* 43(3): 1183–1191.

Qureshi, A.A., Narayanasamy, G. 2005. Residual effect of phosphate rocks on dry matter yield and P uptake of mustard and wheat crops. *Journal of Indian Society of Soil Science* 53(1): 132–134.

Qureshi, A.A., Narayanasamy, G. 1999. Direct effect of rock phosphates and phosphate solubilizers on soybean growth in Typic Ustochrept. *Journal of Indian Society of Soil Science*. 47(3): 475–478.

Raaijmakers, J.M., Paulitz, T.C., Steinberg, C., Alabouvette, C., Moënne-Loccoz, Y. 2009. The rhizosphere: A playground and battlefield for soil borne pathogens and beneficial microorganisms. *Plant Soil* 321: 341–361.

Raaijmakers, J.M., Weller, D.M., Thomashow, L.S. 1997. Frequency of antibiotic-producing *Pseudomonas* spp. in natural environments. *Applied and Environmental Microbiology* 63: 881–887.

Rahman, M., Ishii, Y., Niimi, M., Kawamura, O. 2008. Effect of salinity stress on dry matter yield and oxalate content in Napier grass (*Pennisetum purpureum* Schumach). *Asian Australasian Journal of Animal Science* 21(11): 1599–1603.

Ramans, R., Ramaekers, L., Schelkens, S., Hernandez, G., Garcia, A., Reyes, J.L., Mendez, N., Toscano, V., Mulling, M., Galvez, L., Vanderleyden, J. 2008. Effect of *Rhizobium-Azospirillum* co-inoculation on nitrogen fixation and yield of two contrasting *Phaseolus vulgaris* L. genotypes cultivated across different environments in Cuba. *Plant and Soil* 312: 25–37.

Ramirez, K.S., Craine, J.M., Fierer, N. 2012. Consistent effects of nitrogen amendments on soil microbial communities and processes across biomes. *Global Change Biology* 18: 1918–1927.

Rangeshwaran R., Prasad R.D. 2000. Biological control of sclerotium rot of sunflower. *Indian Phytopathology* 53 (4): 444–449.

Rennie R.J. 1980. Dinitrogen-fixating bacteria: Computer-assisted identification of soil isolates. *Can. J. Microbiol.* 26:1275–1283.

Richardson, A.E., Barea, J., McNeill, A.M., Prigent-Combaret, C. (2009) Acquisition of phosphorus and nitrogen in the rhizosphere and plant growth promotion by microorganisms. *Plant Soil* 321: 305–339.

Rillig M.C., Mummey D.L. 2006. Mycorrhizas and soil structure. *New Phytologist* 171: 41–53.

Rodriguez, R., Redman, R. 2008. More than 400 million years of evolution and some plants still can't make it on their own: Plant stress tolerance via fungal symbiosis. *Journal of Experimental Botany* 59(5): 1109–1114.

Rovira, A.D. 1969. Plant root exudates. *Botancial Review* 35: 35–57.

Ruano-Rosa, D., Mercado-Blanco, J. 2015. Combining biocontrol agents and organics amendments to manage soilborne phytopathogens. In: Organic amendments and soil suppressiveness in plant disease management. Eds. M.K. Meghvansi, A. Varma (Switzerland: Springer Publishing), 457–478.

Rudrappa, T., Czymmek, K.J., Pare, P.W., Bais, H.P. 2008. Root secreted malic acid recruits beneficial soil bacteria. *Plant Physiology* 148: 1547–1556.

Saravanakumar, D., Muthumeena, B., Lavanya, N., Suresh, S., Rajendran, L., Raguchander, T., Samiyappan, R. 2007. *Pseudomonas* induced defense molecules in rice against leaf folder (*Cnephalocrocis medinalis*) pest. *Pest Management Science* 63: 714–721.

Schloss, P.D., Handelsman, J. 2006. Toward a census of bacteria in soil. *PLoS Computation Biology* 2(7): e92.

Schulz, S., Dickschat, J.S. 2007. Bacterial volatiles: The smell of small organisms. *Natural Products Reports* 24(4): 814–842.

Selvakumar, G., Panneerselvam, P., Ganeshamurthy, A.N., Maheshwari, D.K. 2012. Bacterial mediated alleviation of abiotic stress in crops. In: Bacteria in Agrobiology: Stress Management. Ed. D.K. Maheshwari (New York: Springer), 205–224.

Shahid, M., Hameed, S., Imran, A., Ali, S., Van Elsas, J.D. 2012. Root colonization and growth promotion of sunflower (*Helianthus annuus* L.) by phosphate solubilizing Enterobacter sp. Fs-11. *World Journal of Microbiology and Biotechnology* 28: 2749–2758.

Shekhar, N., Bhattacharya, D., Kumar, D., Gupta, R.K. 2006. Biocontrol of wood rotting fungi with *Streptomyces violaceus-niger* XL-2. *Canadian Journal of Microbiology* 52: 805–808.

Singh, N., Yadav, A., Varma, A. 2015. Effect of plant growth promoting activity of rhizobacteria on Cluster bean (*Cyamopsis tetragonoloba* L.) plant growth and biochemical constituents. *International Journal of Current Microbiology and Applied Sciences* 4(5): 1071–1082

Somers, E., Vanderleyden, J., Srinivasan, M. 2004. Rhizosphere bacterial signalling: A love parade beneath our feet. *Critical Reviews in Microbiology* 30: 205–240.

Sommermann, L., Geistlinger, J., Wibberg, D., Deubel, A., Zwanzig, J., Babin, D., Schluter, A., Schellenberg, I. 2018. Fungal community profiles in agricultural soils of a long-term field trial under different tillage, fertilization and crop rotation conditions analyzed by high-throughput ITS-amplicon sequencing. *PLoS ONE* 3(4): e0195345.

Sugiyama, A. 2019. The soybean rhizosphere: Metabolites, microbes, and beyond–A Review. *Journal of Advanced Research* 19: 67–73.

Sugiyama, A., Vivanco, J.M., Jayanty, S.S., Manter, D.K. 2010. Pyrosequencing assessment of soil microbial communities in organic and conventional potato farms. *Plant Diseases* 94: 1329–1335.

Tang, C.S., Cai, W.F., Kohl, K., Nishimoto, R.K. 1995. Plant stress and allelopathy. In: Allelopathy: Organisms, Processes, and Applications. Eds. K.M.M.D. Inderjit, F.A. Einhellig, Acs Symposium Series. Washington: Amer Chemical Soc, 582, 142–157.

Timmusk, S., Nicander, B., Granhall, U., Tillberg, E. 1999. Cytokinin production by *Paenibacillus polymyxa. Soil Biology and Biochemistry* 31(13): 1847–1852.

Tokala, R.K., Strap, J.L., Jung, C.M., Crawford, D.L., Salove, M.H., Deobald, L.A., Bailey, J.F., Morra, M.J. 2002. Novel plant microbe rhizosphere interaction involving *Streptomyces lydicus* WYEC108 and the Pea plant (*Pisum sativum*). *Applied Environmental Microbiology* 68(5): 2161–2171.

Torres, I.F., Bastida, F., Hernandez, T., Garcia, C. 2015. The effects of fresh and stabilized pruning wastes on the biomass, structure and activity of the soil microbial community in a semiarid climate. *Applied Soil Ecology* 89: 1–9.

Uren, N.C. 2007. Types, amounts, and possible functions of compounds released into the rhizosphere by soil-grown plants. In: The Rhizosphere Biochemistry and Organic Substances at the Soil-Plant Interface. Eds. R. Pinton, Z. Varanini, P. Nannipieri, 2nd edn. (Boca Raton: CRC), 1–21.

Venturi, V., Fuqua, C. 2013. Chemical signalling between plants and plant-pathogenic bacteria. *Annual Review of Phytopathology* 51: 17–37.

Von der Weid, I., Duarte, G.F., Van Elsas, J.D., Seldin, L. 2002. *Paenibacillus brasilensis* sp. nov., a novel nitrogen-fixing species isolated from the maize rhizosphere in Brazil. *International Journal of Systematic and Evolutionary Microbiology* 52(6): 2147–2153.

Vorholt, J.A. 2012. Microbial life in the phyllosphere. *Nature Review Microbiology* 10: 828–840.

Wan, T.T., Zhao, H.H., Wang, W. 2017.Effect of biocontrol agent *Bacillus amyloliquefaciens* SN16-1 and plant pathogen *Fusarium oxysporum* on tomato rhizosphere bacterial community composition. *Biological Control* 112: 1–9.

Wang, H., Gu, C., Liu, X., Yang, C., Li, W., Wang, S. 2020. Impact of soybean nodulation phenotypes and nitrogen fertilizer levels on the rhizosphere bacterial community. *Frontiers in Microbiology* 11: 750.

Wang, R., Xiao, Y., Lv, F., Hu, L., Wei, L., Yuan, Z., Lin, H. 2018. Bacterial community structure and functional potential of rhizosphere soils as influenced by nitrogen addition and bacterial wilt disease under continuous sesame cropping. *Applied Soil Ecology* 125: 117–127.

Ward, N., Challacombe, J., Jansses, P.H., Henrissat, B. 2009. Three genomes from the Phylum Acidobacteria provide insight into the lifestyles of these microorganisms in soils. *Applied Environmental. Microbiology* 75(7): 2046–2056.

Whipps, J.M. 1990. Carbon economy. In: The Rhizosphere in Ecological and Applied Microbiology. Ed. J.M. Lynch (Chichester: Wiley series J Wiley), 59–97.

Wu, F., Wan, J.H.C., Wu, S., Wong, M.H. 2012. Effects of earthworms and plant growth promoting rhizobacteria (PGPR) on availability of nitrogen, phosphorus and potassium in soil. *Journal of Plant Nutrition Soil Science* 175 (3): 423–433.

Xiong, W., Li, R., Ren, Y., Liu, C., Zhao, Q., Wu, H., Jousset, A., Shen, Q. 2017. Distinct roles for soil fungal and bacterial communities associated with the suppression of vanilla *Fusarium* wilt disease. *Soil Biology and Biochemistry* 107: 198–207.

Xu, Y., Wang, G., Jin, J., Liu, J., Zhang, Q., Liu, X. 2009. Bacterial communities in soybean rhizosphere in response to soil type, soybean genotype, and their growth stage. *Soil Biology and Biochemistry* 41: 919.

Yaish, M.W. 2015. Proline accumulation is a general response to abiotic stress in the date palm tree (*Phoenix dactylifera* L.). *Genetic and Molecular Research* 14(3): 9943–9950.

Yang, W., Jing, X., Guan, Y., Zhai, C., Wang, T., Shi, D., Sun, W., Gu, S. 2019. Response of fungal communities and co-occurrence network patterns to compost amendment in black soil of Northeast China. *Frontiers in Microbiology* 10: 1562.

Yang, J., Kloepper, J.W., Ryu, C.-M. 2008. Rhizosphere bacteria help plants tolerate abiotic stress. *Trends in Plant Science* 14(1): 1–4.

Zaccardelli, M., De Nicola, F., Villecco, D., Scotti, R. 2013.The development and suppressive activity of soil microbial communities under compost amendment. *Journal of Soil Science and Plant Nutrition* 13: 730–742.

Zahir, A., Mirza, B.S., Mclean, J.E., Yasmin, S., Shah, T.M., Malik, K.A., Mirza, M.S. 2016. Association of plant growth promoting *Serratia* spp. with root nodules of chickpea. *Research in Microbiology* 167(6): 510–520.

Zapata, P.J., Serrano, M., Garcia-Legaz, M.F., Pretel, M.T., Botella, M.A. 2017. Short term effect of salt shock on ethylene and polyamines depends on plant salt sensitivity. *Frontiers in Plant Science* 8: 855. doi: 10.3389/fpls.2017.00855.

Zhang, H., Kim, M.S., Sun, Y., Dowd, S.E., Shi, H., Pare, P.W. 2008. Soil bacteria confer plant salt tolerance by tissue-specific regulation of the sodium transporter HKT1. *Molecular Plant Microbe Interactions* 21: 737–744.

Zhang, M., Li, R., Cao, L., Shi, J., Liu, H., Huang, Y., Shen, Q. 2014. Algal sludge from Taihu Lake can be utilized to create novel PGPR-containing bioorganic fertilizers. *Journal of Environment Management* 132: 230–236.

Zhen, Z., Liu, H.T., Wang, N., Guo, L.Y., Meng, J., Ding, N., Wu, G.L., Jiang, G.M. 2014. Effects of manure compost application on soil microbial community diversity and soil microenvironments in a temperate cropland in China. *PLoS ONE* 9: e108555.

Ziegler, M., Engel, M., Welzl, G., Schloter, M. 2013.Development of a simple root model to study the effects of single exudates on the development of bacterial community structure. *Journal of Microbiological Method* 94: 30–36.

Index

Note: Page numbers followed by *f* indicate figures and *t* indicate tables.

T

U

V

W

X

For Product Safety Concerns and Information please contact our EU representative GPSR@taylorandfrancis.com
Taylor & Francis Verlag GmbH, Kaufingerstraße 24, 80331 München, Germany

www.ingramcontent.com/pod-product-compliance
Lightning Source LLC
LaVergne TN
LVHW010541100826
845148LV00001B/254

* 9 7 8 0 3 6 7 7 0 5 8 6 2 *